Teratocarcinomas and embryonic stem cells

a practical approach

TITLES PUBLISHED IN

THE PRACTICAL APPROACH SERIES

Gel electrophoresis of proteins
Gel electrophoresis of nucleic acids
Iodinated density-gradient media
Centrifugation (2nd Edition)
Transcription and translation
Oligonucleotide synthesis
Microcomputers in biology
Mutagenicity testing
Affinity chromatography
Virology
DNA cloning
Plant cell culture
Immobilised cells and enzymes
Nucleic acid hybridisation
Animal cell culture
Human cytogenetics
Photosynthesis: energy transduction
Drosophila
Human genetic diseases
H.p.l.c. of small molecules
Carbohydrate analysis
Spectrophotometry and spectrofluorimetry
Biochemical toxicology
Electron microscopy in molecular biology
Nucleic acid and protein sequence analysis

Teratocarcinomas and embryonic stem cells
a practical approach

Edited by

E J Robertson

Department of Genetics, University of Cambridge,
Downing Street, Cambridge CB2 3EH, UK

Elizabeth Robertson
CBE FRS FMedSci
DsC MAE
(aka Liz!)

IRL PRESS
Oxford · Washington DC

IRL Press Limited
P.O. Box 1,
Eynsham,
Oxford OX8 1JJ,
England

British Library Cataloguing in Publication Data

Teratocarcinomas and embryonic stem cells : a practical approach.—
(Practical approach series)
1. Teratocarcinoma 2. Cell culture
I. Robertson,E. II. Series
599.03′3 RC260

ISBN 1-85221-005-2 (Hardbound)
ISBN 1-85221-004-4 (Softbound)

Printed in England by Information Printing, Oxford.

Preface

The ability to isolate and grow in tissue culture pluripotential cell lines which possess a set of properties unique to the cells of the early mammalian embryo has provided research workers interested in the study of developmental processes and tumour biology with an invaluable experimental tool. Perhaps the most attractive feature of cultured stem cells, which may be obtained directly from the normal mouse embryo or indirectly from malignant teratocarcinoma tumours, is that they can be variously manipulated to differentiate into a diversity of cell types by altering the growth environment of the cell. Differentiation can be elicited in the culture dish or by using the cells to generate tumours. In addition, some cell lines, notably those derived directly from embryos (ES cell lines), will differentiate normally when they are returned to the embryonic environment to give chimaeric animals. For murine EC and ES cell lines such techniques are well established and with the concurrent development of sophisticated molecular biological techniques, embryonic stem cell lines afford an increasingly important means for the analysis of the complex processes underlying cell and tissue differentiation.

The aim of this book is to provide the basic protocols involved in the derivation of pluripotential cells from tumours and mouse embryos, the routine culture of EC and ES cell lines and methods for inducing their differentiation. In addition, several of the chapters provide extensive protocols covering the more specialist techniques which can be applied to these cell lines such as the production of chimaeras, the introduction of DNA into cells, the isolation and characterization of somatic cell variants and studies on the requirements for cell proliferation *in vitro*. The final chapter concerns human germ cell tumours and provides a useful summary of existing cell lines. Information is given concerning the classification and handling of tumour specimens and comprehensive protocols for the derivation and characterization of tumour-derived cell lines are provided. I would like to express my thanks to all of the authors for their contributions to this volume. Finally, as some of the techniques described in this volume involve the use of live animals, it should be noted that such experiments may be subject to local licensing regulations.

Elizabeth J.Robertson

Contributors

Peter W.Andrews
The Wistar Institute of Anatomy and Biology, 36th Street at Spruce, Philadelphia, PA 19104, USA

Allan Bradley
Department of Genetics, University of Cambridge, Downing Street, Cambridge CB2 3EH, UK

Andrea Damjanov
Department of Pathology, Hahnemann University Medical School, 230 North Broad Street, Philadelphia, PA 19102, USA

Ivan Damjanov
Department of Pathology, Hahnemann University Medical School, 230 North Broad Street, Philadelphia, PA 19102, USA

John K.Heath
Department of Zoology, University of Oxford, South Parks Road, Oxford OX1 3PS, UK

Martin L.Hooper
Department of Pathology, University Medical School, Teviot Place, Edinburgh EH8 9AG, UK

Robin H.Lovell-Badge
MRC Mammalian Development Unit, Wolfson House, 4 Stephensons Way, London NW1 2HE, UK

Michael W.McBurney
Department of Medicine, University of Ottawa, Ottawa, Ontario K1N 6N5, Canada

J.Wolter Oosterhuis
Pathologisch Anatomisch Lab RVG, Oostersingel 63, 9713 EZ Groningen, The Netherlands

Elizabeth J.Robertson
Department of Genetics, University of Cambridge, Downing Street, Cambridge CB2 3EH, UK

Michael K.Rudnicki
Department of Medicine, University of Ottawa, Ottawa, Ontario K1N 6N5, Canada

Davor Solter
The Wistar Institute of Anatomy and Biology, 36th Street at Spruce, Philadelphia, PA 19104, USA

Contents

Abbreviations

AFP	alpha fetoprotein
BRL	Buffalo rat liver
BSA	bovine serum albumin
BPV	bovine papilloma virus
CAT	chloramphenicol acetyltransferase
CFMAM	calcium-free minimal attachment medium
EC	embryonal carcinoma
ECDGF	embryonal carcinoma-derived growth factor
ES	embryonic stem
DMEM	Dulbecco's modified Eagles' medium
DMSO	dimethyl sulphoxide
EDTA	ethylenediamine tetraacetic acid
FACS	fluorescence activated cell sorter
FBS	foetal bovine serum
FITC	fluorocein isothiocyanate
HAT	hypoxanthine/aminopterin/thymidine
HCG	human chorionic gonadotrophin
HDL	high density lipoprotein
Hepes	N-2-hydroxyethylpiperazine-N′-2-ethane sulphonic acid
HPRT	hypoxanthine-guanine phosphoribosyl transferase
ICM	inner cell mass
LTR	long terminal repeats
MEM	minimal essential medium
MNNG	N-methyl-N′-nitro-N-nitrosoguanidine
NCS	newborn calf serum
PBS	phosphate-buffered saline
p.c.	post coitum
PEG	polyethylene glycol
PMSG	pregnant mare serum gonadotrophin
RA	retinoic acid
RIA	radioimmunoassay
RITC	rhodamine isothiocyanate
SDS	sodium dodecylsulphate
SSC	standard saline citrate
TCA	trichloroacetic acid
T-EDTA	trypsin EDTA

CHAPTER 1

Production of teratocarcinomas from embryos transplanted to extra-uterine sites

IVAN DAMJANOV, ANDREA DAMJANOV and DAVOR SOLTER

1. INTRODUCTION

Teratomas are tumours composed of various tissues foreign to their site of origin. Benign teratomas, known simply as *teratomas*, are usually composed of well differentiated somatic tissues which have a rather limited capacity for growth. In contrast, malignant teratomas or *teratocarcinomas*, contain undifferentiated malignant stem cells which have an unlimited proliferative ability and a tendency to metastasize. These tumours, like other biologically malignant neoplasms, are lethal if untreated (1–4).

Benign teratomas are easily generated in several animal species by transplanting embryos, or parts of embryos, to extra-uterine sites (5–8). By contrast, teratocarcinomas can be produced efficiently only in mice (6,7). In this chapter we briefly summarize the technique for the production of teratocarcinomas from mouse embryos transplanted to extra-uterine sites, and the typical features of embryo-derived tumours.

In our experience of the several methods that have been described for producing embryo-derived teratocarcinomas, the method presented here combines technical simplicity with very high reliability. With little practice, the yield of tumours from transferred embryos approaches 100%, thus enabling the investigator to design experiments in which the effect of various factors on tumour yield and on the nature of the tumour (benign versus malignant) can be tested. Experimental teratocarcinomas have been extremely valuable in analysing various aspects of mammalian development and cellular differentiation, but the amount of information potentially available from this system remains enormous.

2. PRODUCTION OF TUMOURS FROM TRANSPLANTED EMBYROS

Embro-derived teratocarcinomas can be produced in most, if not all, inbred mouse strains (9). The technique for the production of malignant tumours is simple to master and easily reproducible. It assumes the ability of the investigator to obtain timed pregnancies, to isolate embryos of a given age and to transplant the embryos under the kidney capsule. Following successful transplantation, no additional treatment is necessary and tumours can be harvested within 6–10 weeks.

2.1 **Equipment**

2.1.1 *Dissecting stereomicroscope*

Any dissecting microscope can be used. The most suitable types are those with adjust-

Table 1. Surgical instruments and supplies for transplanting embryos.

Fine scissors
2 pairs of watchmakers forceps (No. 5)
Blunt-ended forceps
5 inch scissors
Desmarres chalazion forceps
Surgical needle and silk
Wound closing clips
Wound closing forceps
Sterile saline solution
Anaesthetic (Ketamin, Nembutal, Avertin[a])
1 ml hypodermic syringe and 25 gauge needle
Animal balance

[a]Details of the Avertin anaesthetic are given in Chapter 5, Section 8.

able bulb or fibre optic illumination, incandescent light and transmitted light, with a magnification range of 10–40× and a working distance of 10–15 cm measured from the objective [e.g. Nikon SM2-10, Wild M5A, Zeiss (Opton) Jena].

2.1.2 *Surgical instruments and supplies*

A range of standard surgical instruments are required. These are listed in *Table 1*.

2.1.3 *Glassware and plastics*

A normal supply of 35 mm and 100 mm Petri dishes is required.

2.1.4 *Braking pipette*

A 'braking' pipette is used to transfer embryos to the kidney capsule. It consists of a narrow glass tube ('transfer' pipette) attached to a larger diameter cylinder ('holding' pipette). The pipette is illustrated diagrammatically in *Figure 1*. The working end of the transfer pipette should be large enough to accommodate an embryo yet small enough to enable transfer of the embryo with minimal injury to the tissue of the host animal. The opposite end of the transfer pipette is drawn to a very thin diameter capillary and acts as a braking valve for the suction applied through the large cylinder attached to rubber tubing.

Construct a braking pipette by the following procedure.

(i) Hold a 20 cm Pasteur pipette over a Bunsen burner so that the mid-portion of the narrow part is over the flame, rotate until the glass softens, withdraw from the flame and quickly pull the two ends apart in opposite directions. This thins the softened part into a capillary tube with an inside diameter of about 100 μm.

(ii) Allow to cool and break the thin capillary portion, which should be as short as possible, by bending the glass. This produces a glass tube approximately 5 cm long with the original opening (~2 mm diameter) on one end ('working end') and a short, thin capillary section on the other end ('braking end'). Retain the unused part of the Pasteur pipette.

(iii) Hold the tip of the working end over the flame and, using forceps, extend this into a narrower tube (~1 mm inside diameter). Break off the original end of

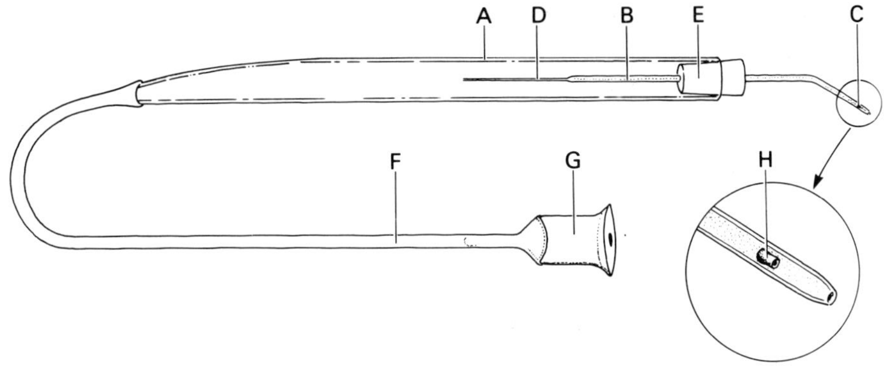

Figure 1. Braking pipette. **A**, holding pipette; **B**, transfer pipette; **C**, working end; **D**, braking end; **E**, rubber stopper or cork; **F**, rubber tubing; **G**, mouth piece; **H**, embryo in the working end of the transfer pipette.

the extended working end and discard. Flame-polish the edges of the working end. While the working end is still warm, introduce a 45° angle bend to enable easier manipulation of the embryos (see *Figure 1*).

(iv) Use the remaining part of the Pasteur pipette to make the holding pipette. Close the wide end with a rubber stopper, or a cork, which contains a central hole large enough to accommodate the transfer pipette.

(v) Pull out the narrow end to give an internal diameter of 2–3 mm and over this fit a piece of rubber tubing approximately 20–25 cm long.

(vi) At the opposite end of the rubber tubing, insert a plastic mouthpiece (like that used in clinical haematology laboratories to aspirate blood into capillary tubes and haemocytometers).

(vii) Assemble the braking pipette by inserting the transfer pipette into the hole of the rubber stopper. Make sure that the thin capillary end does not touch the wall of the holding pipette.

(viii) Test the transfer pipette by immersing the working end into water and exhaling through the mouthpiece. If the braking pipette has been assembled correctly small bubbles should be formed. Upon aspiration through the tubing, water should slowly enter the working end. If the water moves too quickly and cannot be controlled, or if it enters the working end spontaneously due to capillary action, the braking end of the tube is not sufficiently thin. If this occurs re-melt the braking end into a thinner capillary.

Prepare several working pipettes and store with the capillary braking ends carefully protected from damage.

2.2 **Mating of mice**

We prefer to use natural matings. Keep the animals under controlled conditions with a fixed light/dark cycle (19.00 h–5.00 h dark and 5.00 h–19.00 h light is convenient). Under these conditions copulation is assumed to occur, in most cases, at the mid-point of the dark period, i.e. midnight. Keep adult males (6 weeks or older) in individual cages and add two sexually mature (more than 6 weeks old) females per cage in the late afternoon. (Note that females can be housed jointly but males should be housed

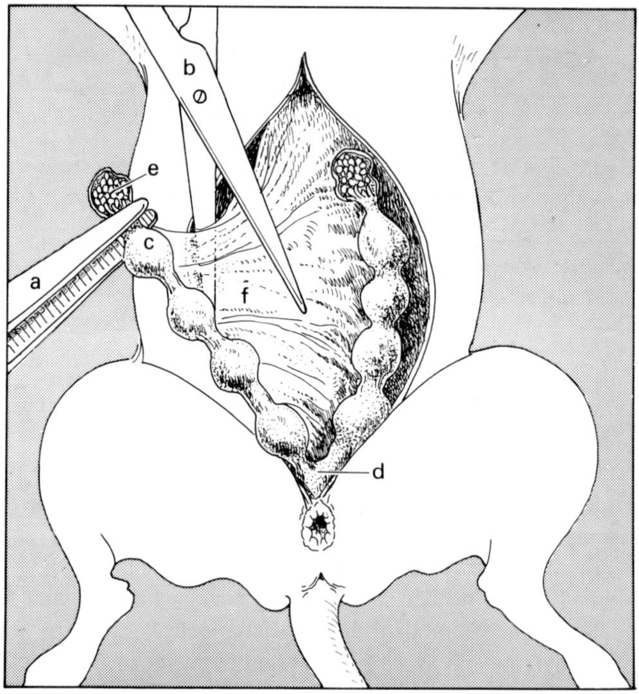

Figure 2. Removal of pregnant uterus. **A**, forceps; **B**, scissors; **C**, pregnant uterus with deciduoma; **D**, vaginal end of the uterus; **E**, ovary; **F**, mesometrium.

individually.) Females enter oestrus and become receptive to males every 4–5 days. Check the females each morning for the presence of vaginal plugs which indicate that copulation has occurred. Immobilize the female face down on a hard surface with the palm or the fifth finger of the hand, and lift the tail with the thumb and second finger. In most instances, the vaginal plug protrudes from the vagina and can be spotted easily. Smaller plugs are located deeper in the vagina and are revealed only when a probe or forceps is used. We routinely insert the tip of blunt forceps into the introitus of the vagina to check for a deeper plug. Since vaginal plugs usually dissolve within 12–14 h, perform the inspection in the early morning, and never later than mid-day, following mating. Keep a record of the mating performance of individual males. We routinely discard males that fail to copulate for 2 weeks when presented with different females. Mark the pregnant females and place them together into a separate cage. Group the non-pregnant females together until the late afternoon. They can then be used again for mating. We find it impractical to keep mating records for individual females, but do keep batches of females of the same age together. Females that have not copulated after a certain period of mating can be used for other purposes, including as recipients of transplanted embryos. Note that in this chapter the age of embryos is given in days, with day 0 corresponding to the day of finding the vaginal plug.

2.3 Recovery of post-implantation embryos

2.3.1 *Recovery of uterine horns*

(i) Sacrifice pregnant females, with 6- or 7-day-old embryos, in the morning (8.00–8.30 h) by cervical dislocation.

(ii) Make a small transverse incision in the abdominal skin (or tear the skin by a light twist of the thumb and second finger of both hands) and pull the skin downward to expose the muscles of the entire abdomen.

(iii) Open the anterior wall of the abdominal cavity and lift the intestines to expose the pregnant uterus. The pregnant uterus is much thicker than the non-pregnant mouse uterus and can be recognized easily. On day 6 or 7 of pregnancy the horns of the uterus are distended by discrete nodules produced by deciduomas. These result from the localized proliferation of decidualized stromal cells surrounding the embryo.

(iv) Using forceps grasp the ovarian end of one of the uterine horns, make a cut between the ovary and uterus, and continue along the mesometrium leaving behind the fat tissue occasionally attached to it (see *Figure 2*). Cut the uterine horn above the vagina, remove from the animal and place into a Petri dish filled with saline. Remove the remaining uterine horn in a similar way. Discard the carcass of the pregnant mouse.

2.3.2 *Isolation of embryos*

(i) Open the pregnant uterine horns by tearing the myometrium with two pairs of watchmakers forceps. This frees the deciduomas. Retain these in the saline-filled Petri dish. All of these manipulations can be completed without the use of a dissecting mciroscope, although the novice may use the microscope when isolating the deciduomas.

(ii) Remove the dish to a dissecting microscope. Deciduomas containing 6- and 7-day-old embryos are pear-shaped in appearance. With reference to *Figure 3* dissect the embryos from the narrow part of this structure using two pairs of watchmakers forceps under saline. Using forceps hold the deciduoma at the broader end and cut the top off the narrow end by closing the second forceps (*Figure 3A*). Insert the second pair of forceps, with tips closed, into the mid-portion of the broad part and allow them to open to expose the implantation chamber (*Figure 3B*). The embryo itself is translucent and hardly visible but the implantation site of the uterus is red and easily identifiable due to extravasated blood. If properly opened, the chamber should contain the egg cylinder attached to the implantation site via the ectoplacental cone. The egg cylinder measures less than 1 mm in length but can be easily discerned with the aid of the microscope even at low magnification.

(iii) Tear the egg cylinder off from the ectoplacental cone by pulling it or by tightly squeezing at the junction of the ectoplacental cone and the extra-embryonic portion of the egg cylinder. Keep the isolated egg cylinder in the same Petri dish while other egg cylinders are collected from the same horn and then transfer them into a fresh dish. Alternatively, the isolated embryos can be transferred

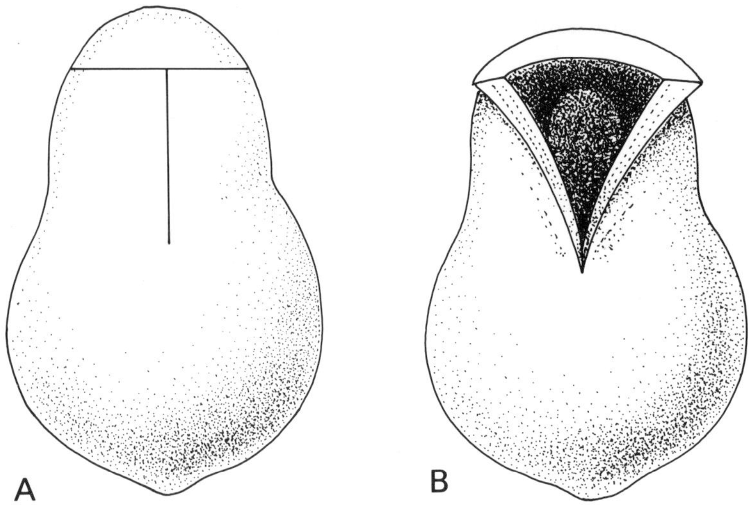

Figure 3. Isolation of embryo from deciduoma. **A**, position of first and second cut is marked; **B**, implantation chamber with embryo inside is exposed.

immediately using a mouth-controlled pipette into a clean dish containing saline or tissue culture medium (see Section 5).

Isolated 6- and 7-day-old embryos have the typical appearance of post-implantational egg cylinders (10). These embryos have an embryonic and extra-embryonic portion, marked by an indentation as shown in *Figure 4*. The 6-day-old egg cylinder has two germ layers; the ectoderm and primitive endoderm, whereas the 7-day egg cylinder, following formation of the primitive streak, has three germ layers in the embryonic portion; an inner layer or ectoderm, an outer layer or proximal (visceral) endoderm and a middle layer or mesoderm. At the junction of the embryonic and extra-embryonic portions in the intra-embryonic cavity is a fold called the amnion. The extra-embryonic portion also consists of ectoderm, endoderm and mesoderm, which are termed extra-embryonic to distinguish them from the embryonic germ layers. Inside the extra-embryonic cavity is a transverse fold called the chorion. The parietal endoderm forms the Reichert's membrane which is tightly bound to trophoblastic cells and deciduoma, thus forming the embryonic portion of the wall of the implantation chamber. This layer is frequently torn during isolation of egg cylinders. The trophoblastic giant cells found external to Reichert's membrane may be partially attached to the embryo. Adhering fragments of Reichert's membrane and trophoblast should be peeled away from the embryo with forceps under the dissecting microscope.

Inspect the embryos to be used for transplantation and select only the morphologically normal embryos. Contorted, haemorrhagic or obviously opaque rather than translucent embryos should be discarded. Also, since teratocarcinomas cannot be obtained from embryos that have undergone neurulation, embryos with an apparent neural fold should not be used. Because development of embryos in a single uterus is not synchronous, some embryos will inevitably be older than 7.5 days. By sacrificing the pregnant mice early in the morning (8.00–9.30 h) the number of older embryos can be minimized.

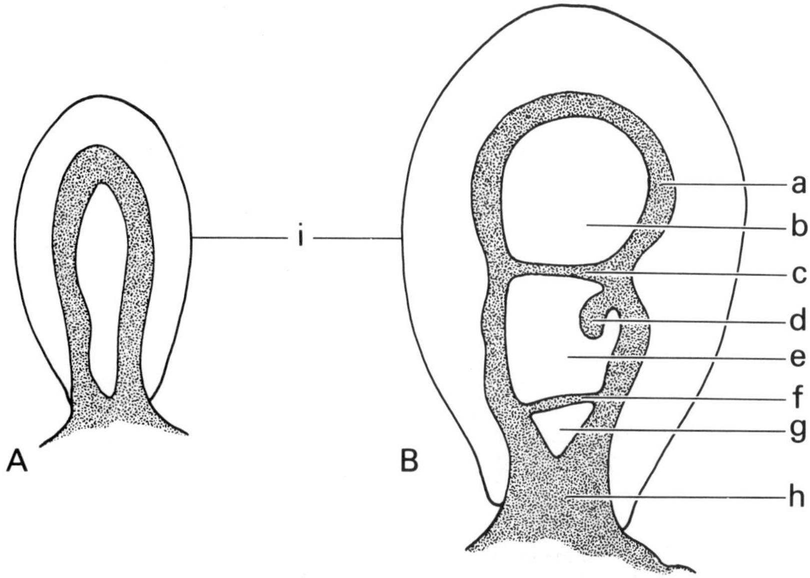

Figure 4. Schematic representation of the 6- **(A)** and 7- **(B)** day-old mouse egg cylinder surrounded by parietal endoderm. **a**, embryonic part (ectoderm, mesoderm, endoderm); **b**, amniotic cavity; **c**, amnion; **d**, allantois; **e**, exocoel; **f**, chorion; **g**, ectoplacental cavity; **h**, ectoplacental cone; **i**, parietal endoderm.

Once isolated, embryos should be transplanted within 1−2 h. Although we have not tested how long embryos can be kept *in vitro* and still yield tumours, we know that 7-day-old embryos incubated in serum-supplemented medium (Section 5) continue to develop and after 24 h do not give rise to teratocarcinomas.

2.4 **Transplantation of embryos**

Embryos freed of Reichert's membrane and attached trophoblastic cells, and severed from the ectoplacental cone, are aspirated into the braking pipette and transferred underneath the kidney capsule of syngeneic or semi-syngeneic recipients. Depending on the experiment, the whole embryo or only a portion is transplanted. In the latter case, the embryo proper should be cut free from the extra-embryonic part (see *Figure 4B*) with a fine iris knife.

Surgical transplantation is carried out as follows. The requirements for the operative procedure are listed in *Table 1*.

(i) Weigh and anaesthetize a recipient and, for convenience, place the mouse into the cover of a 100 mm Petri dish so that it can be moved easily.
(ii) Wipe the back with surgical alcohol and part the hair to allow a clean incision. Make a small longitudinal cut in the skin (1.5 cm) either in the mid-line region or lateral to the spine between the ribs and pelvis.
(iii) Place a pair of closed scissors under the skin and open them to separate the skin from the underlying muscle. Make a 1 cm longitudinal incision in the muscle, parallel to the spine approximately 0.5 cm lower than the last rib. This should reveal the kidney. Either kidney may be exposed.

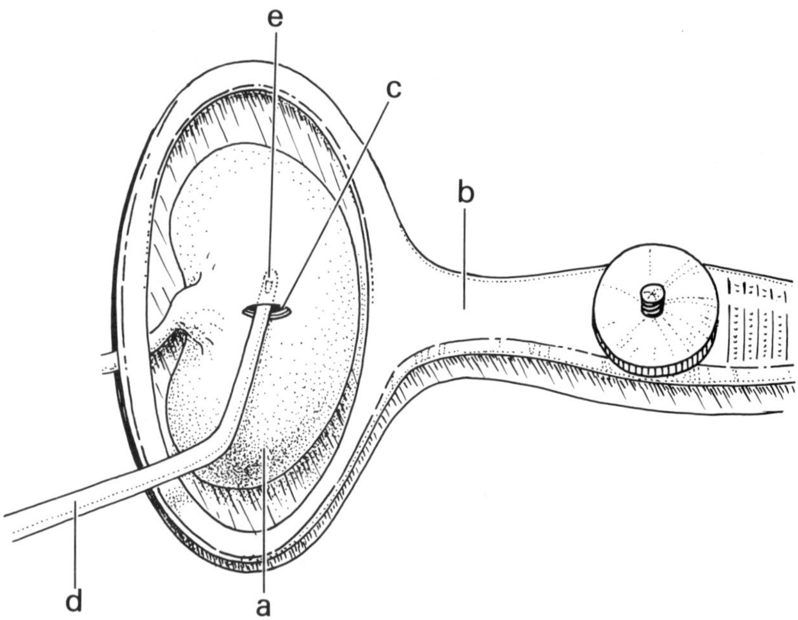

Figure 5. Transfer of egg cylinder under kidney capsule. **A**, kidney; **B**, Desmarres chalazion forceps. **C**, tear in kidney capsule; **D**, transfer pipette; **E**, egg cylinder in the working end of the transfer pipette.

(iv) Express the kidney through this incision by applying pressure to the lateral wall of the abdomen and by pushing the kidney. If the spleen (if working on the left side) or the intestines are accidentally expressed instead, these can be repositioned easily into the abdominal cavity.

(v) Hold the exposed kidney in place with Desmarres chalazion forceps (see *Figure 5*). Leave the exposed kidney outside the body, protruding through the incision, for a few seconds to allow the renal capsule to dry, since a dry capsule is grasped more easily with the watchmakers forceps.

(vi) Transfer the mouse to the stage of a dissecting microscope with good incident lighting. Lift the capsule with forceps at the mid-portion of the kidney and make a 1–2 mm tear in the capsule by pulling it away from the kidney. This may or may not result in some haemorrhage. Once torn, moisten the capsule with saline to prevent the instruments from sticking to it. Insert a prong of a blunt forceps through the tear and under the capsule. By gently pushing the prong of the forceps and moving it from one side of the kidney to the other, separate the capsule from the surface of the kidney to generate a space for the embryo to be deposited. Due to the elasticity of the capsule and the turgor of the renal tissue, this space is compressed, although the forceps and the pipette should be introduced easily without much pressure. Ideally, very little bleeding should occur and the capsule should remain translucent, so that the transfer and placement of the embryo under the capsule can be seen easily.

(vii) Using the braking pipette, aspirate the embryo in approximately 50 μl of saline and position it at 3–4 mm from the tip of the transfer pipette. Make sure there is some saline both above and below the embryo.

(viii) Lift the capsule with watchmakers forceps and insert the tip of the pipette into the hole made in the renal capsule. Gently push the tip toward the pole of the kidney through the previously generated sub-capsular space. When the end has been introduced for a distance of 3−4 mm from the incision, focus on the tip of the pipette and expel the embryo and fluid by gently blowing through the mouthpiece. At this point, withdraw the transfer pipette and, if necessary, use the watchmakers forceps to hold the embryo in place. Usually the elastic capsule of the kidney snaps back and compresses the transferred embryo, fixing it firmly to the surface of the kidney. After withdrawing the transfer pipette, examine the embryo under the dissecting microscope to ensure that it is still under the renal capsule.

(ix) Gently reposition the kidney into the abdominal cavity. Suture the muscle incision (not essential) and clamp the skin with two wound clips.

(x) Allow the mouse to recover from the anaesthesia and transfer it to a properly marked cage. No additional post-operative care is needed.

Note also that reasonably clean laboratory conditions are sufficient for this technique, and none of the instruments used for embryo isolation or transfer need to be sterilized. The procedure is performed on the bench top and there is no need to use tissue culture hoods.

3. ANALYSIS OF TUMOURS

Tumours produced from embryos transplanted to extra-embryonic sites may be harvested at any time after grafting although, during the first 4−5 weeks after grafting, the tumours are small and it is impossible to distinguish teratomas from teratocarcinomas on the basis of histological examination. By 6−7 weeks, the histological differences between teratomas and teratocarcinomas become more obvious. Furthermore, the rate of histogenesis of tumours seems to vary from one mouse strain to another. We sacrifice all tumour-bearing animals at 8 weeks post-transplantation. The reasons for this are:

(i) most benign tumours lack undifferentiated embryonic cells and can be distinguished from malignant teratocarcinomas,

(ii) the nature (malignant versus benign) of most tumours can be predicted from their weight; tumours weighing less than 1 g being invariably benign and those over 2 g malignant, for most mouse strains, and

(iii) tumour-bearing mice are in good health, even with large tumour burden, and do not die from the tumour.

3.1 **Recovery of tumours**

Sacrifice the animals, remove the tumours together with the kidney and weigh the tissues. Small benign tumours are easily separated from the kidney, but larger malignant ones usually penetrate into the parenchyma with the result that it is difficult to completely remove all of the kidney tissue. To obtain quantitative data, weigh the opposite kidney and subtract that from the combined weight of the tumour-bearing kidney. Alternatively, weigh the whole animal before and after removal of the tumour-bearing kidney. Also weigh the liver and the spleen as, if tumour weight is one of the measured parameters, the weight of these organs is useful in standardization and comparison with other animals. Usually it is unnecessary to record all of these measurements.

3.2 Classification of tumours

Although weight is a good indicator of the nature of the tumour, proof that a tumour is either malignant or benign comes from histological examination of properly fixed tissues. In our experience, most tumours (95%) can be definitely classified at 8 weeks post-grafting. A few remain undetermined, especially those generated in some teratocarcinoma non-permissive strains (like C57BL/6), where some large tumours may retain some undifferentiated cells for a long time but do not appear to grow progressively and do not behave like malignant tumours.

Tumour tissue should be fixed, sectioned and stained according to standard histological procedures. For the novice, histological classification of tumours as teratomas or teratocarcinomas is best approached in collaboration with an experienced histologist who can indicate the various components, particularly foci of embryonal carcinoma, from slides stained with haematoxylin and eosin. Initially, it is useful to examine the stained sections with the unaided eye. If the tissue appears cystic (contains holes) and if most of it is stained red with eosin, the tumour is probably benign. On the other hand, if the tumour consists of predominantly solid areas and if these areas are largely stained blue with haematoxylin, the tumour is likely to be malignant. Under the microscope, the bluish areas will reveal any foci of embryonal carcinoma, which are the malignant stem cells of the tumour and the hallmark of a teratocarcinoma. Embryonal carcinoma (EC) cells form small clusters which are usually easily distinguished from the surrounding spindle-shaped connective tissue-like cells (*Figure 6*). EC cells are relatively small

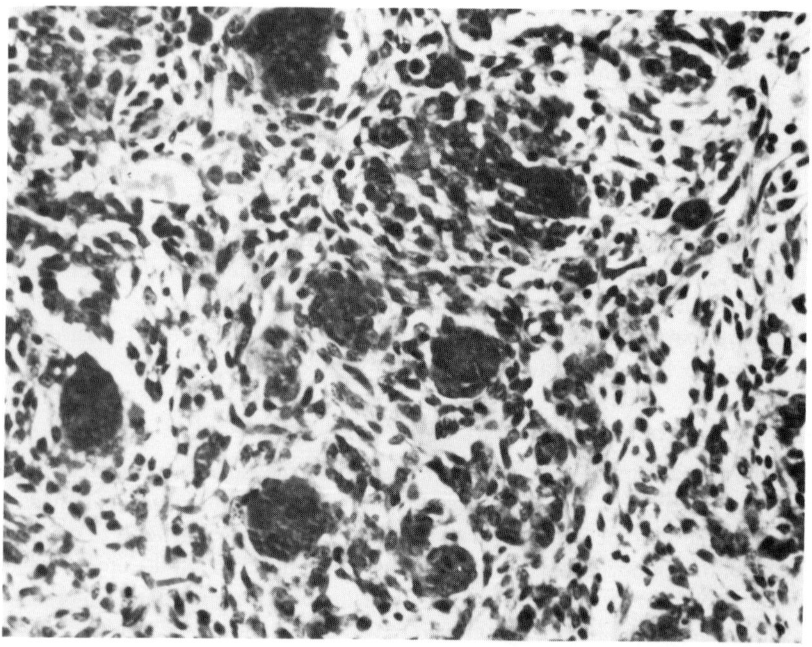

Figure 6. Clusters of embryonal carcinoma (EC) cells within the connective tissue. ×300.

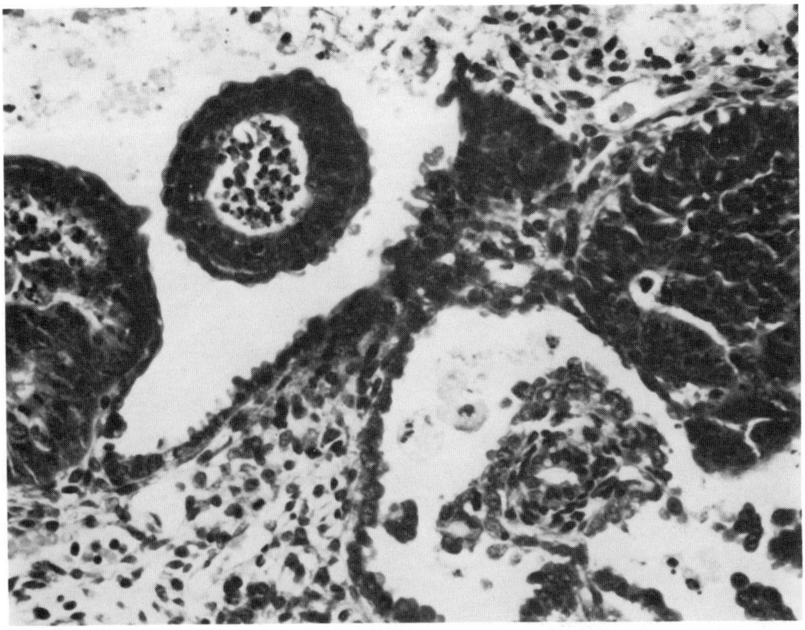

Figure 7. Embryoid bodies with re-transplantable solid teratocarcinoma. Embryoid body in the lumen of the cystic cavity (upper left) is composed of two layers presumably corresponding to ectoderm and endoderm of an egg cylinder. ×300.

because they have very little cytoplasm surrounding a well developed nucleus (*Figure 6*). The nuclei appear finely granular in contrast to the clumped chromatin of some other bluish staining cells, such as those forming neural tubes. Mitoses are easily found. Foci of EC are most easily recognized at low magnification. Examination at a higher magnification is necessary to confirm that the focus is not formed of neuroblasts or condensed connective tissue cells or foci of early squamous differentiation. EC cells usually do not take on any distinct histological structure although sometimes, especially near cavities, they may form so-called embryoid bodies (see *Figure 7*). These structures, when completely developed, resemble early post-implantation embryos and consist of an inner mass of EC cells and a distinctly different layer of cells that have been equated to endoderm. Endodermal cells have more cytoplasm, which appears pink when stained with eosin, and also nuclei polarized to one (basal) side of the cytoplasm. The apical side of the cell cytoplasm may appear scalloped.

Other than the visceral endoderm of embryoid bodies, the only non-somatic components which can be recognized readily are foci of parietal endoderm (yolk sac). Parietal endoderm cells secrete basement membrane-like material corresponding to Reichert's membrane, and this material when present in excess can be recognized easily as an extra-cellular eosinophilic amorphous, hyaline material surrounding individual cells or groups of cells (*Figure 8*). In contrast to human teratocarcinomas, mouse tumours produced from embryos rarely, if ever, contain trophoblastic cells.

In most primary embryo-derived teratocarcinomas, foci of EC cells comprise 20–40% of the entire tumour mass, while the remainder consists of somatic tissues. A wide

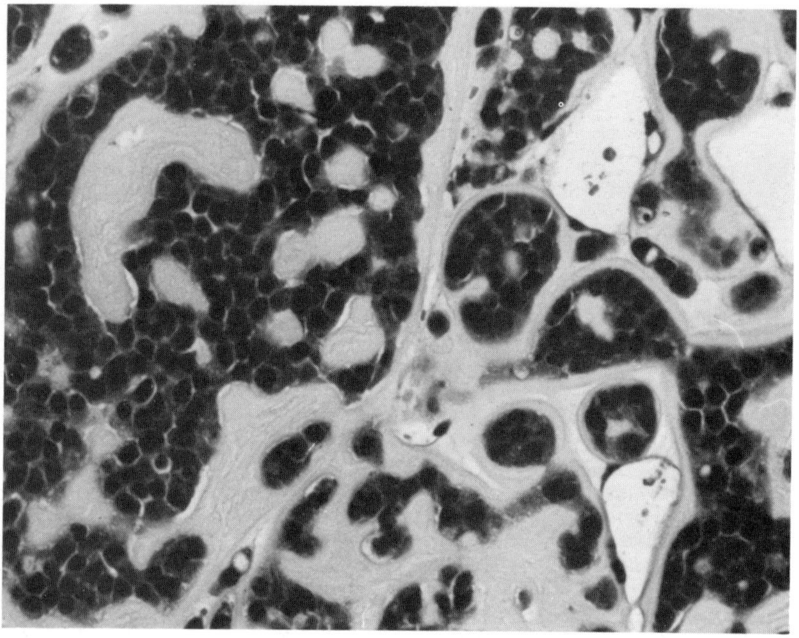

Figure 8. Parietal yolk sac elements within solid teratocarcinoma. Nests of cells are surrounded with amorphous, hyaline material corresponding to Reichert's membrane. ×300.

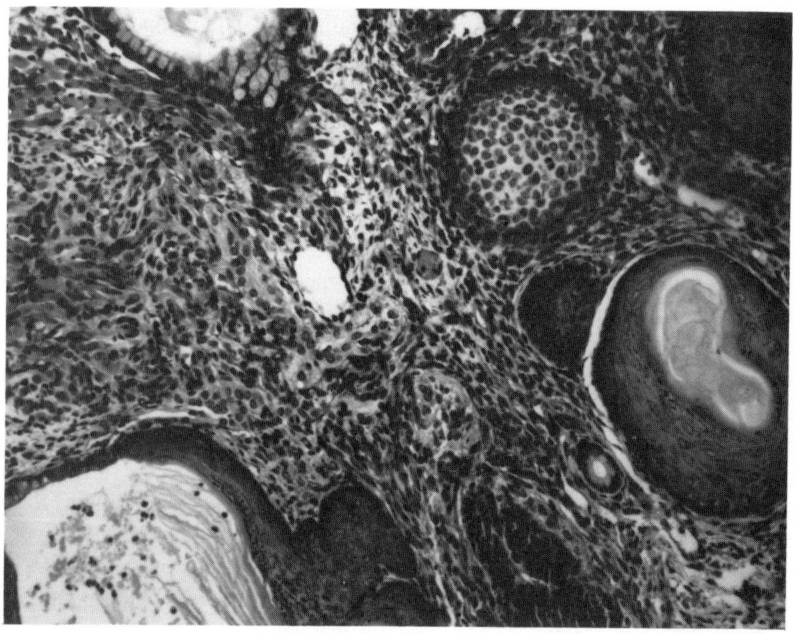

Figure 9. Examples of well differentiated tissues in re-transplantable teratocarcinoma. Notice several areas of immature cartilage (upper right), pearls of keratinizing epithelium (bottom) and glandular epithelium (upper left). ×200.

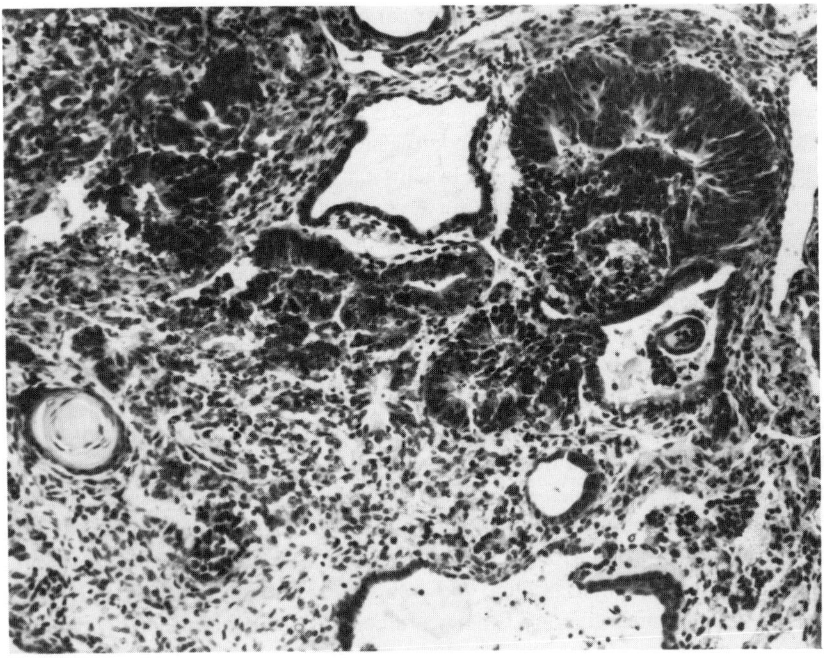

Figure 10. Neural rosettes in solid teratocarcinoma. Such structures can be mistaken for nests of EC cells. ×200.

spectrum of somatic tissues is represented in teratocarcinomas. This includes essentially all epithelia, neural tissue and connective tissues (*Figure 9*). Neural tissue is usually very prominent and consists of neural rosettes and tubes, and brain-like tissue rich in neurophil (*Figure 10*). Specialized neuroectodermal structures such as pigmented retinal epithelium may be recognized easily due to the presence of black pigment, but this is only found in tumours produced in pigmented strains of mice. Epithelial cells usually line small cystic spaces, form glands and solid nests. Epithelium may be low cuboidal, cylindrical, squamous (with or without keratinization) and may contain specialized cells such as ciliated or mucus-secreting cells. Connective tissues that can be recognized easily include smooth, striated and cardiac muscle, cartilage, bone with bone marrow, fat tissue and non-specific fibroblastic stroma. These tissues may be grouped into organ-like aggregates, but well-structured organs which are encountered frequently in human teratomas, such as teeth or skin with hair, are not a feature of mouse teratocarcinomas. Good illustrations of the various components of mouse teratocarcinomas may be found elsewhere (12,13).

4. GENERATION OF RE-TRANSPLANTABLE TUMOURS

Re-transplantable tumours are easily produced from most teratocarcinomas, harvested at 8 weeks after the initial transplantation of 7-day embryos. The tumours are minced into small fragments and pieces measuring 2–3 mm are inserted underneath the skin or into the muscle of histocompatible recipients. Cystic, bony or mucinous parts of the tumour should be avoided, and only the solid areas are used for transplantation.

The tissue can be transplanted using a troacar or using a surgical procedure. We prefer to anaesthetize the recipient, make a 0.5 cm long surgical incision over the proximal part of the hind leg and then make a receptacle in the femoral muscle with the forceps. A watchmakers forceps with closed prongs is inserted into the muscle and the muscle fibres are separated to generate a hole into which the tumour fragment can be inserted. To ensure a better take frequency, we usually transfer tumour pieces bilaterally. Following transfer, the skin is closed with a surgical clip. Tumour outgrowth can be monitored by inspection and periodic palpation. Depending on the malignancy of the primary tumour and the amount of malignant stem cells included in the graft, the re-transplantable tumours will appear within 2–4 weeks.

Re-transplantable tumours can be further passaged by periodic re-transplantation. In most instances, re-transplantation favours the growth of more malignant, i.e. faster growing, clones of stem cells. Therefore, most re-transplantable tumours grow faster than primary embryo-derived teratocarcinomas and contain more stem cells and fewer somatic tissues. With serial transplantation, it is also possible to obtain non-EC malignant stem cells which, when selectively cloned in subsequent passages, give rise to tumours that do not resemble the primary tumour of origin. Such secondary malignant tumours include neuroblastomas, various sarcomas (leiomyosarcoma, fibrosarcoma, osteosarcoma, rhabdomyosarcoma) and yolk sac carcinomas. Theoretically, and from experience with human tumours, adenocarcinomas and squamous cell carcinomas are expected as well; however, we have not so far encountered these forms of secondary malignancy in re-transplantable teratocarcinomas.

Re-transplantable tumours may also be passaged by intraperitoneal injection of finely minced fragments. This procedure usually results in the formation of multiple tumours which grow out from wherever the cells implant on the peritoneum. Occasionally the injected cells continue to grow as ascites. Ascites forms grow either as single cell suspensions or the cells may form aggregates and ‘embryoid bodies’. Embryoid bodies may be small and solid, composed of a core of EC cells and a layer of endoderm-like cells, or be larger and cystic. The latter have a central cavity filled with fluid, and the inner and outer layers are much more complex than the simple solid embryoid bodies. Complex embryoid bodies may contain foci of foetal haematopoiesis beneath the outer layer.

5. ESTABLISHMENT OF TUMOUR CELL LINES

Permanent tumour cell lines can be established from either primary embryo-derived teratocarcinomas or from re-transplantable tumours. Tumours that have been passaged several times *in vivo* (see Section 4) contain a higher proportion of malignant stem cells and are more suitable for this purpose. The success rate for isolating cell lines varies among investigators and there is no single definitive protocol yet available to optimize results. The following protocol has been successful.

(i) Take samples from the solid parts of the tumour and place the tissue fragments into a Petri dish filled with tissue culture medium (e.g. alpha medium; Flow Labs, Gibco).

(ii) Finely mince the tissue with scissors. The fragments generated can be explanted directly into tissue culture (step iii), or can be further disrupted by passing them through a sterile steel mesh. To dissociate the fragments into a single suspension, incubate in 0.25% (w/v) collagenase in Dulbecco's modified Eagle's medium (DMEM) supplemented with 5% calf serum for 1–2 h at 37°C. This incubation can be shortened by vigorously pipetting the suspension to mechanically dissociate the cells.

(iii) Explant the dissociated cells, or small tissue fragments, into one of the standard tissue culture media (e.g. alpha medium, DMEM) supplemented with 1–15% foetal bovine serum. Seed the cells into tissue culture Petri dishes or tissue culture flasks (e.g. Falcon T25) and incubate at 37°C. The use of a feeder layer prepared the previous day from irradiated (6000 rad) WI-38 fibroblasts or STO cells (14) (see Chapter 2, Section 2) facilitates the establishment of cell lines, although the resulting cell lines cannot be easily weaned and may remain feeder layer-dependent. Moreover, cell lines established in the absence of feeder layers seem to be generally more tolerant of alterations to the tissue culture environment.

The procedures for the routine maintenance of EC cell lines are described in Chapter 2. Cell lines derived from teratocarcinomas vary considerably in their biological properties. A comprehensive list of available cell lines derived from mouse teratocarcinomas is presented in ref. 15. Cell lines include: developmentally pluripotent and nullipotent EC cells; parietal, visceral and primitive endodermal (yolk sac) cell lines; and neuroblastic, fibroblastic, leiomyoblastic, osteoblastic and rhabdomyoblastic stem cell lines.

6. ANALYSIS OF FACTORS INFLUENCING THE MALIGNANCY OF TUMOURS

The factors influencing the development of malignant tumours from embryos transplanted underneath the kidney capsule have been reviewed in depth elsewhere (9,16–19). Generally, these factors fall into two categories; embryo-related and host-related. It is possible to experimentally manipulate both the embryo- and host-related factors and further define them as being genetic, metabolic, immunological, etc. Despite considerable data, the factor(s) crucial in determining malignancy of tumours derived from transplanted embryos remain(s) unidentified. The embryo- and host-related factors are discussed here only to the extent that they influence practical aspects of producing embryo-derived teratocarcinomas. Detailed discussion of these factors can be found elsewhere (11).

Table 2. Effect of prior immunization on the incidence of teratoma/teratocarcinoma following embryo transfer.

Strain of the transferred embryo[a]	*Host*	*Incidence*		*Percentage*
		T[b]	*TC*[c]	*TC*
C3H/He	C3H/He	54	49	47
C3H/He	C3H/He immunized F9[d]	9	14	61
C57BL/6	C57BL/6	69	6	8
C57BL/6	C57BL/6 immunized F9[d]	15	9	38
C57BL/6	C57BL/6 bearing F9 tumour	15	15	50

[a]Embryonic part of the 7-day-old egg cylinder was transferred under the kidney capsule.
[b]T, teratoma.
[c]TC, teratocarcinoma.
[d]Animals were immunized with six bi-weekly injections of 10^7 lethally irradiated F9 cells intraperitoneally and subcutaneously. These animals resisted the challenge with 10^6 live F9 cells, which are always tumorigenic in the non-immunized host.

Table 3. Growth of re-transplantable teratocarcinomas in syngeneic, F_1 and allogeneic recipients.

Recipient strains	*Tumours and strains from which transplanted tumours were derived*[a]			
	FA-25 AKR/J	*BMCB-9 C57BL/6J*	*MH-15 BALB/cJ*	*FC-9 C3H/He*
BALB/cJ	−[b]	−	+	−
129/J	−	−	−	−
C3H/He	−	−	−	+
AKR/J	+[c]	−	−	−
C57BL/6	−	+	−	−
C57BL/6 × BALB/c	−	+	+	−
AKR/J × C57BL/6	+	+	−	−
C3H/He × C57BL/6	−	+	−	+

[a]Small pieces of tumours were transferred into thigh muscle on both hind legs. Ten recipients were used for each tumour.
[b]No growth was observed in any recipient up to 6 months after transfer.
[c]Tumours were visible 2−3 weeks after grafting.

Table 4. Tumorigenic potential of several EC cell lines in syngeneic, F_1 and allogeneic recipients.

Recipient strain	*Cell lines and strain designation*[a]			
	PCC4 129/J	*ND-1 129/J*	*FA-25-11-1 AKR/J*	*MH-15-28 BALB/cJ*
129	+[b]	+	−	⊕
C57BL/6J	−[c]	−	−	⊕
BALB/c	−	−	⊕	+
DBA2	−	−	−	⊕
C3H	−	−	⊕	⊕
AKR	−	−	+	⊕
129 × C3H	+	+	⊕	⊕
129 × BALB/c	+	+	⊕	+
129 × C57BL/6	+	+	−	⊕
129 × AKR	+	+	+	⊕
AKR × C57BL/6	⊕[d]	⊕	+	⊕
AKR × C3H	−	−	+	⊕
AKR × BALB/c	−	−	+	+
BALB/c × C57BL/6	⊕	⊕	−	+
BALB/c × C3H	−	−	⊕	+
C3H × C57BL/6	⊕	⊕	⊕	⊕

[a]10^6 cells were injected bilaterally in the groin region. Ten mice were used for each cell line/recipient combination.
[b]Tumours were observed 2−3 weeks after injection.
[c]No tumours were present 6 months after injection.
[d]+ and ⊕ denote growth in syngeneic or semi-syngeneic recipients and in allogeneic recipients, respectively.

6.1 Embryo-related factors

The most important embryo-related factor is the age of the embryo (20). Only embryos 7 days old or less (i.e. before formation of the neural fold) are able to grow into teratocarcinomas. Pre-implantation embryos are capable of developing into teratocarcinomas, but they are more difficult to handle and the success rate is rather low in comparison with that obtained using post-implantation embryos. To date, no differences have been observed in the biological characteristics of teratocarcinomas derived from pre- versus post-implantation embryos.

6.2 **Host-related factors**

In our experience, the strain of mouse used as embryo recipient is the critical host factor (9). Although embryo-derived teratocarcinomas can be produced in many mouse strains (teratocarcinoma-permissive), a few strains, such as C57BL/6 and AKR (teratocarcinoma non-permissive), are resistant. Nevertheless, teratocarcinomas can be generated from embryos of non-permissive strains by using F_1 hosts (18).

Attempts to use immunocompromised mice as recipients of embryo grafts have failed to increase the incidence of teratocarcinomas or to convert non-permissive into permissive strains. Although most mouse tumours and tumour-derived cell lines grow much better in immunodeficient hosts, the incidence of embryo-derived teratocarcinomas is actually decreased (21). Conversely, when mice immunized with teratocarcinoma-derived cells or mice bearing re-transplantable tumours are used as hosts, the incidence of teratocarcinomas is increased (*Table 2*). Thus, it appears that a normal or specifically stimulated immune response is beneficial for development of embryo-derived teratocarcinomas.

It should also be mentioned that the outcome of grafting embryos, or pieces of re-transplantable tumours, differs from that of injecting EC cell lines derived from teratocarcinomas *in vitro*. The fact that neither EC nor cells of early embryos express major histocompatibility antigens (H-2) suggested an approach to both experimentally increase the number of teratocarcinomas and to produce tumours containing only stem cells (EC) through the use of allogeneic recipients. However, all attempts to grow embryos in histoincompatible hosts (even in congenic strains which differ only in H-2) failed. The same is true for re-transplantable tumours (*Table 3*). However, *in vitro*-derived EC cell lines are able to grow and produce tumours in practically any mouse strain, although the extent of permissiveness might vary depending on the cell line (*Table 4*). These results suggest that *in vitro* growth might result in subtle changes in tumour stem cells that affect their tumorigenic potential and ability to grow in an allogeneic host. All of these factors must be considered in comparisons of teratocarcinomas derived directly from embryos and those produced by injection of EC cell lines.

7. ACKNOWLEDGEMENTS

The original work presented in this chapter has been supported by grants CA-23097, CA-38405, HD-12487 and HD-17720 from National Institutes of Health, Bethesda, MD.

8. REFERENCES

1. Stevens,L.C. (1967) *Adv. Morphol.*, **6**, 1.
2. Damjanov,I. and Solter,D. (1974) *Curr. Top. Pathol.*, **59**, 69.
3. Solter,D. and Damjanov,I. (1979) *Methods Cancer Res.*, **18**, 277.
4. Martin,G.R. (1980) *Science*, **209**, 768.
5. Skreb,N., Svajger,A. and Levak-Svajger,B. (1971) *J. Embryol. Exp. Morphol.*, **25**, 47.
6. Stevens,L.C. (1970) *Dev. Biol.*, **21**, 364.
7. Solter,D., Skreb,N. and Damjanov,I. (1970) *Nature*, **227**, 503.

8. Damjanov,I. (1978) *J. Natl. Cancer Inst.*, **61**, 911.
9. Solter,D., Dominis,M. and Damjanov,I. (1979) *Int. J. Cancer*, **24**, 770.
10. Rugh,R. (1967) *The Mouse: Its Reproduction and Development.* Burgess Publishing Company, Minneapolis, p. 84.
11. Damjanov,I., Bagasra,O. and Solter,D. (1983) In *Teratocarcinoma Stem Cells. Cold Spring Harbor Conferences on Cell Proliferation.* Silver,L.M., Martin,G.R. and Strickland,S. (eds), Cold Spring Harbor Laboratory Press, New York, p. 501.
12. Stevens,L.C. (1958) *J. Natl. Cancer Inst.*, **20**, 1257.
13. Damjanov,I., Solter,D. and Skreb,N. (1979) In *Pathology of Tumors in Laboratory Animals. Vol. II. Tumors of the Mouse.* Turosov,V.S. (ed.), World Health Organization, I.A.R.C. Lyon, p. 655.
14. Hogan,B.L.M., Taylor,A., Kurkinen,M. and Couchman,J.R. (1982) *J. Cell Biol.*, **95**, 197.
15. Silver,L.M., Martin,G.R. and Strickland,S., eds. (1983) *Teratocarcinoma Stem Cells. Cold Spring Harbor Conference on Cell Proliferation.* Cold Spring Harbor Laboratory Press, New York, p. 689.
16. Solter,D. and Damjanov,I. (1979) *Nature*, **278**, 554.
17. Solter,D., Dominis,M. and Damjanov,I. (1980) *Int. J. Cancer*, **25**, 341.
18. Solter,D., Dominis,M. and Damjanov,I. (1981) *Int. J. Cancer*, **28**, 479.
19. Damjanov,I. and Solter,D. (1982) *Nature*, **296**, 95.
20. Damjanov,I., Solter,D. and Skreb,N. (1971) *Wilhelm Roux Arch. Entwicklungs. Mech. Org.*, **167**, 288.
21. Damjanov,I., Bagasra,O., Dominis,M. and Solter,D. (1982) *Int. J. Cancer*, **30**, 759.

CHAPTER 2

Cell culture methods and induction of differentiation of embryonal carcinoma cell lines

MICHAEL A. RUDNICKI and MICHAEL W. McBURNEY

1. INTRODUCTION

In the study of development and differentiation of higher eukaryotes, especially mammalia, the complexities involved have led to the use of tissue culture systems in order to facilitate more experimental and analytical approaches. In particular, the culture of embryonal carcinoma (EC) cells, the stem cells of teratocarcinomas, has allowed the analysis of early commitment and differentiation events analogous to those which occur in small numbers of cells in the early mammalian embryo (see *Figure 1*).

Different EC cell lines differ in their predisposition to spontaneous differentiation, the range of cell types capable of being formed and their response to chemical inducers.

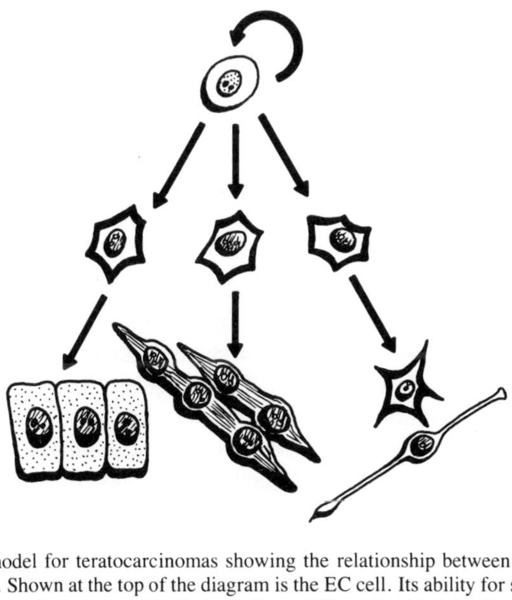

Figure 1. Stem cell model for teratocarcinomas showing the relationship between EC cells and their differentiated derivatives. Shown at the top of the diagram is the EC cell. Its ability for self renewal is indicated by the semicircular line with an arrow. Morphologically undifferentiated but developmentally restricted cells are represented by the second tier of cells and the fully differentiated derivatives of these committed precursors are represented in the bottom row of cells denoting epithelium, skeletal muscle and neurons.

Thus, each biological event may be best approached by exploiting a particular EC cell line. In this chapter, we describe the cell culture techniques which we employ for the growth and differentiation of EC cells (Section 2), we discuss some of the characteristics and potential for differentiation of the various cell lines currently in wide use (Section 3 and Section 4), and we describe some techniques for the characterization of differentiation products (Section 5).

2. CONDITIONS FOR PROLIFERATION OF EC CELLS IN CULTURE

EC cells grow rapidly, with generation times of 14–16 h under conditions normally employed for tissue culture. To maintain cultures of EC cells in their undifferentiated state they must remain in exponential growth phase. Each EC cell line has some likelihood of differentiation particularly under conditions of stress (e.g. low cell density or high cell density). Since some of the differentiated cells are not easily distinguishable morphologically from the EC cells, it is essential to clone cells periodically and to maintain them in exponential growth to ensure that each cell in the population maintains its developmental potential.

We routinely sub-culture cells at intervals of 48 h or less in order to maintain continuous exponential proliferation. Live cultures are inspected daily with the use of an inverted microscope equipped with phase contrast optics and long working distance lenses. Careful records of cell line histories are important and we recommend that cultures be frequently renewed from frozen stocks. Genetic drift may occur in cultures of EC cells, so it is important that the characteristic under study be periodically checked at regular intervals to ensure that this property is not lost from the cell population.

EC cell lines vary widely in conditions required for optimal proliferation. Some cell lines [PSA-1 (1), C145A12 (2), P10 (3)] will grow exponentially only if maintained in the presence of irradiated fibroblast cells. However, most EC cell lines will grow directly on plastic or glass surfaces and can be maintained indefinitely in the undifferentiated state under these conditions.

In general, it is desirable to avoid all potential sources of trauma in dealing with EC cell cultures. We recommend that all media, sera and saline solutions should be stored at 4°C but warmed to 37°C before cells are immersed in these solutions.

2.1 **Materials**

Work with EC cell cultures requires a tissue culture laboratory with a laminar flow hood, mechanical pipetters, sterile pipettes, incubators, low speed centrifuges, an inverted phase contrast microscope and other standard equipment. An autoclave must be used to sterilize materials and liquids. Some liquids, for example growth media, must be sterilized by filtration through a 0.2 μm pore size nitrocellulose filter. All manipulations and waste disposal should be carried out according to the recommendations of the biohazard guidelines. Mouse EC cell lines do not shed infectious virus.

2.1.1 *Growth medium containing serum*

In general, EC cells are grown in liquid medium consisting of a synthetic mixture of salts, amino acids and vitamins supplemented with bovine serum. The medium is generally buffered with bicarbonate (HCO_3^-) so the culture vessels must be incubated in an atmosphere of 5% CO_2 in air to maintain neutral pH. The media may also be buffered with Hepes if the culture dishes are to be incubated in air alone.

Most of the synthetic media used for growing EC cells are derivatives of Eagle's minimal essential medium (MEM). We routinely culture EC cells in the α-modified MEM (4). Provided the medium is supplemented with serum, most EC cells grow well in the other commonly-used synthetic media which are commercially available (e.g. Flow Labs, Gibco). The serum used to supplement the growth medium is normally bovine (newborn calf) or foetal bovine serum. The serum comprises 5–10% of the volume of the growth mixture. We supplement the medium with 7.5% (v/v) bovine serum and 2.5% (v/v) foetal bovine serum. Batches of serum vary widely in their growth-promoting characteristics and should be tested for their ability to support the clonal proliferation of the EC cell line under study (see Section 2.3 and Chapter 4, Section 2). The serum should be maintained frozen and heat-inactivated by heating to 56°C for 30 min prior to use. It may be stored at 4°C following heat inactivation. The heat inactivation removes complement activity which, along with natural heterophile antibodies, may be toxic for EC cells. The heating apparently also destroys various bovine mycoplasma which may also be present within untested batches of sera.

2.1.2 *Serum-free medium*

Some cell lines [F9 (5) and PC13 (6)] are able to proliferate in the absence of sera provided the medium is supplemented with 1 μg/ml insulin and 5 μg/ml transferrin as the only macromolecular growth factors (7,8). The growth of the cells in this medium can be improved by supplementing the medium with a source of lipoprotein such as purified high density lipoprotein (HDL). Unfortunately, the two cell lines able to grow in this defined medium also have very limited developmental potential. Exposure of most pluripotent EC cells to the same serum-free medium results in their differentiation and/or death (see Section 3.3). The culture of EC cell lines in serum-free medium is discussed at length in Chapter 7, Section 3.

With the exception of cells growing in serum-free medium, most EC lines will proliferate directly on the tissue culture plastic surfaces. In serum-free media, cells will not efficiently attach to plastic unless it has previously been coated with collagen followed by fibronectin (alternatively the medium can be supplemented with 1 μg/ml fibronectin). The coating of the dishes can be achieved by wetting the surface of a plastic dish with a solution of 0.1% (w/v) collagen dissolved in 1.0% (v/v) acetic acid followed by drying. Before use, the plates are then briefly covered with a solution of 1 μg/ml fibronectin in sterile phosphate-buffered saline (PBS, Section 2.2.1). The cells may then be grown directly on this surface without drying the plates. Certain cells, such as F9, attach poorly to plastic even in the presence of serum. The attachment properties can be improved by pre-coating the plastic surface with a solution of 0.1% (w/v) gelatin in sterile water and drying the glass or plastic surface before use.

2.1.3 *Supplements and antibiotics*

Some medium components are unstable and should be added to the medium just prior to its use. Glutamine is one such component, as are some of the antibiotics frequently used to supplement the culture medium. Streptomycin and penicillin (50 μg/ml each) are often used to control bacterial contamination. However, both antibiotics, particularly penicillin, are unstable and may be totally inactive by the time the liquid medium is used. We recommend that no antibiotic should be used to supplement the media and that antibiotic and anti-fungal agents should be used only in cultures which are irreplaceable. In our experience, gentamycin (100 μg/ml) and Fungizone (amphotericin B, 0.5 μg/ml, Flow Labs) have been the most useful agents for controlling bacterial and fungal (mould and yeast) contaminations, respectively. However, the use of Fungizone is not without effect on the growth rate of EC cell cultures and therefore should not be used prophylactically in all growth media.

2.2 **Routine culture**

2.2.1 *Removal of EC cells from plastic surfaces*

(i) To facilitate the removal of EC cells from the solid surfaces on which they grow, remove the growth medium by aspiration and wash the culture with calcium and magnesium-free PBS. PBS contains 0.8% NaCl, 0.02% KCl, 0.02% KH_2PO_4 and 0.115% Na_2HPO_4 (pH 7.4).

(ii) Flood the culture with a small volume of trypsin EDTA (T-EDTA) consisting of 1 mM EDTA and 0.025% (w/v) trypsin in PBS. Generally, 0.5 ml and 1.0 ml of this T-EDTA solution are sufficient to cover the surface of 60 mm and 100 mm diameter tissue culture dishes, respectively.

(iii) Following incubation for 10 min at 37°C the cells will have lifted from the plastic surface. Disperse the cells into a single cell suspension by vigorously pipetting up and down with the aid of a narrow bore Pasteur pipette. It is generally very difficult to obtain a good single cell suspension from most EC cell cultures. If small clumps of cells remain undispersed, incubate the culture for an additional 5–10 min and then proceed with another round of vigorous pipetting. However, note that this additional T-EDTA exposure does result in some loss of cell viability.

2.2.2 *Passaging mass EC cell cultures*

(i) Dilute the suspended cells into serum-containing medium and count with the aid of an electronic particle counter (e.g. a Coulter Counter, Coulter Electronics Inc.) or a haemocytometer.

(ii) Tissue culture plastic dishes are most convenient for culturing EC cells. The most common sizes are 35, 60 and 100 mm diameter. These dishes differ by approximately 3-fold in their surface areas and should be seeded with the cells dispersed in 2, 5 and 15 ml of medium, respectively. Cultures should be seeded with no less than 10^5 cells/ml. It is generally not necessary to wash the cells following dispersal in T-EDTA (Section 2.2.1) because there is sufficient trypsin inhibitor in the sera supplementing the medium to inhibit this enzyme and the EDTA is normally diluted to levels much lower than those of the medium divalent cations.

(iii) Following introduction of the cells into the tissue culture dish, care should be taken to disperse the cells uniformly over the surface. This can be done by gently rocking the dish back and forth in several orientations. EC cells require several hours to attach to plastic surfaces, so it is important that the dish is not incubated in a vibrating or moving condition.

(iv) Cultures seeded at 10^5 cells/ml may be incubated for 48 h before further attention. By that time, they should have undergone a 10-fold increase in number.

Large quantities of EC cells may be grown up in roller bottles, provided the bottles are seeded at very low initial rotation velocities.

2.3 **Plating efficiency**

It is frequently desirable to test the viability of a cell population following exposure to a toxin or chemical inducer of differentiation and this can be most readily achieved by measuring the culture plating efficiency. In addition, the plating efficiency test is used to test the efficacy of various batches of bovine sera for their abilities to promote growth at low cell density (see Chapter 4, Section 2).

(i) For testing plating efficiency it is essential that the culture be dispersed into a single cell suspension. Prepare this as described in Section 2.2.1.

(ii) Prepare serial dilutions of the suspension in growth medium so that 60 or 100 mm tissue culture dishes can be seeded with 100–200 viable cells. If the cell viability is unknown, one can increase the number of seeded cells by factors of three or ten. For plating experiments, the medium should be supplemented with 0.1 mM 2-mercaptoethanol. A 100 times concentrated stock solution is made by aseptically adding 7 μl of pure 2-mercaptoethanol to 10 ml of sterile medium. The stock solution should be made the day it is used.

(iii) Following the addition of the cells and 2-mercaptoethanol to the medium, carefully rotate the dishes to ensure uniform dispersal of the cells and incubate the dishes for 6–9 days without disturbance.

(iv) Following incubation, aspirate the medium and replace it with 95% ethanol containing 2% methylene blue to fix and stain the colonies.

(v) Pour the solution off after 10 min and rinse the plates carefully in water to remove the excess stain.

(vi) The attached colonies retain the blue dye and can now be counted.

2.4 **Cloning EC cells**

It is impossible to eliminate all the small clumps of two or three cells in a dispersed population of EC cells. Since the plating efficiency of a small clump is very much larger than that of a single EC cell, the conventional limiting dilution assays do not guarantee that an EC cell clone has been obtained. Two more rigorous cloning procedures are described below.

2.4.1 *Cloning by mechanically isolating single cells*

Rigorously cloned cultures can be derived by seeding individual cultures with single cells. We use either 10 μl cultures in individual wells of Terasaki plates (Falcon) or 2 ml cultures in 35 mm tissue culture dishes containing medium supplemented with 2-mercaptoethanol to a final concentration of 0.1 mM.

(i) Suspend the cells to be cloned at a concentration of approximately 10^4 cells/ml in PBS containing 10% calf serum.

(ii) Pour 5 ml of this mixture into a 100 mm diameter Petri dish (bacteriological grade) and allow 10 min for the cells to settle to the bottom.
(iii) A finely drawn glass capillary is used to pick individual cells from this culture. To prepare this hold the narrow portion of a Pasteur pipette over a flame until it is soft, then remove the pipette from the flame and pull the glass rapidly. It is important that the diameter of the capillary created is approximately 20 μm. This allows sufficient room for accommodating one cell and accurate control of the flow-rate by mouth pressure. Attach the Pasteur pipette via a length of flexible rubber tubing and a cotton wool filter to a mouth piece. Using suction fill the end of the Pasteur pipette with the PBS–serum solution. Using mouth control and a dissecting or inverted microscope, individual cells can now, with practice, be picked from the large dish and placed into the individual culture dishes. This procedure should be monitored by continuous examination under the microscope.
(iv) Confirm that a single cell is present in the culture by allowing 10–15 min before re-examining each of the cultures under the microscope.

For most EC cells the plating efficiency from this procedure is 30–75%.

2.4.2 *Isolating single cells with cloning cylinders*

The general procedure is as follows.

(i) Plate out a dispersed population of cells (Section 2.2.1) at a density of 100–200 cells per 100 mm tissue culture dish in medium containing 0.1 mM 2-mercapto-ethanol.
(ii) After 4–6 h of incubation to allow the cells to firmly attach to the plastic, examine the cultures with an inverted microscope to identify individual cells well separated from all others.
(iii) While continuously monitoring under the microscope, surround an individual cell with a cloning cylinder (consisting of a plastic or glass cylinder of about 1 cm internal diameter with silicon grease on the end) by applying the cloning cylinder (greased end down) to the surface of the culture dish. The height of the cloning cylinder must exceed the level of the medium by several millimeters, but may not be so high as to come in contact with the top of the tissue culture dish.
(iv) Return the culture dish to the incubator for an additional 8–9 days. It is a good practice to check the individual cloned cultures 3–4 days after the cloning procedure to ensure that viable colonies are developing and that only one colony is developing within each culture.
(v) After 8–9 days incubation, each colony should consist of several hundred cells. Aspirate the medium from these colonies and disperse the cells within each colony using the procedure outlined in Section 2.2.1. New pipettes must be used for the manipulation of each clone to avoid cross-contamination of clones.
(vi) Seed the dispersed clonal cultures initially in individual wells of 96-well Linbro dishes (Flow Labs) in 0.2 ml of culture medium containing 0.1 mM 2-mercapto-ethanol (Section 2.2.3).
(vii) As the cultures grow up, seed them into progressively larger dishes and omit the 2-mercaptoethanol from the culture medium.

2.5 Long-term storage of EC cells

Most tissue culture cells may be successfully recovered following freezing in the presence of protein and a cryopreservative such as dimethyl sulphoxide (DMSO) or glycerol. Successful recovery depends upon the rates of freezing and thawing. In general, freezing should be relatively slow and thawing rapid. Devices are available for controlled freezing of tissue culture cells, but the procedure outlined below requires no such apparatus and has been uniformly successful for all tissue culture cells investigated.

2.5.1 *Freezing EC cells*

(i) Disperse exponentially growing cultures of EC cells and suspend these at a concentration of $1-5 \times 10^6$ cells/ml in a mixture consisting of 90% (v/v) calf serum and 10% (v/v) DMSO.

(ii) Place 1 ml aliquots of this mixture into 1 ml screw cap plastic vials (Flow Labs, Gibco).

(iii) Place the vials into a small cardboard box at room temperature and then place the box into an ultra-low freezer operating at −75°C.

(iv) After 2−3 h, transfer the vials into an appropriate inventory control system in the ultra-low freezer or into liquid nitrogen storage tanks.

2.5.2 *Thawing EC cells*

(i) To recover the frozen cells, remove the vial from the freezer or nitrogen tank and immediately immerse it directly into water at 37°C with rapid agitation until the contents of the vial have thawed.

(ii) Transfer the contents to a 15 ml tube and slowly add 10 ml of pre-warmed (37°C) medium.

(iii) Centrifuge the diluted suspension at low speed in a clinical centrifuge (300 r.p.m., 5 min).

(iv) Aspirate the supernatant and resuspend the pellet of cells in fresh medium and plate it into a 60 mm tissue culture dish. In general, recovery of viable cells should exceed 80%.

Due to the unavoidable possibilities of power failure and/or mechanical failure of ultra-low freezers, important stocks of cells should be maintained in both ultra-low electric and liquid nitrogen freezers. The viability of cells in ultra-low freezers is excellent for several years provided the temperature is maintained below −70°C. Loss of viability does occur usually as the result of the vials warming during 'searching' of inventory boxes. We recommend, therefore, that a careful inventory book of frozen stocks be maintained and updated so that retrieval of frozen vials may be achieved without exposure of the neighbouring vials to room temperature for more than a few seconds.

2.6 Culturing feeder-dependent EC cell lines

Some EC cells die and/or differentiate spontaneously at a very high rate when cultured in the absence of other cells. These so-called feeder-dependent cell lines [for example, PSA-1 (1), C145A12 (2) and P10 (3)] must be maintained in the presence of mitotically inactivated feeder fibroblasts. The source of the fibroblasts is not critical, except that they must remain metabolically active following the treatment which eradicates their ability to proliferate. Mouse embryo fibroblasts, mouse 3T3 cells (9) and mouse STO cells (10) all make excellent feeder layers. The STO cell is useful because it maintains viability for long periods (up to 2 or 3 weeks) following treatment with irreversible anti-mitotic agents. Embryo fibroblasts and 3T3 cells often lyse within 5–7 days of drug or radiation treatment.

Cultures of the feeder cells must be maintained continuously if one wishes to maintain the feeder-dependent EC cell line. The fibroblasts are rendered incapable of proliferation by treatment with ionizing radiation, X-rays or gamma rays with a total dose greater than 50 Gy (5000 krad), or by treatment with fresh mitomycin C. Feeder cell cultures should be exposed to mitomycin C at a concentration of 10 μg/ml in medium for 2 h followed by extensive washings in drug-free medium. The mitomycin C is obtained as a powder in an injection vial (Sigma Chemical Co.). It should be dissolved in sterile water to make a 2 mg/ml solution. The mitomycin C is stored in the dark at 4°C for no more than 1 week as the drug is unstable in solution. Following drug exposure, the feeder cells can then be stored for several days at 8°C. The killed feeder fibroblasts should be seeded into dishes at 3×10^5 cells/ml or sufficient density to form an almost confluent monolayer. The function of the feeder fibroblasts is to maintain EC cells in the undifferentiated state. Conditioned medium from these feeder cells cannot be used as a substitute. This suggests that cell contact or some other mechanical event(s) between the feeder cell and the EC cell is responsible for preventing differentiation and thus allowing proliferation of the EC cell.

For many experiments, it is desirable to remove the feeder cells. This can be easily achieved because the EC cells rapidly overgrow these feeder cell populations. Therefore one or two subcultures in the absence of feeder cells is usually sufficient to eliminate these almost entirely from the population. Feeder-dependent EC cells will tolerate at least 48 h in the absence of feeder cells before significant loss of viability and/or differentiation occurs. The 3T3 cells generally lyse 4–5 days after irradiation or mitomycin C exposure and this lysis is useful for obtaining feeder-free EC populations. On the other hand, irradiated STO cells remain viable for much longer and they are, therefore, more difficult to eliminate.

Feeder-dependent EC lines can be cloned using the single cell procedure or the cloning cylinder procedure (Section 2.4.2) provided the cultures are supplemented with feeders after confirmation that a single EC cell is present within each culture. The plating efficiencies of feeder-dependent EC cell lines is generally somewhat lower than that of feeder-independent lines, usually in the range of 10–40%.

2.7 Karyotype analysis

Unlike various continuous mouse cell lines (L cells, 3T3 cells, etc.) mouse EC cell lines have stable and sometimes euploid karyotypes. It is frequently desirable to examine the chromosome integrity and the replication kinetics of various parts of these chromosomes. Because EC cells proliferate rapidly in culture, it is very easy to obtain large numbers of mitotic figures from the culture and therefore performing karyotopic analysis of these cells is straightforward (11). The techniques are described below.

2.7.1 *Mitotic arrest by treatment with Colcemid*

To obtain a growing culture of EC cells appropriate for chromosomal analysis, seed a culture at 1–2 × 10^6 cells/60 mm dish and incubate overnight. The next morning, add Colcemid (Sigma Chemical Co.) to a final concentration of 0.06 μg/ml and incubate the culture for an additional 30–60 min. Feeders do not have to be eliminated from the populations for mitotic analysis because most treated feeder cells do not attempt mitosis and those that do give rise to aberrant mitotic figures which can be easily distinguished from those of the EC cells. If one is interested only in counting the chromosome number, incubate the culture for 3–4 h in Colcemid in order to accumulate very large numbers of mitotic figures. The chromosomes in those mitotic figures will be short and unsuitable for G-banding or most other selective chromosome staining procedures.

2.7.2 *Preparing chromosome spreads*

(i) Following Colcemid exposure, wash the cultures with PBS being careful not to remove the loosely attached mitotic cells. Disperse the treated cells by incubating for 5 min in 1 ml of T-EDTA then pipetting up and down through a Pasteur pipette.
(ii) To this suspension add 10 ml of fresh 0.56% (w/v) KCl. The cells swell in this hypertonic solution. Incubate this cell suspension for 7–8 min at room temperature.
(iii) Transfer the suspension into a 15 ml centrifuge tube and pellet the cells by centrifuging in a clinical centrifuge for 5 min at 300 r.p.m.
(iv) Remove all but 100 μl of the supernatant by aspiration. Disperse the pelleted cells into the remaining supernatant by gentle agitation.
(v) Fix the cells by adding drop-wise 5–10 ml of *fresh* fixative consisting of methanol: glacial acetic acid (3:1 v/v). Allow this mixture to stand for 10 min at room temperature or overnight at 4°C.
(vi) Centrifuge at 300 r.p.m. for 5 min, then remove the supernatant by aspiration.
(vii) Resuspend the cell pellet in 10 ml of fresh fixative. Centrifuge the cells as before then resuspend in fixative such that the concentration of cells is approximately 10^6 cells/ml.
(viii) Drop 1–5 drops of the cell suspension onto a clean microscope slide. Briefly heat the slides in the heat of a 100 W light bulb to dry any remaining fixative. Make at least five slides in this way.
(ix) The slides may be stored at room temperature in a slide box for several weeks.

2.7.3 *Solid staining of chromosome spreads*

(i) Immerse the slides (prepared as described above) in a freshly made solution of 2% (v/v) Giemsa stain (Fisher Scientific Co.) in 50 mM phosphate buffer pH 7.0 for 5 min.
(ii) Rinse the slides with water and allow to air dry. The slides can now be examined directly under a microscope without the necessity of a coverslip.

2.7.4 *G-banding chromosomes*

(i) Prior to G-banding, the slides must be aged at least 3 h following their preparation.
(ii) After ageing, immerse the slides for 30–180 sec in 0.25% trypsin in 0.15 M NaCl.

(iii) Rinse the slides in 0.15 M NaCl solution (without trypsin) then stain for 3 min in fresh 2% Giemsa (v/v) in 50 mM phosphate buffer pH 7.0.

Note: In order to ensure proper G-banding, it is necessary to determine empirically the optimum times of exposure to the trypsin solution and Giemsa stain. In general, it is advisable to prepare several slides and expose them to the trypsin solution for various lengths of time to ensure that a successful preparation is obtained.

2.8 Mycoplasma

The bane of all cell culturists is mycoplasma. These are intracellular parasites whose presence is usually not obvious upon inspection of live cultures. Procedures do exist for the elimination of mycoplasma contamination from cell cultures (12) but these procedures are time consuming, not always efficient and represent a potential source of contamination. Mycoplasma contamination can arise from cross-contamination from other contaminated cultures, from the human experimentalist or from contaminated sera. Once infected, a culture should be immediately discarded unless it is irreplaceable.

A number of commercial kits are available for microbiological and enzymatic detection of mycoplasma. The simplest procedure (13) involves staining the mycoplasma DNA with the fluorescent dye Hoechst 33258 (Flow Labs).

(i) Grow the cells overnight, or longer, on coverslips then fix them with fresh methanol:glacial acetic acid (3:1, v/v) and air dry.
(ii) Stain the coverslips with 0.05 μg/ml Hoechst 33258 in PBS for 10 min.
(iii) Wash the coverslips extensively in water and mount upside down on a glass slide in 50% glycerol.
(iv) Examine the cells under a fluorescence microscope equipped with filters appropriate for u.v. excitation and blue light emission. Mycoplasma-contaminated cultures are characterized by flecks of fluorescent material in the cytoplasm. In uncontaminated cultures the nuclear fluorescence is not accompanied by any cytoplasmic fluorescence.

3. INDUCTION OF DIFFERENTIATION *IN VITRO*

EC cell differentiation consists of at least two separable events. The first is the induction of differentiation *per se*, that is the irreversible withdrawal from continuous proliferation and the loss of tumourigenicity. The second event, or series of events, are those which direct each cell into a particular developmental lineage. The EC cell culture system is unique in the variety of lineages which are possible.

All EC cell lines have an inherent predisposition to differentiate, but the extent of this predisposition varies widely between different EC cell lines. Furthermore, different EC cell lines differ in the cell types which develop following differentiation. The choice as to which EC cell line to use for any particular line of investigation must, therefore, be based upon the scientific question to be addressed. Differentiation of EC cells can be obtained in many different ways. EC cells will differentiate *in vivo* when forming a tumour (see Section 4). Differentiation *in vitro* however has proven to be much more amenable to experimental analysis because of the inherent benefits and control afforded by tissue culture. Differentiation can be produced *in vitro* by various physical manipulations of culture conditions as well as by the addition of chemicals to the media.

A number of procedures have been described in which a culture consisting predominantly of EC cells develops into a culture consisting primarily or exclusively of differentiated benign cell types. Achieving this fully differentiated status may be the result of *induction* of differentiation, *selection* in favour of the differentiated cell types pre-existing in the original population or both. Differentiated derivatives of EC cells become abundant in cultures following certain physical manipulations of the culture conditions or the addition of chemical inducers of differentiation or both. Here we review some of the methods used to produce differentiation *in vitro*.

3.1 **Spontaneous differentiation of EC cells**

3.1.1 *Long-term culture*

The rate of cell differentiation appears to be increased under conditions in which the EC cells are maintained at high cell density. This can be achieved by maintaining cells in confluent cultures without subculture for several weeks (16). Under these conditions, differentiated cells appear and undifferentiated cells become less prominent in the population probably as a result of both differentiation induction and strong selection against the EC cells. The same effect can be achieved by cloning cells on fibroblast feeder layers (11). The clones initially grow up almost exclusively as EC cells, but if the cultures are maintained undisturbed for 2–3 weeks in the presence of feeder cells, the EC cells pile up and differentiated cells become more and more abundant in the clonal population.

3.1.2 *Low density plating*

Cultures of feeder-dependent EC cells develop into differentiated cells upon removal of feeders. In an apparently similar fashion, certain feeder-dependent cell lines will also differentiate if plated at low cell density in media insufficiently rich to allow for colony formation [e.g. medium supplemented with 5% (v/v) horse serum or 2% (v/v) bovine serum]. Both of the above procedures are of relatively little utility because the efficiency of differentiation is dependent on cell density and because a great deal of cell death usually occurs in these cultures.

3.1.3 *Formation of aggregates in suspension cultures*

The most efficient and reproducible way to achieve the high cell density required to induce spontaneous differentiation is to allow the EC cells to aggregate into solid spheres of cells (*Figure 2*). EC cells stick to each other naturally and aggregates can be conveniently constructed by plating cells at 10^5 cells/ml onto siliconized glass surfaces or non-tissue culture grade plastic surfaces (1). Provided the EC cells do not stick to the plastic or glass, they will stick to each other and form small aggregates. These aggregates can be maintained suspended for prolonged periods provided the culture medium is replaced at regular 48 h intervals. Cells within the aggregates proliferate at slightly reduced rates and, depending upon the cell line, differentiated cells develop. Initially, the outside cells become morphologically and biochemically altered to resemble cells of the extra-embryonic endoderm. This developmental step occurs within 2–4 days with feeder-dependent cell lines. Feeder-independent lines develop endoderm only after incubation of the aggregates for periods in excess of 7 days.

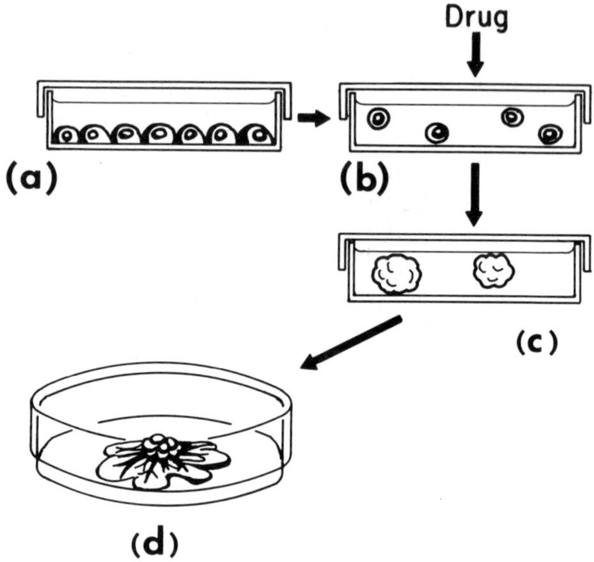

Figure 2. Differentiation is facilitated by the formation of aggregates **(a)** EC cells are dispersed then transferred into bacteriological grade Petri dishes. **(b)** The cells are unable to to attach to the surface of the Petri dish and aggregate with each other. **(c)** Chemicals such as retinoic acid, when added to aggregates, induce the differentiation of cells within the aggregate. **(d)** By transferring the aggregates to tissue culture dishes the products of differentiation can be more easily viewed because the cells migrate onto the plastic surface.

3.1.4 *Controlling aggregate size*

When allowed to aggregate spontaneously, the number of cells incorporated into each aggregate is variable but can be controlled to some extent by the cell density; the lower the cell density the smaller the average aggregate size. To more accurately control the number of cells in each aggregate, one can use the following hanging drop method (*Figure 3*).

(i) Suspend dispersed EC cells in growth medium at a concentration of 10^4 cells/ml. The cell concentration can be varied in order to achieve the desired number of cells per aggregate.
(ii) Place aliquots of 10 μl (containing ~100 cells) of this suspension onto the lids of bacteriological grade Petri dishes in regular arrays (see *Figure 3a*). Approximately 50 drops can be easily fitted on the lid of a 60 mm Petri dish.
(iii) Invert the lid and place over the bottom of a Petri dish which is filled with 5 ml of sterile water, PBS or medium (see *Figure 3b*). On average, about 100 cells will be present in each 10 μl drop. When the top is inverted, each drop hangs and the cells fall to the bottom of the drop where they aggregate into a single clump (see *Figure 3c*). Aggregation is usually complete within 24–48 h.
(vi) After 48–72 h, remove the lid, invert it and flood it with 3 ml of growth medium.
(v) Transfer the suspension of uniform aggregates into bacteriological grade Petri dishes. Add fresh medium to make the volume up to 5 ml/60 mm dish.

The number of cells aggregated into each of the drops can be controlled by varying the number of cells in the initial cell suspension. The sterile PBS, water or medium

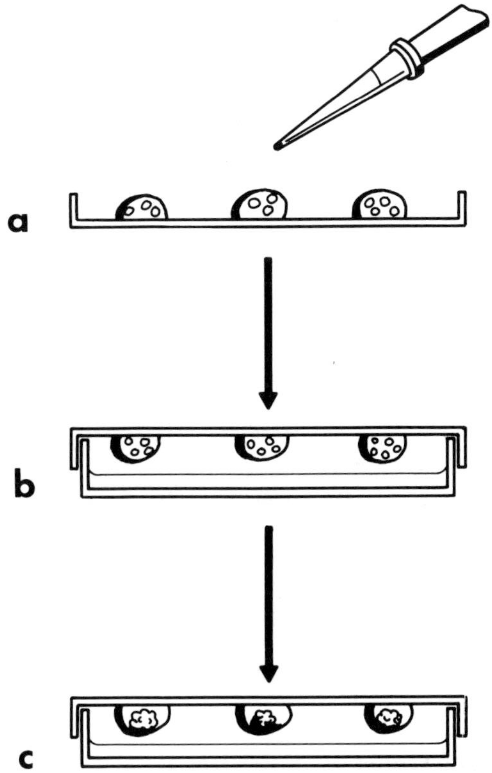

Figure 3. Aggregates composed of pre-determined numbers of cells can be produced by suspending cells in hanging drops as described in the text.

Table 1. Preparation of a methylcellulose stock solution.

1.	Dissolve 120 g of solid methylcellulose powder (Fisher Scientific Co.) into 500 ml of medium (without serum) to give a 2.4% (w/v) stock solution.
2.	Stir overnight at 4°C with a magnetic stir bar and an electric stirrer, or use a mechanical shaker.
3.	Sterilize the solution by autoclaving and store it at −20°C. Keep a working stock at 4°C.

Because it is viscous, this 2.4% methylcellulose solution must be manipulated by pouring or with a syringe. An accurate volume cannot be delivered with a pipette.

placed in the Petri dish is essential to prevent evaporation from the hanging drops. It also appears to be necessary to use bacteriological grade plastic to prevent excessive wetting of the plasting surface since otherwise the formation of excessively wide bases for each drop results in the fusion of adjacent drops.

3.1.5 *Forming aggregates of clonal origin*

Aggregates of clonal origin can also be constructed relatively easily in the following way.

(i) Add 1 ml of a single cell suspension containing 10^3-10^4 cells/ml to 1.7 ml of medium, 0.5 ml of serum and 1.7 ml of 2.4% methylcellulose in medium (see *Table 1*). The final concentration of methylcellulose is 0.8% (w/v). 0.1 mM 2-Mercaptoethanol is added to the medium to improve the cloning efficiency.

The solution should be mixed with a vortex mixer and then allowed to stand for a few minutes to allow the release of air bubbles.

(ii) With a syringe and needle, transfer the 5 ml cell suspension into a 60 mm bacteriological grade Petri dish and incubate for 7–8 days at 37°C.

(iii) The resulting colonies which develop from each viable cell are tight clusters or aggregates which can be removed individually with a Pasteur pipette. Aggregates can be removed *en masse* by dilution of the methylcellulose. Add 10–15 ml of medium to the viscous methylcellulose-containing medium and agitate to disperse the methylcellulose. Allow the aggregates to settle for 10 min, remove the supernatant by aspiration then wash the aggregates twice in PBS to remove the remnants of the methylcellulose.

(iv) Transfer the aggregates either into bacteriological grade Petri dishes to maintain as aggregates, or into tissue culture dishes to allow for attachment. It is undesirable to maintain the aggregates for periods longer than 8–9 days in methylcellulose because the medium becomes exhausted and the viscous nature of the medium does not allow for its replacement.

Methylcellulose offers advantages over agar and agarose because the former is merely viscous and can be made less viscous by dilution. In addition, it is not necessary to maintain the methylcellulose at elevated temperatures prior to plating as is necessary with agar and agarose. Frequently with agar and agarose, the cells are exposed to temperatures higher than 37°C and this may have a deleterious effect on their plating efficiencies.

3.2 **Drug-induced differentiation of EC cells**

The spontaneous differentiation which can be achieved with the procedures outlined in Section 3.1 occurs with relatively little synchrony. In addition, the cell types which develop are highly heterogeneous because, in addition to the extra-embryonic endoderm which normally surrounds each aggregate, the cells internal to the endoderm differentiate into a wide spectrum of cell types. The problems of asynchrony and, to a lesser extent, heterogeneity can be largely alleviated with the use of certain chemicals to induce differentiation, notably retinoic acid and its derivatives. The drugs are used at concentrations which are not toxic so that one is seeing the effect of differentiation induction rather than selection of the minor proportion of differentiated cells pre-existing within the population. Different cell lines respond differently to the same chemical inducers. It is, therefore, important to choose a cell line and a drug appropriate for the study of the process of interest. With some cell lines which are in wide use today, such as the F9 line, there appear to be considerable differences from laboratory to laboratory in the spectrum of cell types which result from a drug exposure.

The two classes of drugs most useful for inducing EC cell differentiation are the retinoids [the most potent being all-*trans* retinoic acid (RA) (17)] and bipolar compounds such as dimethyl sulphoxide (DMSO) and hexamethylene bis-acetamide (HMBA) (18–20). The stock solution of RA (Eastman Kodak or Sigma Chemical Co.) is made up as a 0.01 M solution in DMSO or a 1 mM solution in 95% ethanol. No sterilization of the solutions is necessary. They should be stored at 4°C and protected from light because RA is photosensitive. Stock solutions of RA have a finite shelf life; we use a solution for 4–6 weeks before replacing it. Normally, the RA is diluted to 10 μM in growth medium prior to use. The DMSO or ethanol solvents become sufficiently diluted as to have no effects on biological systems.

3.2.1 *Differentiation of F9 cells with RA*

(i) Seed F9 cells at a density of 10^5 cells/ml in medium containing 0.1 μM RA. F9 cells are usually grown on tissue culture dishes pre-coated with a solution of 0.1% gelatin.
(ii) After 2 days, remove the medium by aspiration and replace with fresh medium containing RA.
(iii) By 4 days the culture should be confluent. Disperse the cells by treatment with T-EDTA and re-seed at 3×10^5 cells/ml in RA-containing medium in fresh gelatin-coated dishes.

Within 7 days of initial exposure to RA, the cultures will have differentiated into a population consisting of cells resembling parietal extra-embryonic endoderm (17). These F9-derived parietal endoderm cells can be induced to more closely resemble parietal endoderm both morphologically and biochemically by the addition to the culture medium of 1 mM dibutyryl cAMP (21).

3.2.2 *Differentiation of F9 cell aggregates with RA*

RA-treated F9 cells have also been shown to differentiate into cells resembling visceral extra-embryonic endoderm (22). This cell type is characterized by the synthesis of alpha-foetoprotein (AFP). Visceral endoderm-like cells develop in RA-treated F9 cultures provided the cells are aggregated and exposed to low doses of RA. This can be achieved by seeding cells at 10^5 cells/ml in bacteriological grade Petri dishes in the presence of 0.01 μM RA. The drug-containing medium should be replaced at 48 h intervals and the formation of visceral endoderm-like cells around the outside of the F9 aggregates becomes maximal at about 14 days of incubation.

In addition to cells resembling extra-embryonic endoderm, RA-treated F9 cultures also contain cells which respond to nerve growth factor by differentiation into adrenergic neurons (23). It is not yet clear whether different sub-clones of F9 cells differ in their abilities to differentiate into neurons nor is it clear what the relationship is between the endodermal and neuronal derivatives of F9 cells.

3.2.3 *Differentiation of P19 cells in RA*

When cultured in the presence of 0.3 μM RA while attached to plastic surfaces, P19 cells differentiate into cells resembling fibroblasts which do not express any of the markers characteristic of derivatives of the extra-embryonic endoderm. The nature of these fibroblast-like cells is not yet clear. When aggregated and exposed to 0.3 μM RA, the P19 cells differentiate and develop large numbers of neurons and astrocytes in addition to relatively small numbers of fibroblast-like cells (24,25) (*Figures 4* and *5*).

(i) Monodisperse the P19 EC cells and seed at a density of 10^5 cells/ml in medium containing 0.3 μM RA in 60 mm bacteriological grade Petri dishes.
(ii) After 48 h transfer the aggregates into 10 ml tubes, allow the aggregates to settle, then remove the supernatant by aspiration.

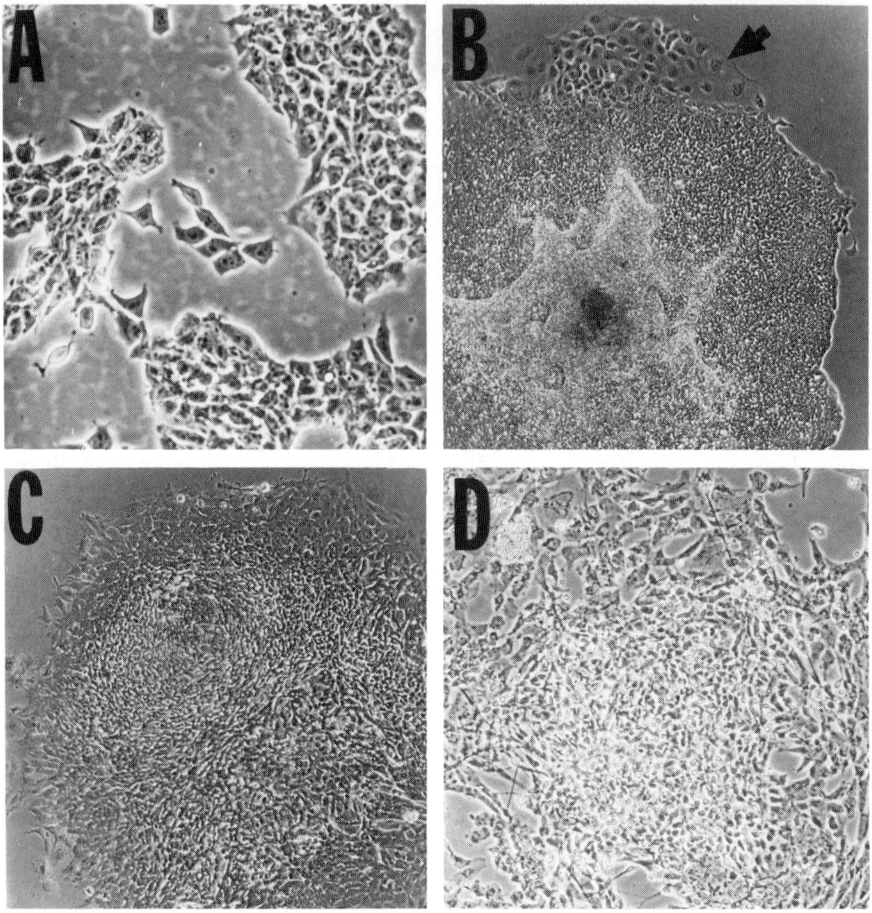

Figure 4. Phase-contrast photomicrographs of live cultures of P19 cells before and after aggregation and exposure to different inducers of differentiation. **(A)** P19 EC cells growing exponentially in the undifferentiated state. This photograph is twice the magnification of those in the other panels. **(B)** Cells were aggregated for 4 days without drug, transferred to a tissue culture surface then photographed 3 days later. Most of the cells remained undifferentiated; however a small amount of extra-embryonic endoderm was formed (arrow). **(C)** An aggregate after attachment to a tissue culture dish following treatment with 1% DMSO. These cultures contain between 10 and 15% muscle cells. **(D)** An aggregate after treatment with 0.5 μM retinoic acid. This treatment produces large numbers of neurons and astroglia cells.

(iii) Resuspend the aggregates in fresh medium containing 0.3 μM RA and transfer into 100 mm bacteriological grade Petri dishes.
(iv) After an additional 48 h, collect the aggregates as described in step (ii) and transfer between 100 and 200 aggregates into each 100 mm tissue culture grade Petri dish in fresh medium without RA.
(v) Change the medium every second day.

The aggregates attach to the tissue culture surface and the cells grow out onto the surface. Cells with long processes become evident 1 or 2 days after the aggregates are plated onto the tissue culture surface. These cells have been identified as cholinergic

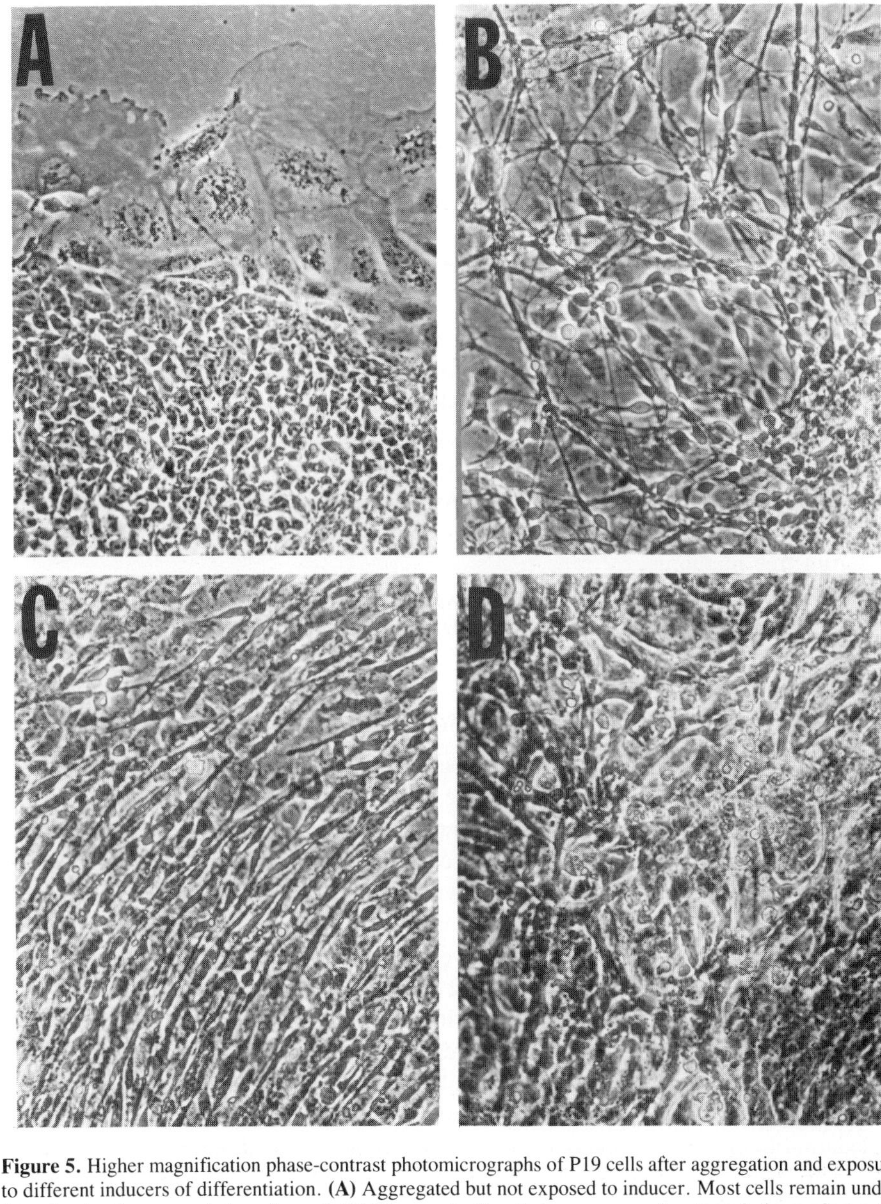

Figure 5. Higher magnification phase-contrast photomicrographs of P19 cells after aggregation and exposure to different inducers of differentiation. **(A)** Aggregated but not exposed to inducer. Most cells remain undifferentiated with some endoderm formed. **(B)** Neurons (cells with long processes) on top of a layer of astroglia produced by aggregation and exposure of the cells to 0.5 μM retinoic acid. Embryonic skeletal muscle **(C)** and cardiac muscle **(D)** produced by aggregation and exposure of P19 to 1% DMSO.

neurons. The proportion of neurons to non-neuronal cells is initially very high, but the non-neuronal cells proliferate and eventually become more numerous than the neurons. Amongst these non-neuronal cells are the precursors to astrocytes. These astrocyte precursors proliferate and spontaneously differentiate into astrocytes by 8–10 days after the initiation of the experiment. The fibroblast-like cells also proliferate, but eventually senesce following one or two subcultures.

The culture conditions can be modified to enrich for cells of interest. For example, to produce a population consisting primarily (greater than 90%) of neurons, we use the following protocol.

(i) Take the 4-day-old aggregates (step iv above) and disperse the aggregates by treatment with T-EDTA.

(ii) Plate the resulting cell suspension into tissue culture dishes in fresh medium without RA.

(iii) One to 2 days later aspirate the medium and replace it with fresh medium containing 5 μg/ml cytosine arabinoside.

This treatment kills the proliferating non-neuronal cells and a neuron-enriched population is obtained by day 8 or day 9. Alternatively, selective inhibition of the growth of non-neuronal cells can be achieved by plating dispersed aggregates into serum-free medium (Section 2.1.2). In this case, the tissue culture surface must first be coated with collagen (0.1% solution in 1% acetic acid) and fibronectin (1 μg/ml). The medium must be supplemented with 5 μg/ml transferrin and 1 μg/ml insulin. However, while the neurons survive and differentiate under these conditions, they do not survive well in serum-free medium.

P19 cells respond to aggregation and exposure to *low doses* (1–10 μM) of RA in a very different way. The spectrum of cells produced includes embryonic cardiac and skeletal muscle (26).

3.2.4 *Differentiation of P19 cells in DMSO*

Aggregates of P19 cells formed as described in Section 3.2.3 but exposed to medium containing 0.5–1.0% (v/v) DMSO for the 4-day culture in the aggregated state, differentiate into a spectrum of cells which include cardiac and skeletal muscle (20) (*Figures 4* and *5*). Cultures are initiated by plating cells into bacteriological grade Petri dishes at 10^5 cells/ml in medium supplemented with 0.5–1.0% DMSO. After 2 days incubation the medium must be replenished with fresh drug-containing medium. After 4 or 5 days in suspension the aggregates are plated onto tissue culture grade surfaces in medium lacking the drug. Medium must be replenished at 2-day intervals. By 5–6 days after the initiation of the experiment, rhythmically contracting cardiac muscle cells appear in these cultures. Several days later, bipolar skeletal myoblasts also appear. These DMSO-treated cultures appear similar to cultures derived from aggregates of feeder-dependent EC cells in that they consist of an external layer of extra-embryonic endoderm-like cells and an internal core of morphologically undifferentiated cells, some of which develop into muscle. The muscle cells comprise 10–30% of the derivatives of the DMSO-treated P19 cells. Various other cell types appear in DMSO-treated P19 cultures, but these do not include neurons or astrocytes. Most of these other cell types, due to their lack of lineage-specific markers, have not yet been identified.

3.3 **Differentiation of EC cells in serum-free medium**

Most EC cell lines do not grow continuously in a defined serum-free medium. Some EC cell lines, such as OC15S1 and 1003, differentiate when cultured in serum-free medium (27,28). Cells must be plated onto collagen and fibronectin (or laminin) coated tissue culture surfaces in order to become attached in serum-free medium (Section 2.1.2). The extent of cell death in these cultures appears to be significant, but the surviving cells consist of differentiated derivatives which may be largely neuronal as in the case of the EC cell 1003 or endodermal as in the EC line OC15S1. In at least some cases, the exposure to fibronectin-coated plastic surfaces is itself sufficient to induce differentiation even in the presence of serum. This suggests that the fibronectin may be the biologically important inducer molecule and the serum-free medium may simply serve to select against the EC cells by encouraging their death.

4. DIFFERENTIATION OF EC CELLS *IN VIVO*

EC cells are derived from teratocarcinomas and will form teratocarcinomas upon introduction into appropriate animal hosts (see Chapter 1 and refs. 14 and 15). The advantage of using EC-derived teratocarcinomas for analysis of developmental potential is that histological procedures are well developed for identifying cell types in sectioned material. A variety of staining techniques are available to be used with the standard procedure of paraffin embedding and sectioning. The identification of particular tissues can be made easily by comparison with tissues present in normal embryos. The other advantage of this methodology is that the nutritional milieu of the animals allows for the terminal differentiation of all cell types where the *in vitro* conditions do not allow for the terminal differentiation of certain cell types (for example, bone matures efficiently in tumours but not *in vitro*).

(i) Remove the EC cells to be analysed from the culture dish by treatment with T-EDTA and suspend them in medium at a concentration of 10^6-10^7 cells/ml.
(ii) With the aid of a syringe, inject 0.1 ml of the suspension subcutaneously into the flank of a histocompatible or immune-compromised mouse.
(iii) When the tumour reaches a diameter of 1–2 cm (usually 3–6 weeks), sacrifice the animal and surgically remove the tumour. The tumour may then be fixed, embedded, sectioned and stained according to standard histological procedures.

Identification of the various cell types is based upon their morphology, their staining characteristics and the staining characteristics of the extracellular material surrounding the cells of interest (see Chapter 1).

5. IDENTIFYING THE DIFFERENTIATED DERIVATIVES OF EC CELLS

Following their differentiation either *in vitro* or *in vivo*, EC cells can develop into a very large spectrum of differentiated cell types which resemble cells present in the extra-embryonic and embryonic portions of developing mouse foetuses. The cell types most frequently encountered include: parietal extra-embryonic endoderm, visceral extra-embryonic endoderm, neurons, astroglia, melanocytes, cartilage, bone (seen only *in vivo* and then frequently with marrow), muscle (skeletal, cardiac and smooth) and epithelia (squamous, columnar, cuboidal, keratinized, secretory). EC cells differentiating *in vivo* or *in vitro* normally do not develop into tissues whose normal development requires the interaction between tissues of different origin. One does not, therefore, find EC cells differentiating into liver, kidney or lymphoid organs, although the potential to develop into these and all other tissue types is apparently retained in EC cells and can be demonstrated by blastocyst injection experiments (see Chapter 5).

Studies both *in vivo* and *in vitro* strongly suggest that the developmental sequence of events which occurs in differentiating populations of EC cells resembles the analogous events occurring in the developing embryo. To identify the cell types developing from populations of EC cells, it is necessary to make use of the morphological and biochemical properties of embryonic tissues to provide the means of discriminating between the various cell types present. In deciding which lineage-specific markers are appropriate, two important considerations must be made.

(i) The cell types which develop from populations of EC cells resemble differentiated *embryonic* tissues and often do *not* express the genes whose products are characteristic of a lineage in the adult organism.

(ii) Many differentiated cell types do not have a single biochemical marker which can be used reliably to indicate the presence of that lineage. One must, therefore, use a *variety* of cellular properties to identify the presence of a particular lineage. Ideally, it should be demonstrated that the same cell type contains all of the characteristic structures or molecules under study.

In general, the identification of a particular cell type can be made using morphological and/or biochemical criteria. Conventional histological techniques can be applied to the differentiated products of EC cultures. The procedures and mode of tissue identification for tumour samples are described in Chapter 1. Cells differentiating *in vitro* can be scraped off the culture dishes, fixed and embedded for histological processing in the same way. These procedures are, however, laborious and rely upon being able to identify groups of similar cell types within a single section. This situation is not always possible in differentiated cultures.

Electron microscopy of differentiated cell types can be valuable for the identification of organelles or other intracellular and extracellular structures characteristic of particular cell lineages. Once again, the embryonic nature of the differentiated cells often limits the value of electron microscopy analysis.

A few cell types can be unequivocally identified by their morphology in live cultures viewed with phase contrast optics. Skeletal muscle has a characteristic multinuclear bipolar structure which can generally be easily distinguished from all other cell types (*Figure 5C*), including multinuclear cells which often develop in differentiated cell populations apparently by certain cells failing to undergo cytokinesis following nuclear division. Cardiac muscle cells can frequently be unequivocally identified by their rhythmic contraction (*Figure 5D*). Unfortunately, the initiation of contraction does not occur in all cardiac muscle cells and even in those cardiac muscle cells which do rhythmically contract, they do so for only 3 or 4 days before contraction ceases. Neurons formed in differentiating EC populations have small nerve cell bodies and long branching processes (*Figure 5B*). The processes can be very long indeed (up to several millimetres) and such neurons can be easily distinguished from other cell types in the population which might develop processes as a consequence of their terminal differentiation or state of ill health.

Although morphology can be frequently highly suggestive of the nature of a particular cell type, confirmation normally requires the identification in that cell type of one or more macromolecular components of restricted distribution. Many of the biochemical procedures used to identify particular macromolecules require analysis which destroys the cell population. For example, the identification of proteins by one- or two-dimensional gel electrophoresis or the identification of a particular mRNA species by Northern blot analysis or nuclease S1 mapping allows one to conclude that the gene product under investigation is present, but such analyses do not indicate the extent of differentiation into the lineage for which the macromolecule is characteristic or the state of maturation of the cells within that lineage.

The more valuable means of making use of the lineage-specific macromolecules is to use them in a way that allows the identification of positive cells without destroying the architecture of the culture. Such *in situ* identification procedures allow one to confirm suspicions based upon morphology and allow one simultaneously to compare the distribution of markers of the same or different lineages.

In situ staining procedures for identifying the presence of particular mRNA species should prove valuable in the future. At the present time, the most valuable markers for lineage identification are monospecific antibodies which are reactive with lineage-specific proteins or other macromolecules. The most useful lineage-specific proteins are those which comprise the intermediate filament network present within most, if not all, cell types (29). Monoclonal antibodies have been isolated which react specifically with each of the components of each of the five major types of intermediate filaments. These antibodies are generally not species-specific and are commercially available (e.g. Boehringer). By using antibodies reactive with the five classes of intermediate filaments, one can identify a particular cell type as being derived from one of five lineages: neurons (neuro-filaments), astroglial (glial fibrillar acidic protein), epithelial (cytokeratin), muscle (desmin) and mesoderm (vimentin) (for examples see *Figure 6*). Since EC cells and other cell types growing in cell culture express vimentin, this marker is the least valuable of the five.

Once identification of a particular lineage has been made, one can make use of markers more restricted within the identified lineage to further define the cell type under investigation. For example, the presence of desmin may indicate skeletal, cardiac or some types of smooth muscle. These different muscle cell types can be distinguished by the isoforms of the actins and myosins present within them. Such identification may take the form of immunofluorescence experiments using specific antibodies, biochemical experiments using two-dimensional gel electrophoresis or protein isolation and identification of peptide fragments.

The use of cell surface antigens as markers for different functional populations of lymphoid cells and other haematopoietic derivatives has found widespread application recently. Haematopoietic cells have only infrequently been identified within differentiating populations of EC cells and the value of cell surface antigens has been limited primarily to their use in identifying the undifferentiated EC cells themselves. A number of the lineage-specific cell surface antigens appear to be species-specific. Some plant lectins have been used as cell surface probes and these, particularly in combination with other markers, may be valuable in identifying certain lineages. Our experience, however, is that intracellular antigens appear to be more reliable lineage-specific markers than cell surface antigens.

Because of their value in identifying cell types derived from differentiating EC populations, we describe below the procedures we use for immunofluorescence using antibodies to specific cell structural components. The procedures of indirect

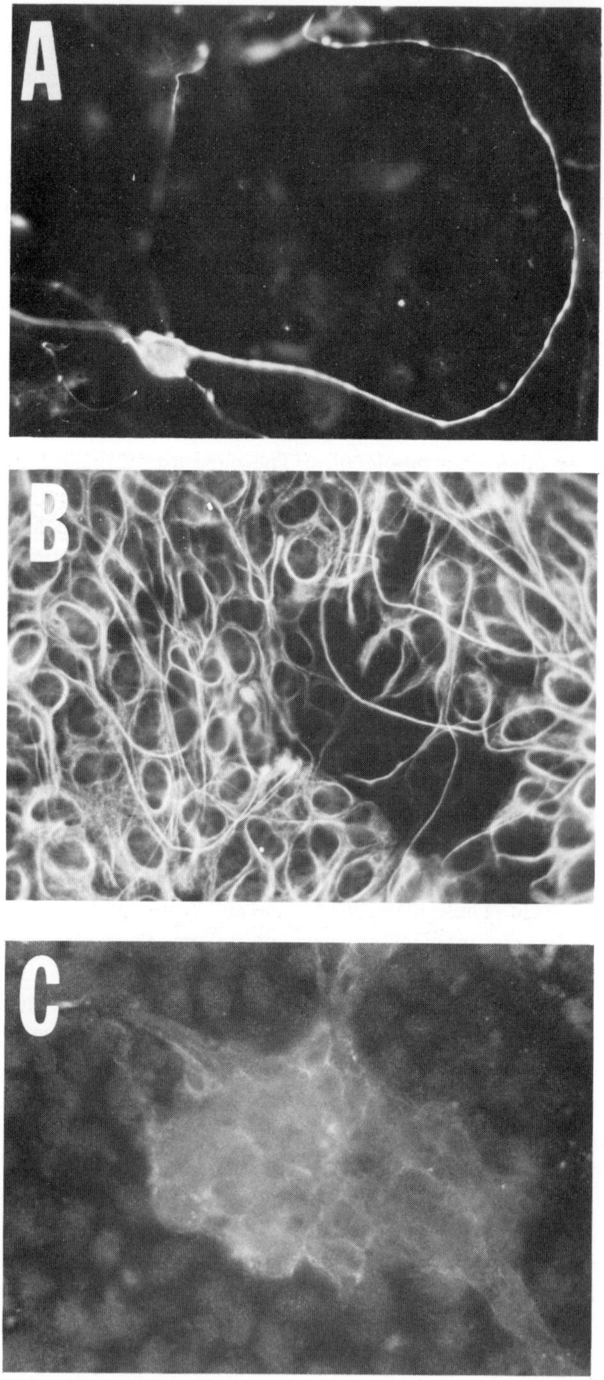
A
B
C

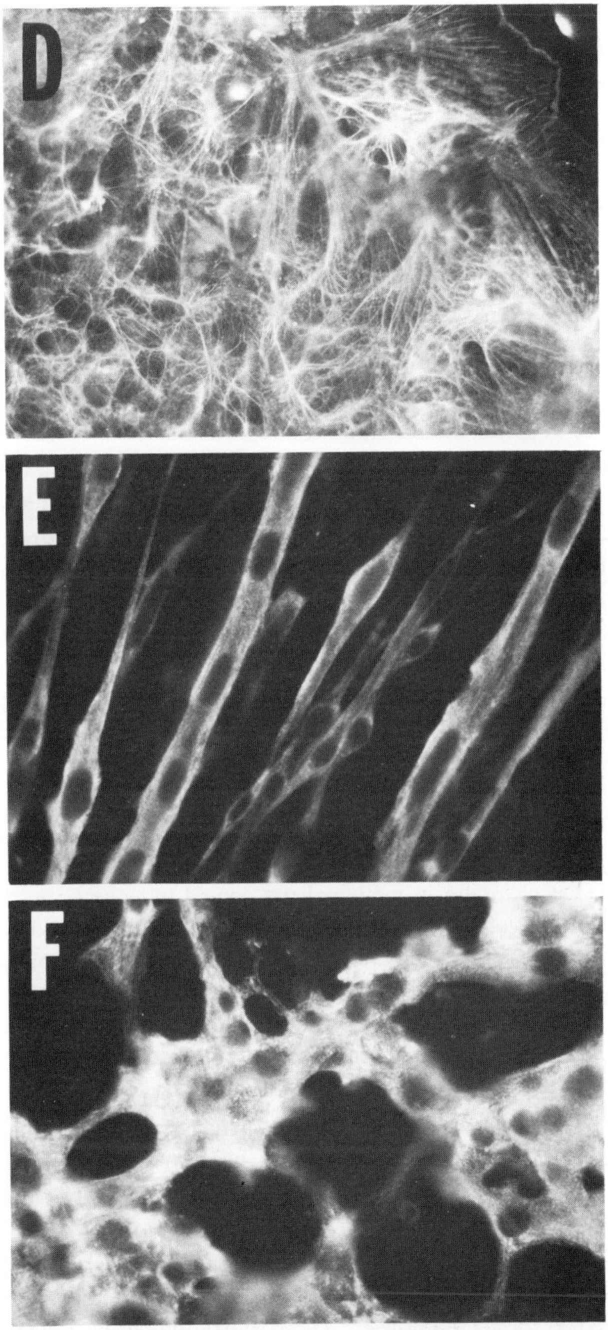

Figure 6. Differentiated cells derived from P19 EC cells identified by staining of lineage-restricted markers by indirect immunofluorescence microscopy. **(A)** Neuron stained with anti-neurofilament antisera. **(B)** Glia cells stained with anti-glial fibrillar protein antisera. **(C)** Muscle cells stained with anti-sarcomeric α-actin antisera. **(D)** Epithelial cells stained with anti-keratin antisera, Troma-I (30). Skeletal muscle **(E)** and cardiac muscle **(F)** stained with anti-muscle myosin monoclonal antibody, MF20 (31).

immunofluorescence microscopy require a light microscope equipped with epifluorescence optics with filter systems suitable for excitation and visualization of fluorescein, rhodamine or various other possible fluorochromes.

5.1 **Preparation of samples for immunofluorescence**

All of the washing, permeabilization and fixation steps can be done by immersing the coverslip in a 35 × 35 mm plastic weigh boat. Handle the coverslip with fine tweezers. Between each step of the procedure the coverslip should have the excess liquid removed by touching the edge of the coverslip to tissue paper. Unless otherwise stated, do not allow the coverslip to dry at any time throughout the procedure and carry out all manipulations at room temperature.

5.1.1 *Growth of cells for in situ staining*

Cells are grown on 22 × 22 mm coverslips placed individually into 35 mm diameter tissue culture dishes. The coverslips should be pre-coated with 0.1% gelatin in water then air dried prior to use to facilitate attachment of cells or aggregates.

Alternatively, cells may be cultured on plastic tissue culture surfaces for differentiation then dispersed with T-EDTA before being transferred onto a coverslip. The cell population is suspended in growth medium and plated onto coverslips coated with 1 μg/ml of poly-L-lysine immediately prior to use. Within 30 min the cells attach to the coverslip and are ready for fixation and staining. The dispersed populations are valuable for determining the proportion of cells in a population containing a particular antigen(s). However, cellular morphology is lost during the dispersal of the culture.

5.1.2 *Fixation method for cytoskeletal antigens*

(i) Wash the coverslip twice for 30 sec in PBS.
(ii) Incubate the coverslip in methanol pre-cooled to −20°C for 4 min.
(iii) Immerse the coverslip in acetone at −20°C for 2 min (optional). [Note: acetone destroys some antigens (notably cell surface antigens) but allows greater permeability of the antibodies into the fixed cells.]
(iv) Allow the coverslip to air dry then rehydrate the sample by incubating the coverslip in PBS for at least 15 min at room temperature.
(v) The coverslip can now be stained with antibody.

For slightly better preservation of the cell structure, one can use the following procedure provided the antigen is not destroyed by formaldehyde and glutaraldehyde treatments.

(i) Wash the coverslip twice for 30 sec in two changes of PBS at room temperature.
(ii) Immerse the coverslip in 3% paraformaldehyde (see *Table 2*) for 30 sec.
(iii) Permeabilize the cells in 1% Triton X-100 in stabilization buffer (see *Table 2*) for 60 min at room temperature. The stabilization buffer preserves cytoskeletal structures (32).
(iv) Wash the coverslip in stabilization buffer twice for 30 sec each time.

Table 2. Reagents for immunofluorescence.

Phosphate-buffered saline (PBS)

Prepare a 10× PBS stock as follows:

NaCl	76.5 g
Na_2HPO_4 (anhydrous weight)	7.25 g
KH_2PO_4	2.12 g
Distilled water	to 1 litre final volume

Adjust the pH to 7.0. Check the pH after diluting to the working concentrations.

Paraformaldehyde

Prepare fixative fresh on the day it is going to be used. To 90 ml of boiling water add 3.0 g paraformaldehyde (reagent grade) and 100 μl 1 M NaOH. Stir on a hot plate until dissolved then cool to room temperature. Add 10 ml of 10× PBS, adjust the pH to 7.0 and the volume to 100 ml.

Stabilization buffer

Prepare a 2× stock as follows:

Imidazole	3.4 g
KCl	3.73 g
$MgCl_2$-$6H_2O$	0.10 g
EGTA	0.38 g
EDTA	0.04 g
2-Mercaptoethanol	70 μl
Glycerol	368 g
Water	to 500 ml final volume

Adjust the pH to 6.7.

Sodium borohydride ($NaBH_4$)

Prepare 1.0 mg/ml $NaBH_4$ in PBS within 10 min of use. This solution is active when bubbles are present.

Glutaraldehyde

Use electron microscope grade glutaraldehyde stored in sealed vials. Open a fresh vial each week.

(v) Fix the cells in 1.0% glutaraldehyde (see *Table 2*) in stabilization buffer for 10 min.
(vi) Reduce the free aldehyde groups formed during fixation by immersing in sodium borohydride (see *Table 2*) three times for 4 min each at room temperature.
(vii) Wash the coverslip for 4 min each time in PBS.
(viii) The cells can now be stained with antibody.

5.1.3 *Procedure for the fixation of soluble intracellular antigens*

(i) Wash the coverslip twice for 30 sec each time in PBS at room temperature.
(ii) Fix in 3% paraformaldehyde (see *Table 2*) for 20 min.
(iii) Wash the coverslip twice in PBS for 4 min each time.
(iv) Permeabilize the cells by immersing the coverslips in two changes of 0.2% Triton X-100 in PBS, for 2 min each time.
(v) Immerse the coverslip in three changes of sodium borohydride (see *Table 2*) for 4 min each time.

(vi) Wash the coverslip in four changes of PBS for 4 min each time.
(vii) The cells can now be stained with antibody.

Note: the antigenic determinants of many antigens may be lost when fixed in paraformaldehyde. An alternate fixation method for such antigens is to immerse the coverslip in methanol (pre-cooled to −20°C) for 20 min, air dry, then rehydrate the cells by incubating in PBS for 15 min.

5.2 **Procedures for detecting antigens by indirect immunofluorescence**

5.2.1 *Intracellular antigens*

(i) Fix the cell preparation, either grown on or attached to a coverslip, in one of the ways described in Section 5.1.
(ii) Dilute the primary antibody (usually between 1:5 and 1:100) into 60 μl of PBS per coverslip (the optimal dilution must be determined empirically).
(iii) In a humidity chamber, cover the coverslip with diluted primary antibody and incubate for 45 min at room temperature. Note that some monoclonal antibodies, particularly those of the IgM class, are unstable at room temperature and in these cases this incubation should be carried out at 4°C instead. A humidity chamber is a small air-tight plastic box with a raised shelf where the coverslip can rest. The bottom of the chamber is filled with water to minimize evaporation of the solution over the coverslip.
(iv) Wash the coverslip three times for 5 min each time in 3 ml of PBS in small plastic weigh boats.
(v) Incubate the coverslip with either fluorescein- or rhodamine-conjugated secondary antibody directed against the primary antisera. The secondary antibody is usually diluted 1:50 (v/v) into 60 μl of PBS per coverslip. Incubate the coverslip covered in antibody for 45 min at room temperature in the humidity chamber.
(vi) Wash the coverslip three times for 5 min each time in PBS as in step (iv).
(vii) Mount the coverslip onto a slide as described in Section 5.2.7.

5.2.2 *Double staining two intracellular antigens*

This procedure requires a species or an immunoglobulin class difference between the two primary antibodies. The two secondary antibodies must be labelled with different fluorochromes and must be specifically reactive with only one of the two primary antibodies. In general, one secondary antibody is fluorescein-conjugated (FITC) and the other rhodamine-conjugated (RITC). In this way both antigens can be visualized in the same field.

(i) Prepare the sample on a coverslip in one of the ways described in Section 5.2.1.
(ii) Dilute both primary antibodies in PBS, allowing a final volume of 60 μl per coverslip.
(iii) In a humidity chamber, cover the coverslip with the primary antibodies and incubate for 45 min at room temperature.
(iv) Wash the coverslip three times for 5 min each time in PBS in a small plastic weigh boat.
(v) Incubate the coverslip with both secondary antibodies [diluted (usually 1:50) into 60 μl of PBS per coverslip] in a humidity chamber as in step (iii).

(vi) Wash the coverslip three times for 5 min each time as in step (iv).
(vii) Mount the coverslip onto a slide as described in Section 5.2.7.

5.2.3 *Procedure for staining cell surface markers*

This procedure differs from that described in Section 5.2.1 in that fixation follows the exposure of the live cells to the antibody.

(i) Wash the coverslip twice for 30 sec each time in a solution of 1:1 PBS and alpha-MEM (PBS – MEM) or Hepes-buffered medium. The buffered medium allows the live cells to remain attached to the coverslip.
(ii) Incubate the coverslip with primary antibody diluted into 60 μl of PBS – MEM per coverslip for 45 min at room temperature in a humidity chamber as described in Section 5.2.2.
(iii) Wash the coverslip three times for 5 min each time in 3 ml of PBS – MEM.
(iv) Incubate the coverslip with secondary antibody diluted into 60 μl of PBS – MEM per coverslip for 45 min at room temperature in a humidity chamber as in step (ii).
(v) Wash the coverslip three times for 5 min each time as above.
(vi) Fix the cells by immersing the coverslip in freshly-prepared methanol:acetic acid (95:5, v/v) for 15 min at −20°C.
(vii) Air-dry the coverslip then rehydrate the cells under PBS for 15 min.
(vii) Mount the coverslip onto a slide as described in Section 5.2.7.

5.2.4 *Double staining against a cell surface ligand and an intracellular antigen*

This procedure is used to stain both the outside of the cell with antibodies or lectins reactive with cell surface molecules and intracellular antigens. The following is an example describing the detection of tetanus toxin binding to neurons and concurrent staining with antibodies against neurofilament protein by indirect immunofluorescence (*Figure 7*).

(i) Rinse the coverslip with attached live cells twice in PBS – MEM for 30 sec each time then incubate in a humidity chamber under 60 μl of tetanus toxin (e.g. Connaught Research Labs), diluted 1:20 with PBS – MEM, for 30 min at room temperature.
(ii) Wash the coverslip six times for 5 min each time in 3 ml of PBS – MEM in a small plastic weigh boat.
(iii) Incubate the coverslip with 60 μl of horse anti-toxin antisera (e.g. Connaught Research Labs), diluted 1:50 in PBS – MEM, in a humidity chamber for 45 min at room temperature.
(iv) Wash the coverslip three times for 5 min each time as in step (ii).
(v) Incubate the coverslip under 60 μl of rhodamine-conjugated anti-horse IgG (e.g. Cappel Labs) for 45 min at room temperature in a humidity chamber.
(vi) Wash the coverslip three times for 5 min each time as in step (ii).
(vii) Fix the sample by immersing the coverslip under freshly-prepared methanol:acetic acid (95:5, v/v) for 15 min at −20°C.
(viii) Air-dry the coverslip then rehydrate the cells under PBS for 15 min.
(ix) Dilute the second primary antibody (mouse anti-neurofilament antisera) between

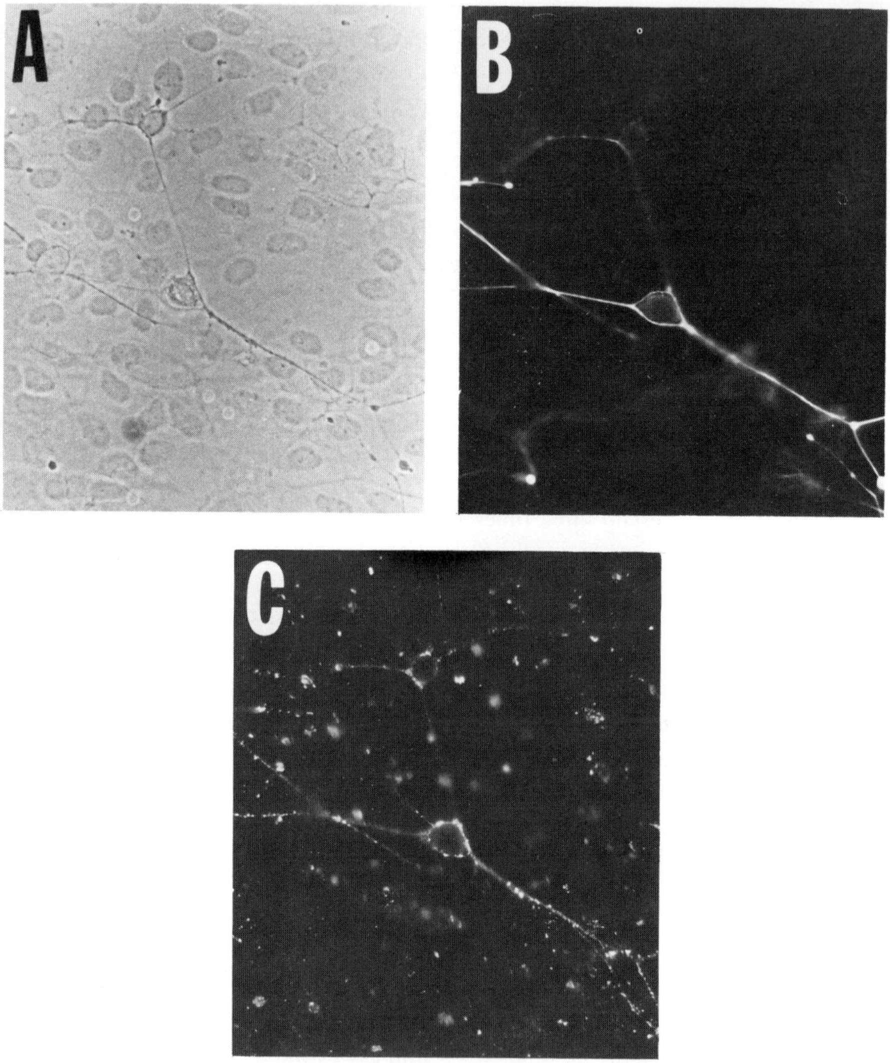

Figure 7. Double staining of neurons but not glia with tetanus toxin and anti-neurofilament antisera. **(A)** Phase-contrast photomicrograph of P19 cells following aggregation and treatment with 0.5 μM retinoic acid. **(B)** Tetanus toxin staining of the same field as photographed through a rhodamine filter. **(C)** Anti-neurofilament staining as photographed through a fluorescein filter. Both the cell surface marker, tetanus toxin receptor, and the cytoplasmic marker, neurofilament protein, are present in the cells with neuronal cell morphology.

1:5 and 1:100 into 60 μl of PBS per coverslip (optimal dilution must be determined empirically).

(x) Cover the coverslip with diluted second primary antibody and incubate for 45 min at room temperature in a humidity chamber.

(xi) Wash the coverslip three times for 5 min each time in 3 ml of PBS in a small plastic weigh boat.

(xii) Incubate the coverslip with fluorescein-conjugated secondary antibody (anti-mouse antisera). The secondary antibody is usually diluted 1:50 (v/v) into 60 μl of PBS per coverslip. Incubate the coverslip covered in antibody for 45 min at room temperature in the humidity chamber.

(xiii) Wash the coverslip three times for 5 min each time in PBS as in step (xi).

(xiv) Mount the coverslip onto a slide as described in Section 5.2.7.

Tetanus toxin is a 150 000 molecular weight holoprotein produced by *Clostridium tetani* which specifically binds neuronal gangliosides and can therefore be used as a cell surface marker of neurons in culture (33). Tetanus toxin specifically binds the morphologically distinct class of neuron-like cells in RA-treated cultures of P19 (25). The tetanus toxoid should be dialysed overnight against PBS (pH 7.4) in order to eliminate non-specific binding. The toxin must be stored at 4°C since freezing will cause degradation of the protein. As a safety precaution it is suggested that workers ensure that they are immunized against tetanus toxin.

5.2.5 *Absorption of antisera*

Many preparations of polyclonal antisera contain varying amounts of non-specific reactivity. The following is an absorption procedure to remove this activity from antisera used with methanol-fixed preparations.

(i) Disperse 10^7 cells which do *not* express the antigen of interest with 1.0 ml of T-EDTA.

(ii) Add to the cell suspension 5 ml of complete growth medium. Centrifuge at 300 r.p.m. for 5 min.

(iii) Resuspend the cell pellet in 5 ml of PBS and centrifuge as in step (ii).

(iv) Resuspend the cells in 5 ml of methanol pre-cooled to -20°C. After 5 min, centrifuge as in step (ii). Wash the cells twice more by resuspending the cells in PBS and re-centrifuging (step ii).

(v) Mix equal volumes of packed cells and concentrated antisera in a 1.5 ml microfuge tube. Incubate overnight on ice.

(vi) Pellet the cells by centrifuging for 10 min at 300 r.p.m. The antiserum contained in the supernatant fraction is then ready for use.

The cells used for antisera absorption should be fixed in the same way as the cells used for immunofluorescence. Live cells are used for absorption of antisera against cell surface antigens. Absorption of secondary antibodies sometimes will reduce background staining but usually this is not necessary.

5.2.6 *Counter-staining*

On occasions it is necessary to counter-stain cells with a flurochrome. For example when counting dispersed cells containing a particular antigen, it is useful for the negative cells to be counter-stained for easier visualization. At the end of the immunofluorescence procedures, immerse the coverslip into PBS containing 0.015 μg/ml of Hoechst 33258 for 2 min. Wash the coverslip three times for 5 min each time in PBS then mount it as described in Section 5.2.7.

Table 3. Preparation of mounting medium.

The mounting medium contains paraphenylene diamine in order to retard the bleaching of fluorescein and permit photography of the dye with minimal decay. The paraphenylene diamine needs to be recrystallized before use. Alternatively, photobleaching of both fluorescein and rhodamine protein conjugates can be simultaneously reduced by the presence in the mounting medium of 0.1–0.25 M n-propyl gallate[a] in glycerol.

Recrystallization of paraphenyl diamine[b]

1. Dissolve 3 g of paraphenylene diamine (e.g. Fisher Scientific Co. or Sigma Chemical Co.) in 50 ml of HCl at 25°C while stirring slowly.
2. Add 1 g of activated charcoal and stir for 2 h.
3. Filter through Whatman paper and cool the filtrate on ice.
4. Add 35 ml of ice-cold concentrated HCl and allow the mixture to crystallize on ice for 6 h.
5. Collect the crystals in a Buchner funnel and store them in the dark at −20°C.

Preparation of mounting medium

1. Make a 1.0% (w/v) aqueous solution with the recrystallized paraphenylene diamine. Adjust the pH to 7.0 with NaOH.
2. To prepare 10 ml of mounting medium, mix 1 ml of 1.0% paraphenylene diamine with 1 ml of 10 times concentrated PBS (*Table 2*), 3 ml of water and 5 ml of glycerol.
3. Store the mounting medium in 1 ml aliquots in the dark at −20°C. Discard the solution when it turns brown.

[a]See ref. 34.
[b]See ref. 35.

5.2.7 *Mounting the coverslip*

Place a drop of mounting medium (see *Table 3*) onto a clean slide. Gently lower the coverslip *upside down* onto the drop ensuring that no bubbles are trapped. Gently wipe away excess mounting fluid with tissue paper. Seal around the perimeter of the coverslip with clear nail polish. The slide can now be inspected. Photographs should be taken as soon as possible since the fluorochromes fade noticeably within 1 week. Store slides in the dark at 4°C.

5.2.8 *Viewing samples stained by immunofluorescence*

The stained preparations can be examined in a microscope equipped with epifluorescence optics. A combination of barrier and excitation filters appropriate for the fluorochromes used must be employed. Fluorescence objectives are available to permit high fluorescent light intensities. It is convenient to have a low-power lens (20×) for scanning and higher-power lens (40× to 63×) for detailed work. For black and white photography one can use Kodak Tri-X Pan (ASA 400) or Ilford XPI 400 film, whereas for colour photography use Kodak Ektachrome 400 or Fujichrome 400 colour slide film.

Some empirical measurements may be necessary in order to obtain the appropriate exposure times. Exposure meters are useful for phase-contrast or bright-field photomicrography. However, for photography of fluorescent samples exposure meters are of much less utility. Reasonably intense samples such as those shown in *Figures 6* and *7* can be photographed using exposure times of between 5 and 20 sec using ASA 400 film.

6. ACKNOWLEDGEMENTS

The work describing the P19 EC cell system was supported by grants from the National Cancer Institute of Canada, the Medical Research Council of Canada and the Muscular Dystrophy Association of Canada. M.A.R. holds a research studentship of the National Cancer Institute and M.W.McB. is a National Cancer Institute research associate.

7. REFERENCES

1. Martin,G.R. and Evans,M.J. (1975) *Proc. Natl. Acad. Sci. USA*, **72**, 1441.
2. McBurney,M.W. and Strutt,B. (1979) *Exp. Cell Res.*, **124**, 171.
3. McBurney,M.W. and Strutt,B.J. (1980) *Cell*, **21**, 357.
4. Stanners,C.P., Elicieri,G. and Green,H. (1971) *Nature New Biol.*, **230**, 52.
5. Bernstein,E.G., Hooper,M.L., Grandchamp,S. and Ephrussi,B. (1973) *Proc. Natl. Acad. Sci. USA*, **70**, 3899.
6. Hooper,M.L. and Slack,C. (1977) *Dev. Biol.*, **55**, 271.
7. Rizzino,A. and Crowley,C. (1980) *Proc. Natl. Acad. Sci. USA*, **77**, 457.
8. Heath,J.K. and Deller,M.J. (1983) *Exp. Cell Res.*, **104**, 255.
9. Todaro,G.J. and Green,H. (1963) *J. Cell Biol.*, **17**, 299.
10. Ware,L.M. and Axelrad,A.A. (1972) *Virology*, **50**, 339.
11. McBurney,M.W. (1976) *J. Cell Physiol.*, **89**, 441.
12. Schmidt,J. and Eufle,V. (1984) *Exp. Cell Res.*, **152**, 565.
13. Chen,T.R. (1977) *Exp. Cell Res.*, **104**, 255.
14. Stevens,L.C. (1960) *Dev. Biol.*, **2**, 285.
15. Pierce,G.B., Dixon,F.J. and Verney,E.L. (1960) *Lab. Invest.*, **9**, 583.
16. Nicholas,J.F., Dubois,P., Jakob,H., Gaillard,J. and Jacob,J. (1975) *Ann. Microbiol.*, **126A**, 3.
17. Strickland,S. and Mahdavi,V. (1978) *Cell*, **15**, 393.
18. Jakob,H., Dubois,P., Eisen,H. and Jacob,J. (1978) *C.R. Acad. Sci.*, **286**, 109.
19. Speers,W.C., Birdwell,C.R. and Dixon,F.J. (1979) *Am. J. Pathol.*, **97**, 563.
20. McBurney,M.W., Jones-Villeneuve,E.M.V., Edwards,M.K.S. and Anderson,P.J. (1982) *Nature*, **299**, 165.
21. Strickland,S., Smith,K.K. and Marotti,K.R. (1980) *Cell*, **21**, 347.
22. Hogan,B.L.M., Taylor,A. and Adamson,E.D. (1981) *Nature*, **291**, 235.
23. Wartiovaara,J., Liesi,P. and Rechardt,L. (1984) *Cell Differ.*, **15**, 125.
24. Jones-Villeneuve,E.M.V., McBurney,M.W., Rogers,K.A. and Kalnins,V.I. (1982) *J. Cell Biol.*, **97**, 253.
25. Jones-Villeneuve,E.M.V., Rudnicki,M.A., Harris,F. and McBurney,M.W. (1983) *Dev. Biol.*, **3**, 2271.
26. Edwards,M.K.S. and McBurney,M.W. (1983) *Dev. Biol.*, **98**, 187.
27. Rizzino,A. (1983) *Dev. Biol.*, **95**, 126.
28. Darmon,M.J., Bottenstein,J. and Sato,G. (1985) *Dev. Biol.*, **85**, 463.
29. Lazarides,E. (1982) *Ann. Biochem.*, **51**, 219.
30. Kemler,R., Brulet,P., Schenebelin,M.T., Gaillard,J. and Jacob,F. (1981) *J. Embryol. Exp. Morphol.*, **64**, 45.
31. Bader,D., Masaki,T. and Fischman,D.A. (1982) *J. Cell Biol.*, **95**, 763.
32. Berchadsky,A.D., Gelsand,V.I., Svitkina,T.M. and Tint,I.S. (1978) *Cell Biol. Int. Rep.*, **2**, 425.
33. Mirsky,R., Wendon,L.M.B., Black,P., Stolkin,C. and Bray,D. (1978) *Brain Res.*, **148**, 57.
34. Giloh,H. and Sedat,J.W. (1983) *Science*, **217**, 1252.
35. Johnson,G.D. and de C.N.Aaujo,G.M. (1981) *J. Immunol. Methods*, **43**,349.

CHAPTER 3

Isolation of genetic variants and fusion hybrids from embryonal carcinoma cell lines

MARTIN L.HOOPER

1. INTRODUCTION

Application of the techniques of somatic cell genetics to embryonal carcinoma (EC) or embryonic stem cell lines opens the way to two novel experimental strategies. Firstly, by studying the effects of a genetic change on the developmental capacity of these cells in culture, in tumours and in the somatic tissues of chimaeras one can obtain information about the developmental role of the altered function. Secondly, by colonizing the germ line of chimaeric mice it should be possible to produce new animal genotypes. The principal tools at the disposal of the somatic cell geneticist are the isolation of genetic variants, the production of fusion hybrids and the introduction of defined exogenous genes into cells. This chapter deals with the first two of these topics, while the third is the subject of Chapter 6. The term 'variant' is used to designate a cell line with a stably inherited change of phenotype. The term thus includes both true genetic mutants and cell lines with phenotypic changes of unknown mechanistic basis. In addition to the isolation of variants of intrinsic developmental interest, it is possible to introduce markers of general utility in somatic cell genetic manipulations, such as deficiency for the enzyme hypoxanthine-guanine phosphoribosyl transferase ($HGPRT^-$) and ouabain-resistance, without interfering with developmental capacity. The production of fusion hybrids is of interest in studying the regulation of gene expression and of pluripotency as well as in the analysis of variants by dominance and complementation studies.

As a route for the introduction of genetic markers into the mouse germ line the use of these cell lines is less direct than microinjection of DNA into the mouse egg. The advantage of the cell line approach, however, lies in the range of genetic techniques made available by the possibility of using selective isolation techniques.

2. THE DESIGN OF ISOLATION PROCEDURES

2.1 General considerations

Both variants and hybrids are initially generated as rare events in a cell population and therefore require selective techniques for their isolation. These are commonly based on selective survival, although physical separation and rapid screening techniques can also be used (reviewed in ref. 1). All the methods presented in this chapter rely on selective survival, but may be subdivided into those involving continued growth in the presence of the selective agent and those in which exposure to the selective conditions is followed by a period of recovery under non-selective conditions.

The isolation of $HGPRT^-$ variants by growth in the presence of the purine analogue 6-thioguanine (Section 4) is described as an example of a procedure involving continuous exposure. Before the advent of high-efficiency cloning techniques, selection for resistance to such toxic anti-metabolites was carried out by 'training' procedures in which a starting culture was split into a number of subcultures which were then exposed to different concentrations of the anti-metabolite. Because of the lack of a 'feeder' effect at low cell density, survival tended to be all or none. The culture exposed to the highest concentration allowing survival was then subcultured and the process repeated, and in this way by a multistep procedure it was possible to 'train' cells to grow at progressively higher concentrations of anti-metabolite. However, rather than progressively enriching for stable mutant phenotypes, this method tended to give cell lines which had obtained their resistance in a series of increments, and whose resistance was unstable in the absence of continued selective pressure. Now that high cloning efficiencies can be obtained routinely it is preferable to select for the desired degree of resistance in a single step, using a concentration of anti-metabolite which will kill all the sensitive cells and allow resistant cells to form colonies. Under these conditions one can usually assume that any survivors differ in some way from the starting material, so that characterization of the survivors will involve merely discriminating between possible mechanisms of resistance to the selective agent.

In contrast, procedures involving recovery under non-selective conditions, although they may enrich for the phenotype of interest, commonly lead to the survival of a substantial proportion of cells with unchanged properties, so that several rounds of selection are usually necessary to achieve isolation of cells with the phenotype of interest. The isolation of metabolic cooperation-defective variants (Section 6) involves a selection procedure of this type. Unless a rapid screening technique is available to allow a large number of colonies to be scored for the phenotype of interest without undue commitment of time and effort, it is usually preferable to pool survivors and screen for the presence of altered cells in this pooled population. If present at an acceptably high frequency these can then be isolated by cloning and their characterization completed. Selections of this type involve a compromise so that one obtains enrichment for the desired cells together with acceptable recovery. Where several rounds of selection are to be used it may be advantageous to use different conditions in different rounds of selection. In the first round of selection the priority is to achieve high recovery even if the enrichment factor is relatively modest, otherwise all the desired cells may be lost. In subsequent rounds a lower recovery is tolerable and one may thereby achieve a higher enrichment factor. In optimizing conditions it is useful to have available a cell line of the phenotype to be isolated so that one can perform reconstruction selections by adding a small number of these cells to the population on which the selection is to be performed.

The frequency of variants in a population may be increased by the use of physical or chemical mutagens. A number of factors influence the decision as to whether such a mutagen should be used in any particular isolation procedure. If the frequency with which a class of variant is generated spontaneously is very low, it may not be possible to isolate such a variant at all without the use of a mutagen. In other instances it may be desirable to produce variants by a particular mechanism (single-base transition, frameshift, large deletion, etc.) or the broadest possible spectrum of variants, and here the chemical specificity of mutagens may be advantageous. However, the use of a mutagen

Table 1. Mutagenesis with N-methyl-N′-nitro-N-nitrosoguanidine (MNNG).

1. Prepare a stock solution of 10 mM MNNG in water, sterilize by filtration through a membrane filter of pore size 0.2 μm, aliquot and store at −70°C. **(Caution — MNNG is carcinogenic and should be handled with appropriate precautions. Wear gloves at all times during handling and a face-mask when weighing out the solid. Destroy residues and decontaminate glassware by incubation overnight in 0.1 M HCl in a fume hood, checking that the pH remains below 5 during this procedure.)** Each aliquot should be thawed once only.
2. Set up the cells to be subjected to mutagenesis in one or more 175 cm^2 tissue culture flasks using the growth conditions appropriate to the particular cell line.
3. When the cells are in late exponential phase, replace the growth medium with growth medium lacking β-mercaptoethanol, supplemented with 50 μM MNNG from the frozen stock solution immediately prior to use. Incubate for 2 h at 37°C.
4. Remove the mutagen-containing medium.
5. Wash the cells twice with normal growth medium and incubate for 24 h at 37°C in normal growth medium.
6. Trypsinize each culture and re-seed into a flask of the same area. This speeds up the elimination of dead and dying cells from the monolayer. Incubate at 37°C and change the medium daily. The cell density will initially fall because of the killing effect of the mutagen but after 5−10 days it will regain its original density. At this point the cells are ready to be seeded out under selective conditions. It is important for the viability of the cells to be high at this stage. If there are doubts about this, due for instance to the cell density rising too high, re-seed under non-selective conditions at a split ratio of 1:4 and seed out for selection when late exponential phase is reached.

increases the risk of second site mutational events at irrelevant loci which may cloud the interpretation of the effects of the lesion of interest. This is a particularly important problem for embryonal cell lines whose developmental potency depends on the coherent functioning of a large number of genes. Certainly where isolation of a class of variant is readily achieved without the use of a mutagen, as is the case with $HGPRT^-$ variants, the use of a chemical mutagen is to be avoided. *Table 1* gives a procedure for mutagenesis with N-methyl-N′-nitro-N-nitrosoguanidine (MNNG) which is one of the most efficient mutagens currently available for use with cultured cells. Other mutagens which have been shown to be effective in tissue culture lines are discussed elsewhere (1). Although MNNG is an efficient mutagen there are drawbacks in its use. Firstly, it causes mutational events preferentially at the replication fork of DNA, commonly resulting in multiple, closely-linked mutations which may be undesirable. Secondly, it has been shown to produce shifts in the state of differentiation of cells by a mechanism not involving mutagenesis but possibly mediated by changes in DNA methylation (ref. 2 and references cited therein). Whether this also occurs with other mutagens which exert their effect by the alkylation of DNA remains to be established. The mutagenicity of MNNG is enhanced by high intracellular thiol concentrations. As it is rapidly broken down in thiol-containing media it seems likely that the product of the reaction with thiol is an intermediate in the mutagenic process. However until more is known about the stability of this intermediate and its ability to enter cells it is probably better to allow the reaction to occur within the cell rather than to add thiol to the medium during mutagenesis. The yield of mutants is a function of the expression time, i.e. the time allowed to elapse between the end of the mutagen treatment and the application of selective conditions (discussed in ref. 1). The expression time allowed in the MNNG mutagenesis procedure (*Table 1*) is governed by the important practical consideration that because selective procedures impose limitations on the initial cell density (see Section 2.2 and Section 4) one aims to avoid the presence of a large proportion of non-viable cells in the population seeded out for selection. Another important consideration is that the number of cells subjected to mutagenesis must be sufficiently large to avoid loss of mutants by sampling, which can arise as a result of the killing effect of the mutagen or as a result of inappropriate subculturing procedures. There is little point in using a mutagen to boost a mutation frequency to 1 in 10^7 if one is seeding into the selection the progeny of only 10^6 cells. This means that, in general, it is not worthwhile to make frozen stocks of mutagen-treated cells before applying selective conditions but preferable to use freshly treated cells on each occasion.

By proper design of the isolation procedure it is possible to obtain a number of independently arising variants from a single wild-type clonal line. It is important to realize that variants are generated spontaneously not at the moment that selective conditions

Table 2. Preparation of media used in selection procedures.

Culture medium (CM)

Supplement the Glasgow modification of Eagle's medium (Flow Labs, Gibco) with the following reagents (final concentrations given).

Non-essential amino acids:	
Glycine, L-alanine, L-aspartic acid, L-asparagine and L-glutamic acid	each 0.1 mM
L-Proline and L-serine	each 0.2 mM
Sodium pyruvate	1.0 mM
Serum (foetal or newborn calf, selected batches)	10% (v/v)

Culture medium β (CMβ)

Supplement CM (see above) with 0.1 mM β-mercaptoethanol.

Feeder-conditioned medium

1. Prepare feeder layers in 175 cm² tissue culture flasks from STO fibroblast cells by mitomycin treatment (see Chapter 4, Section 2).
2. Incubate each feeder layer with 30 ml of CM[a] for 7 days at 37°C (discard the feeder layer after one incubation).
3. Remove the medium from the feeder layer and pass it through a membrane filter of pore size 0.8 μm or smaller to remove cells and cell debris.
4. Supplement with 0.1 mM β-mercaptoethanol and an additional 2 mM glutamine and 10% (v/v) foetal calf serum. Use immediately.

BRL-conditioned medium

1. Seed 175 cm² tissue culture flasks with Buffalo rat liver (BRL) cells and grow to confluence in CM (see above).
2. Incubate each monolayer with 30 ml of CM[a] for 3 days at 37°C. Monolayers may be used repeatedly for up to 2 weeks.
3. Remove the medium from the monolayer and pass it through a membrane filter of pore size 0.8 μm or smaller. Store filtered medium at −20°C until required. This is designated '100% BRL-conditioned medium'.
4. Immediately prior to use mix 6 volumes of 100% BRL-conditioned medium with 4 volumes of fresh CM[a] and add β-mercaptoethanol to a final concentration of 0.1 mM. The resulting medium is designated '60% BRL-conditioned medium'.

[a]The use of foetal calf serum (selected batches) rather than newborn calf serum is recommended for optimal results.

are applied but throughout the lifetime of the cell line. The only method which ensures independent origin of variants is to subclone the starting line and isolate a single variant from each. An alternative approach which gives a high probability (but not certainty) of independent origin is to use a physical or chemical mutagen which appreciably increases the yield of variants over that obtained spontaneously, and, after mutagenesis for less than one cell generation, to divide the culture into subcultures, and isolate a single variant from each subculture (or, alternatively, to mutagenize several cultures in parallel and isolate a single variant from each). It is a more straightforward matter to guarantee independence of origin of fusion hybrids; it suffices merely to set up a series of parallel co-cultures, expose each separately to a chemical fusogen such as polyethylene glycol, and isolate a single hybrid from each.

A further important variable is the size of the population subjected to selection. $HGPRT^-$ variants arise at high frequency because, being X-linked, the gene is hemizygous or functionally so. A typical mutation frequency even in a culture not treated with a chemical mutagen is of the order of 1 in 10^6 so that an initial population of 1 or 2×10^7 suffices. This is readily achieved with standard tissue culture flasks or dishes. Autosomal recessive lesions, however, arise at a much lower frequency and scaling-up the initial population to 10^9 or even 10^{10} cells may be desirable. Because of limitations on initial cell density imposed by selection techniques this becomes an unwieldy procedure if standard tissue culture plasticware is used. The use of microcarrier culture techniques (3), which allow the achievement of very high surface areas within a manageable culture volume, is then worthy of consideration.

Because selective techniques require the growth of rare survivors into colonies, it is important to carry out selections under conditions where, but for the selective agent, a high cloning efficiency would be achieved. Recipes for the preparation of tissue culture media used in selective procedures are given in *Table 2*. Supplementation of media with non-essential amino acids and pyruvate (*Table 2*) increases the cloning efficiency in most cell types since, although cells are capable of synthesizing these compounds, they are readily lost to the external medium if the concentration gradient across the cell membrane is high.

2.2 Embryonal carcinoma (EC) and embryonic stem cell lines

Many EC and embryonic stem cell lines require the presence of a feeder layer of mitomycin C-treated fibroblast cells for growth as an undifferentiated population (Chapter 2, Section 2, and Chapter 4, Section 2). This complicates the design of selection procedures for a number of reasons. Firstly, many toxic compounds used as selective agents cause feeder cells to detach, resulting in the loss of potentially resistant but feeder-dependent cells. Secondly, loss of potentially resistant cells may arise because toxic metabolites are transferred to them by metabolic cooperation either from feeder cells or from wild-type cells via the feeder cells. An example of this occurs in selection for thioguanine resistance. In feeder-independent lines, $HGPRT^-$ variants, which in pure culture are resistant to 6-thioguanine, become sensitive if they are co-cultured with wild-type cells at high density. This is a result of transfer of the metabolite 6-thio-GMP from the wild-type cells via gap junctions (4). This leads to a loss of potentially resistant variants if selection is carried out at high cell density. This problem can be overcome in the case of feeder-independent cell lines by reducing the cell density below that where appreciable cell contact occurs. However in feeder-dependent lines, wild-type and variant cells, even if themselves seeded at low cell density, remain in indirect contact via bridging feeder cells. Thus although the feeder can be prepared from STO cells, which are themselves $HGPRT^-$, transfer of thio-GMP from wild-type cells causes both detachment of the feeder cells and killing of variant cells. Conversely, metabolic cooperation can lead to survival of cells under conditions where they would normally be killed, thereby rendering a selective system useless. For instance, ouabain-sensitive EC cells are rescued from ouabain toxicity by virtue of the passage of sodium ions through gap junctions formed with ouabain-resistant STO feeder cells (4).

We have sought ways of overcoming these problems and a number of different approaches have been fruitful. In the case of selection for $HGPRT^-$ variants, the problem can be reduced in magnitude by an appropriate choice of feeder cell and EC cell densities so that feeder cells do not detach and recovery of variants is still possible (see Section 4). In the case of selection for metabolic cooperation-defective cell lines, the problem of feeder detachment can be overcome by repeated re-seeding of feeders, while loss of EC cells by virtue of metabolic cooperation is a desired component of the selection system. More generally applicable solutions have, however, become available with the demonstration first that the effect of feeder layers can be substituted by feeder cell-conditioned medium provided that β-mercaptoethanol is present (5) and recently that medium conditioned by Buffalo rat liver (BRL) cells functions similarly to feeder cell-conditioned medium but gives superior results (6). While growth in feeder cell-conditioned medium is rather variable and usually accompanied by some differentiation, reproducible growth without differentiation is readily achieved in medium conditioned by BRL cells, provided that β-mercaptoethanol is present and foetal calf serum is used (newborn calf serum gives inferior results). Procedures for the preparation of these conditioned media are given in *Table 2*. Both are used in conjunction with gelatin-coated culture surfaces (see Chapter 4, Section 2). Large volumes of BRL-conditioned medium are conveniently prepared using microcarrier culture techniques. We have successfully used feeder cell-conditioned medium to allow isolation of hybrids between feeder-dependent and feeder-independent cell lines (Section 7).

An entirely separate consideration which influences the design of selection procedures involving EC cells is that the overall capacity to differentiate tends to decrease with increase in passage number. This effect is probably due to selection for faster-growing variants, resulting in the gradual accumulation of genetically altered cells in the population. It should come as no surprise that developmental capacity, which depends upon the coordinated functioning of a large number of genes, is more unstable in the face of such changes than, for instance, the activity of an enzyme. Therefore if it is desired either to introduce a marker mutation without change in developmental capacity or to examine the effect of a mutation on developmental capacity, it is important that selection procedures should be so designed as to be completed within the minimum possible number of cell multiplications.

The procedures detailed in this chapter are those which we have successfully used to isolate variant and hybrid cell lines. As a result of our more recent work it seems likely that the use of 60% BRL-conditioned medium to substitute for feeder layers may give better results than the already established methodology. However, as this has yet to be formally established the original methods have been detailed and, at the appropriate points in the text, possible improvements are indicated.

3. THE CHARACTERIZATION OF SELECTED CELLS

It cannot be overemphasized that no selective system specifically yields the desired variant or hybrid and that the characterization of the products of selection is equally as important as the design of the selective system. Where selection is not applied continuously the surviving population usually contains cells of unaltered phenotype and a screening procedure is necessary to identify populations and clones worthy of more detailed characterization. An example of such a screening procedure is the ouabain rescue assay for metabolic cooperation described in Section 6. Once it is clear that the products of selection differ in some way from the starting cells, a complete characterization is required in order to pin down the exact nature of the lesion. This may involve the application of a number of different techniques and is frequently the most time-consuming part of an isolation procedure. These technqiues are, however, unique to each individual class of variant. Rather than detailed protocols for the steps involved in characterizing the selection products described in the following sections, which are chosen to illustrate selective techniques of broader applicability, an outline of desirable characterization strategies is provided with references to further reading where appropriate.

4. ISOLATION OF HGPRT$^-$ VARIANTS

Variants defective in HGPRT are selected on the basis of their resistance to the toxic analogue 6-thioguanine. Since this enzyme is coded by a gene on the X-chromosome, variants arise spontaneously at high frequency and mutagenesis is unnecessary for their isolation.

A procedure for the isolation of HGPRT$^-$ variants from feeder-dependent cell lines is given in *Table 3*. In this protocol, colonies are picked from the plates using drawn-out Pasteur pipettes. Preparation and use of these pipettes is as follows.

(i) Heat a sterile plugged Pasteur pipette in a small Bunsen flame (the pilot light of a 'Touch-o-Matic' burner is ideal) at a point close to the taper, rotating the pipette in the flame to ensure uniform heating (*Figure 1a*).

(ii) When the tubing in the flame reaches red-heat, remove from the flame and, with the indicated motion (*Figure 1b*), draw out the tubing and introduce a bend into it. It is not necessary to draw out the tubing very finely; a diameter of around 0.5 mm will be found quite satisfactory.

(iii) With sterile forceps break off the pipette at the point shown in *Figure 1c* to give an approximately square end. Uneven breaks make it difficult to hold a drop of medium at the tip in the following operations, but can usually be rectified by breaking away glass from the tip with the forceps. It is not necessary to flame-polish the tip.

(iv) Fit a rubber bulb (*Figure 1d*). To operate the pipette, hold the shaft of the pipette in the fingers and use the thumb to operate the rubber bulb. With a little practice fine control of both the movement of the tip and the liquid level in the pipette can be achieved.

(v) To pick up individual colonies, remove most of the medium from the dish leaving

Table 3. Isolation of HGPRT$^-$ variants.

1. Prepare a sterile stock solution of 10 mg/ml of 6-thioguanine as follows: weigh out 50 mg of 6-thioguanine (2-amino-6-mercaptopurine), add 4 ml of water and then 10 M NaOH dropwise with constant stirring until solution is complete (small amounts of impurity may remain undissolved at the endpoint). Make up to 5 ml with water and sterilize by filtration through a membrane filter of pore size 0.2 μm. Store at -20°C in 1 ml aliquots. Each aliquot may be thawed and re-frozen four or five times but should then be discarded.
2. On day 1, trypsinize a 25 cm^2 culture of EC cells or embryonic stem cells and, after neutralization of the trypsin, dissociate the culture, as far as possible by pipetting alone, into a *single* cell suspension. Inoculate the entire suspension on to a 175 cm^2 feeder layer in 50 ml CMβ (see *Table 2*).
3. On day 2
   (i) Change the medium of the above culture, replacing with 50 ml of fresh CMβ;
   (ii) Prepare feeder layers in twenty 100 mm tissue culture dishes using mitomycin-treated STO fibroblast cells (as described in Chapter 4, Section 4). Seed out 3×10^6 feeder cells per dish in 10 ml of CM.
4. On day 3
   (i) Trypsinize the 175 cm^2 flask and, after neutralization of the trypsin, prepare a suspension of single cells by pipetting. If cell clumps are difficult to dissociate, pass the suspension twice through a 23-gauge needle[a].
   (ii) Count the suspension and dilute 8.4×10^6 cells to a final volume of 210 ml in CMβ.
   (iii) Remove the medium from the feeder layers prepared on day 2 and add 10 ml of the diluted suspension just prepared (i.e. 4×10^5 cells) to each dish. Discard the remaining suspension.
5. On day 4, replace the medium in all dishes with CMβ supplemented with 10 μg/ml of 6-thioguanine from sterile stock solution.
6. Replace the medium in all dishes twice weekly with CMβ + 10 μg/ml of 6-thioguanine, freshly prepared from sterile stock solution. Killing of wild-type cells should be complete within 1 week, but surviving colonies are usually not visible for at least 2 weeks. Continue to replace medium with fresh selective medium until colonies are easily visible with the naked eye (2$-$3 mm in diameter). This stage is normally reached at between 2 and 3 weeks. Do not allow colonies to become larger, as otherwise substantial cell death occurs in the centre of the colonies.
7. Prepare feeder layers in 24-well multi-dishes (well diameter 17 mm), seeding 10^5 feeder cells per well. Prepare feeder layers between 1 and 7 days prior to picking colonies from selection plates. Pick the colonies using drawn-out Pasteur pipettes as described in Section 4. Seed each colony into a separate feeder well in fresh CMβ (without 6-thioguanine).
8. Monitor growth in the wells and passage by trypsinization when required (this will be dictated initially by the size of the clumps and the need to avoid cell death in the centre). Expand each culture in stages to a near-confluent 175 cm^2 flask for preparation of frozen stocks and further characterization.

[a] This step should be avoided unless absolutely necessary since it does result in some loss in viability, and is usually not necessary where the cells have been subcultured only 2 days previously.

Table 4. Characterization of HGPRT$^-$ variants.

1. Test for stability of resistance to 6-thioguanine by seeding the cells into various concentrations of 6-thioguanine after growth under non-selective conditions.
2. Test for incorporation of [^{3}H]hypoxanthine into trichloroacetic acid-precipitable material by intact cells. HGPRT$^-$ variants fail to incorporate.
3. Test for growth in HAT medium (medium supplemented with 0.1 mM hypoxanthine, 0.8 μM aminopterin, 20 μM thymidine; see Section 7). HGPRT$^-$ variants die in this medium.
4. Assay HGPRT enzyme in cell extracts[a].

[a]For procedure see ref. 7.

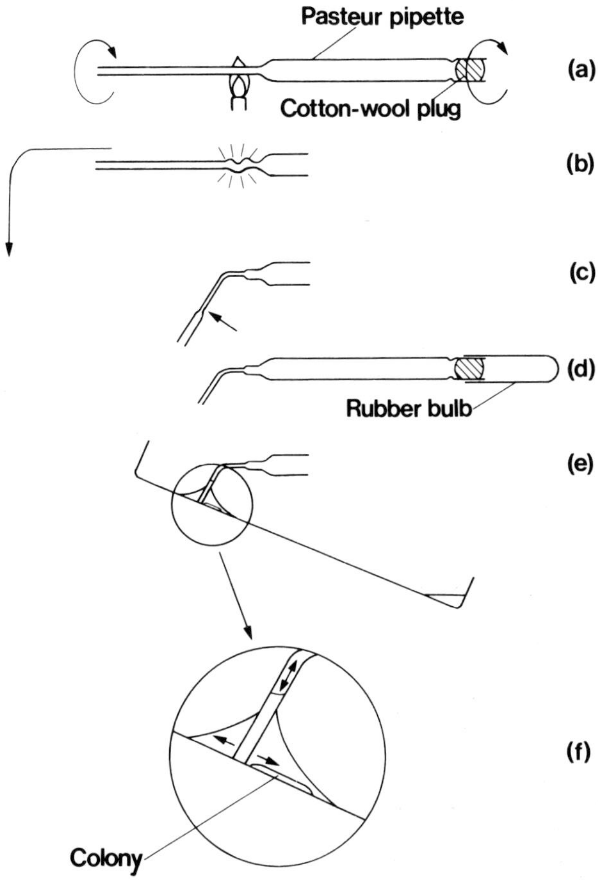

Figure 1. Preparation and use of pipettes for colony picking. Details of the procedures are described in Section 4 of the text.

only about 1 ml to cover the cells. Take up enough fresh CMβ (*Table 2*) into the picking pipette to fill the drawn-out section. Tilt the dish so that the colony of interest is drained of medium. It is important to be able to see the colony easily at this stage and therefore viewing the dish against a dark background may be helpful. Place the tip of the pipette near to the colony and expel a small volume of medium sufficient to cover the colony (*Figure 1e*).

(vi) Scratch across the colony so as to dislodge it, while at the same time taking the medium in and out of the pipette using the rubber bulb (*Figure 1f*). With a little practice the colony can be broken up into several large clumps of cells and these clumps taken up in the pipette. Do not attempt to break up the colony into single cells at this stage.

Putative HGPRT$^-$ variants are verified according to procedures described in *Table 4*. An important consideration in the design of selection procedures is to avoid loss of

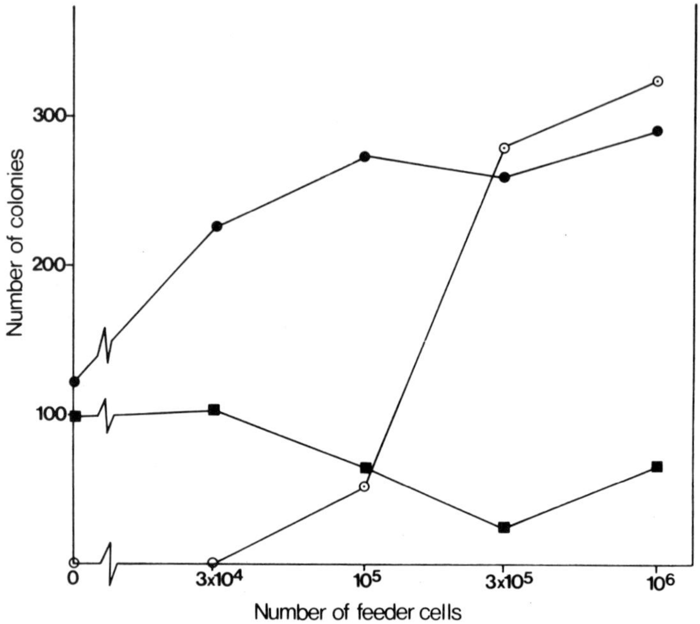

Figure 2. Effect of feeder cells on recovery of HGPRT$^-$ variants. 60 mm dishes were inoculated with mitomycin C-treated STO cells at various cell numbers per dish (abscissa). Reconstruction selections were then performed by seeding 500 cells per dish of the feeder-independent HGPRT$^-$ EC cell variant PC13TG8 either alone (●) or together with 10^5 cells per dish of its wild-type parent PC13 (■). Medium containing 10 μg/ml of 6-thioguanine was applied the following day and renewed twice weekly, and dishes stained for colony counting at 10 days. Dishes seeded with 10^5 PC13 cells alone gave no surviving colonies. To investigate the dependence of the survival of feeder-dependent cells on feeder density under non-selective conditions, 1000 cells of the feeder-dependent EC line PSA4 were seeded into similar feeder dishes, maintained in non-selective medium for a week with a medium change after 4 days, and then stained for colony counting (○). Note that the plates used in this experiment have one-third of the area of the plates used in the protocol of *Table 3*.

mutant cells by metabolic cooperation with wild-type cells (see Section 2.2). In the case of feeder-independent cells this can be achieved by limiting the seeding density to 4×10^5 cells per 100 mm dish but, in the case of feeder-dependent cells, metabolic cooperation via the feeder layer introduces additional problems (see Section 2.2). The extent of the effect on mutant recovery can be investigated by carrying out reconstruction selections with mixtures of wild-type and HGPRT$^-$ feeder-independent cells at various feeder densities. Typical results from such an exercise are given in *Figure 2*. Relative to the survival of HGPRT$^-$ cells alone at the same feeder density, their survival in the presence of wild-type cells is progressively reduced with increase in feeder cell density up to 3×10^5 cells per 60 mm dish but, if the feeder cell density is increased to 10^6 cells per 60 mm dish, survival increases again. This effect occurs reproducibly. At first sight this appears to be paradoxical, but it can be rationalized as follows. At 3×10^5 feeder cells per dish, most HGPRT$^-$ cells are in indirect contact with wild-type cells through a thinly spread monolayer of feeder cells, so that further increase in the number of feeder cells does not increase the number of interconnecting pathways but merely interposes more cell junctions in between the wild-type and HGPRT$^-$ cells, resulting in less transfer of 6-thioGMP. The best compromise between mutant recovery and cloning efficiency of feeder-dependent cells is thus fortuitously achieved at the feeder cell density which is routinely used. Although clearly there are losses occurring under these conditions, the recovery of HGPRT$^-$ cells is above 10%, making the method a viable one. This is the basis of the protocol of *Table 3*. In theory the recovery could be increased by increasing the feeder cell density still further. However, this leads to practical problems in that more dense feeder layers are mechanically unstable and tend to peel from the dish as an intact sheet.

We have successfully used this method to isolate thioguanine-resistant variants of both the EC line PSA4 and the embryonic stem cell line B2B2. Now that we can reproducibly achieve feeder-free growth of these cells using 60% BRL-conditioned medium, however, there is little doubt that this will provide us in the future with an improved procedure for the isolation of thioguanine-resistant variants.

Variants resistant to 6-thioguanine are unusual in that, because of the high frequency at which HGPRT$^-$ variants are generated, the overwhelming majority of 6-thioguanine-resistant variants are HGPRT$^-$ Other analogues such as 8-azaguanine have yielded variants of other types in addition to HGPRT$^-$ but it is not clear whether there is a real difference between the different analogues in this respect or whether the different behaviour reflects other differences in the design of selection procedures. Strictly speaking the characterization of a HGPRT$^-$ variant should include all four steps of *Table 4* but, in practice, depending upon the intended use of the variant, a limited characterization will often suffice. If the variant cell line is to be used for the isolation of metabolic cooperation variants (Section 6), stable 6-thioguanine resistance and lack of incorporation of [^{3}H]hypoxanthine are the crucial requirements, while if it is to be used for the selection of fusion hybrids (Section 7) inability to grow in HAT medium is the most important property. In most applications the requirement will be for a variant line whose developmental properties are unchanged compared with the wild-type parent, and in such cases characterization should include analysis of karyotype and developmental capacity, either *in vitro*, in tumours or in chimaeras, depending upon the intended application. As genetic changes at other loci can and do occur it is a good idea to pick and screen a number of surviving colonies in order to eliminate those with undesirable features. However, note that the method of colony picking shown in *Figure 1* is not satisfactory for the isolation of several colonies from a single dish, since once one colony has been picked there is an appreciable chance that some of its cells will be dispersed in the dish and will contaminate other colonies. Isolation of several colonies from one dish is in any case undesirable as some colonies may have arisen as satellites of others. It is therefore preferable to pick a single colony from each dish, although even this does not, of course, guarantee independence of origin. If truly independent variants are sought one of the measures outlined in Section 2.1 should be adopted. Only where the original selection plate contains exactly one colony should the picked population be regarded as clonal. Cultures established from colonies picked from dishes in which additional colonies were present should be subjected to re-cloning.

5. ISOLATION OF OUABAIN-RESISTANT VARIANTS

Similar considerations apply to the isolation of ouabain-resistant variants, which are of interest because of the use of the ouabain-resistance marker in the selection of fusion hybrids (Section 7). Ouabain-resistant variants are probably heterozygous for an autosomal-dominant mutation (see ref. 1 for a discussion of the experimental evidence on this topic) and can be obtained without mutagenesis at manageable frequencies, albeit somewhat lower than those obtainable for HGPRT$^-$ variants. Rodent cells are several orders of magnitude more resistant than primate cells. Isolation of resistant variants in mouse cell lines requires selection in 3 mM ouabain, which is close to the limit of solubility. To date we have isolated ouabain-resistant variants from feeder-independent cells only. On STO feeder layers ouabain-sensitive cells are rescued from ouabain toxicity by metabolic cooperation, but isolation of ouabain-resistant feeder-dependent cells should be readily achieved by the use of 60% BRL-conditioned medium.

6. ISOLATION OF METABOLIC COOPERATION-DEFECTIVE VARIANTS

Variant cell lines deficient in metabolic cooperation, that is the passage of low molecular weight metabolites between cells via gap junctions, can be selected using a procedure which has been termed the 'kiss of death'. This requires the prior presence of the HGPRT$^-$ marker in the cell line on which the selection is to be performed. This can be introduced if necessary by selection as described in Section 4 (*Table 3*). HGPRT$^-$ cells, normally resistant to 6-thioguanine, become sensitive when co-cultured with wild-type (donor) cells due to the passage of 6-thioGMP from wild-type to HGPRT$^-$ cells via gap junctions. Survivors are therefore enriched for cooperation-defective cells (4). Because of the death of the donor cells, selective conditions cannot be continuously maintained and the survivors of a single round of selection usually contain an overwhelming predominance of non-defective cells. The strategy outlined in *Figure 3* is therefore adopted. The pooled survivors of the first round of selection are screened for the presence of an appreciable proportion of defective cells. If such cells are detected, the population is cloned by dilution (see Chapter 2, Section 2) and the individual clones are then screened and, where the screening procedure indicates different behaviour from the parental cells, complete characterization is carried out. If the pooled survivors behave similarly to the parental cells in the screening procedure they are recycled through further rounds of selection until differences are revealed.

The 'kiss of death' procedure is described in detail in *Table 5*. It is advantageous to use an EC cell line as the HGPRT$^+$ donor since metabolic cooperation between EC cells is highly efficient. There is no particular advantage, however, in using a feeder-dependent cell line as donor even when the HGPRT$^-$ parental cells are feeder-dependent, and it is more convenient to use a feeder-independent cell line such as PC13.5. This also has the advantage that PC13.5 carries a metacentric marker chromosome, so that in the unlikely event of derivatives of the donor cells surviving the selection their presence would be detectable. In practice we have never found this to be a problem, presumably because a minimum of two genetic changes would be required for wild-type cells to acquire resistance to the selection conditions (i.e. an HGPRT$^-$ mutation and a second lesion causing resistance to the 'kiss of death' effect). The total cell density has quite a profound effect on the efficiency of the kill, while the ratio of donor cells to HGPRT$^-$ cells is less critical provided that donor cells are in the majority. Because of logistic problems in coordinating the growth of several cell lines it is often difficult to achieve exactly the same conditions in each round of selection and we have successfully isolated metabolic cooperation-defective variants from selection procedures

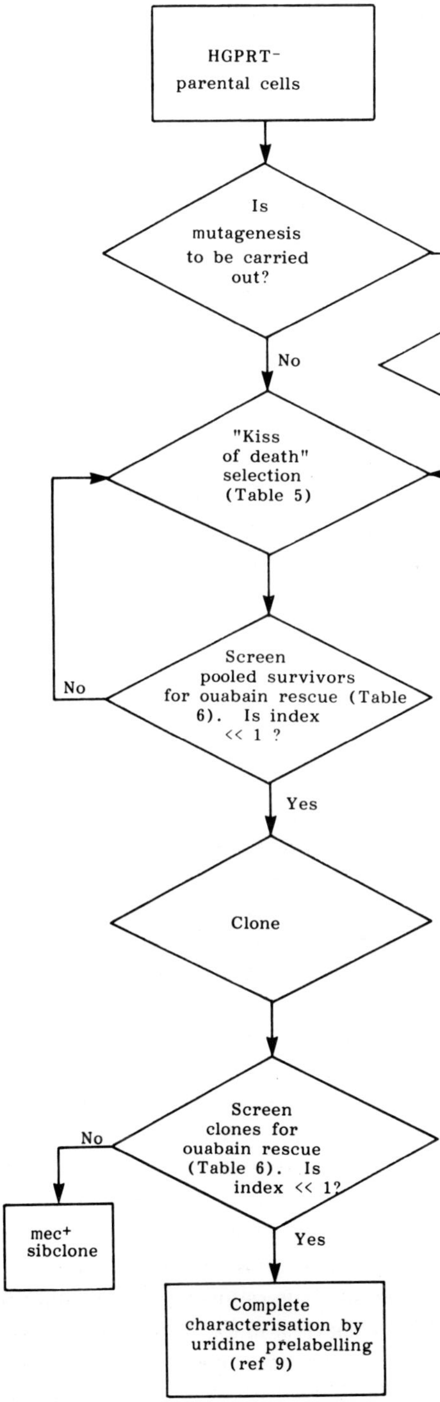

Figure 3. Outline of a procedure for the isolation of variants defective in metabolic cooperation.

Table 5. Thioguanine 'kiss of death' enrichment for metabolic cooperation-defective variants of an $HGPRT^-$ feeder-dependent cell line.

1. On day 1, prepare feeder layers in twenty 100 mm tissue culture dishes from STO cells by mitomycin treatment (as described in Chapter 4, Section 2). Seed out 3×10^6 feeder cells per dish in 10 ml CM (*Table 2*).
2. On day 2
   (i) Trypsinize two 175 cm^2 late exponential phase cultures of the $HGPRT^-$ parental cell line (or survivors from previous rounds of selection, as appropriate) and five 175 cm^2 late exponential phase cultures of $HGPRT^+$ EC cells.
   (ii) Prepare single-cell suspensions from each as in *Table 3*.
   (iii) Count the suspensions and prepare a mixture of 4.2×10^7 $HGPRT^-$ cells and 1.05×10^8 $HGPRT^+$ cells in a total volume of 210 ml CMβ (*Table 2*).
   (iv) Remove the medium from the feeder layers prepared on day 1 and add 10 ml of the mixed cell suspension just prepared (i.e. 2×10^6 $HGPRT^-$ cells and 5×10^6 $HGPRT^+$ cells) to each dish. Discard the remaining suspension.
3. On day 3, replace the medium in all dishes with CMβ supplemented with 10 μg/ml of 6-thioguanine from stock solution prepared as in *Table 3*.
4. Replace the medium every 2 days with fresh CMβ + 10 μg/ml of 6-thioguanine. Monitor the progress of cell killing. If detachment of feeders begins to occur then seed on a further 3×10^6 freshly mitomycin C-treated STO cells per dish. A frequent problem is that developing colonies retain cells killed by the thioguanine which remain attached to the substratum and restrict the growth of the survivors. If this occurs, trypsinize the culture and, after neutralizating the trypsin, pellet the cells by low-speed centrifugation ($100 \times g$, 3 min), resuspend in fresh CMβ + 10 μg/ml of 6-thioguanine and re-seed into another dish of the same size. Feeder cells will re-attach but may require supplementation with freshly mitomycin C-treated STO cells as described above.
5. When killing is complete, allow the survivors to grow up until colonies are 2–3 mm in diameter. Trypsinize, pool the survivors and expand the culture for the subsequent operations which are outlined in *Figure 3*.

in which the ratio of donor cells to $HGPRT^-$ cells was allowed to vary between 1:1 and 6:1 with the availability of the cells (8).

The method used for the screening of survivors is rescue from ouabain toxicity. The protocol is detailed in *Table 6*. Cells defective in metabolic cooperation show a reduction in ouabain rescue index because transfer of sodium ions from the test cells to the ouabain-resistant STO feeder cells is required for survival (4). For this assay the STO cells are treated with mitomycin C *in situ* so that the test cells can be seeded on to them on the same day, thus avoiding any problems associated with ageing of the cells after mitomycin treatment. There are advantages and disadvantages to using the ouabain rescue assay as a screening procedure rather than a quantitative assay based on survival in the thioguanine 'kiss of death' procedure used for selection. The advantages are 2-fold.

(i) Because the ouabain rescue assay depends upon the transfer of a different molecule to that important in the thioguanine 'kiss of death' (sodium ions as opposed to nucleotides) it will not be influenced by alterations other than at the level of the gap junction which could facilitate survival in the 'kiss of death' selection (e.g. increased pool sizes of nucleotides competing with 6-thioGMP for incorporation into nucleic acid).

(ii) The requirement for feeder cells makes quantification of 'kiss of death' procedures difficult.

Table 6. Screening of selected populations by assay of rescue from ouabain toxicity.

1. On day 1, seed out a number of 60 mm dishes with 5×10^5 live STO cells (i.e. not mitomycin C-treated) in 4 ml of CM. Four dishes are required for each population under test and four for the parental cell control.
2. On day 2
   (i) Replace the medium with 4 ml of CM + 10 μg/ml of mitomycin C prepared less than 1 week previously.
   (ii) Incubate for 2 h at 37°C.
   (iii) Wash twice with fresh CM and incubate for 3 h in CM at 37°C.
   (iv) Trypsinize the cultures of cells under test and of HGPRT$^-$ parental cells (control). Prepare single cell suspensions, count, and by serial dilution prepare suspension of each cell type at 250 cells/ml in 20 ml of CMβ.
   (v) Remove the medium from mitomycin C-treated STO dishes and replace with 4 ml of cell suspension (i.e. 10^3 cells), each cell suspension being used to inoculate four dishes. The diluted cell suspensions should be allowed to stand for the minimum possible time before plating to avoid loss of viability and/or losses due to attachment to glassware.
3. On day 3 prepare CMβ + 3 mM ouabain as follows. Weigh out ouabain octahydrate and dissolve in serum-free CMβ to give 2.43 mg/ml. **(Caution — ouabain is highly toxic. Wear gloves and a face-mask when handling solid.)** Warming to 37°C may be necessary to achieve complete solution. When solution is complete, sterilize by filtration through a membrane of pore size 0.2 μm. Add serum to 10% (v/v) after filtration (this reduces the probability of clogging the filter). Replace the medium in two dishes of each group of four with CMβ + 3 mM ouabain and in the remaining two with CMβ.
5. On day 6, replace the medium in all dishes by medium of the same composition. (Note: ouabain-containing medium should be freshly prepared.)
6. On day 9
   (i) Remove the medium and replace with 2 ml of Leishman's stain (1.5 g/l in absolute methanol, filtered immediately before use). Leave for at least 5 min to allow fixation.
   (ii) Add 2 ml of distilled water to each dish and leave for a further 5 min.
   (iii) Rinse and allow to dry. Count the colonies on each dish.
7. Calculate the ouabain rescue index as follows. Let
   mean number of colonies per dish given by selected cells without ouabain = a
   mean number of colonies per dish given by selected cells with ouabain = b
   mean number of colonies per dish given by parental cells without ouabain = c
   mean number of colonies per dish given by parental cells with ouabain = d.
   Ouabain rescue index = $\frac{bc}{ad}$.

The disadvantage is that since wild-type cells score more highly than variant cells in the ouabain rescue assay it is not sensitive to the presence of a small proportion of variant cells. Clearly there is some degree of subjective judgement associated with the decision whether, in a single assay, deviations in ouabain rescue index from unity are sufficiently large to be worthy of further investigation. As a rough guide, values of 0.5 or below usually fall into this category while higher values may merely differ from 1 by virtue of experiment error. In borderline cases cloning may be undertaken in parallel with a repeat assay on the uncloned population.

Because passaging of EC cell lines over extended periods is often associated with loss of developmental capacity (Section 2.2), it is important to minimize the amount of multiplication between rounds of selection. It is therefore not desirable to passage the survivors while awaiting the results of screening. The most convenient solution to this problem is to carry out screening in parallel with the preparation of a frozen stock and a subsequent round of selection, so that if screening suggests the presence of variant cells the frozen stock may be thawed for cloning.

A number of improvements to this isolation procedure are possible. The use of 60% BRL-conditioned medium to replace feeder cells in the 'kiss of death' selection should allow problems of feeder detachment to be circumvented (note, however, that conditioned medium cannot replace the mitomycin-treated STO cells in the ouabain rescue assay where they have an entirely separate function in addition to the normal one of inhibiting differentiation). It would also enable the design of a quantitative screen for ability to resist the selective conditions although, as discussed above, this would have both advantages and disadvantages. In the case of feeder-independent EC cells, the efficiency of killing in 'kiss of death' selections can be increased by re-seeding fresh donor cells at 3-day intervals. This approach was not adopted in the presence of feeders because it exacerbates the problem of feeder detachment. However, it could be used to increase the killing efficiency in conditioned medium. Reconstruction selections on mixtures of wild-type and variant cells could be carried out under different conditions in an attempt to achieve predictable enrichment factors and efficiencies of recovery. As indicated in Section 2.1, this information may then be used to tailor the stringency of selection to suit the immediate requirements. It may also be used to set a limit to the number of rounds of selection which can usefully be applied in succession. Clearly after four rounds of selection each giving 100-fold enrichment, for instance, any variant present in an initial population of 4×10^7 cells should dominate the culture. If no variant has been detected it is preferable to start again with a fresh population of cells rather than to continue recycling the survivors. Increasing the scale of the first round of selection may also be profitable to allow the isolation of variants present at low frequency. In pursuit of this aim we have found that satisfactory large-scale culture of EC cells can be achieved on Cytodex 3 microcarriers (Pharmacia) which have a covalently bound collagen coating and provide a surface similar to that of gelatinized tissue culture plasticware. Growth is nevertheless not as good as under standard conditions and the formation of multilayered clumps tends to occur leading to difficulties in trypsinization, and so repeated passaging on microcarriers is not recommended. It is not possible at this stage to give a clear answer regarding whether chemical mutagenesis is desirable in the isolation of metabolic cooperation-defective variants. While mutagenesis can lead to loss of developmental capacity (possibly compounded for mutagens such as MNNG by side-effects — see Section 2.1), so can passaging the cells, and more rounds of selection are likely to be necessary on cells not subjected to mutagenesis.

Clonal derivatives of selected populations giving ouabain rescue indices appreciably less than unity are subjected to further characterization. Subclones with ouabain rescue indices of 1 (mec^+ subclones) are frequently useful as controls as they have been subjected to the same selective conditions as cooperation-defective variants isolated during the same cloning. Many of the techniques used to assay for metabolic cooperation are indirect in that they quantify not the immediate products of transfer but some consequence of their transfer such as incorporation into nucleic acid or cell killing. The most direct technique is that of dye-coupling, but this has other drawbacks in that it is intrusive, demanding of specialized equipment and expertise, and can only be applied to a small number of cell contacts in any one experiment (4). The solution which we have adopted is to use the ouabain rescue assay (*Table 6*) in conjunction with the uridine pre-labelling assay, in which donor cells are incubated with [^{3}H]uridine and thereby acquire labelled pyrimidine nucleotide pools. These donor cells are then co-cultured with unlabelled recipient cells and transfer of labelled nucleotide to the recipient cells leads to its incorporation into nucleic acid where it can be detected autoradiographically (9). Quantitative indices can be extracted from grain count data in a number of ways (4) and compared with the ouabain rescue index. If both differ significantly from the corresponding values for the parental cells then, since the two assays depend on the transfer of different molecules, it is reasonable to conclude that the selected cells differ from the parental cells in their capacity for metabolic cooperation.

Because of the possibility that developmentally relevant intercellular signalling may involve metabolic cooperation we are investigating the question of whether metabolic cooperation deficiency is associated with reduction in developmental capacity. Interpretation of the data we have obtained to date is, however, not straightforward because of the presence of additional unrelated changes in our selected metabolic cooperation-defective cell lines and the question must still be regarded as an open one.

7. ISOLATION OF FUSION HYBRIDS OF EC CELLS

The spontaneous frequency of production of fusion hybrids in cell co-cultures is very low but higher frequencies can be obtained by exposure to a variety of agents. In early work, u.v.-inactivated Sendai virus was employed as a fusogen but it has now been largely superceded by polyethylene glycol (PEG) which is simpler to use and has a wider spectrum of application (10). A protocol for PEG-mediated fusion is given in *Table 7*. PEG is an efficient fusogen only in a narrow concentration range near 50% w/w (11). For this reason, careful removal of wash medium prior to addition of the PEG is important to avoid dilution. Cytotoxic effects become significant at these concentrations but may be reduced by omitting Ca^{2+} ions during the fusion and immediately afterward. Fusion is more efficient when the PEG solution is prepared in Hepes buffer

Table 7. Isolation of fusion hybrids between PSA4 and R5/3OA EC cell lines.

1. Prepare the following solutions.

   Polyethylene glycol (PEG). Melt PEG (molecular weight 1000) in a 60°C water bath and add to an equal volume of 0.15 M Hepes buffer, pH 7, containing 3.3 μM phenol red indicator. Adjust the pH to 7.55 with 12 M HCl or 10 M NaOH, taking care that the HCl or NaOH is fully dissolved in the viscous PEG solution prior to taking the final pH reading. Sterilize by filtration through a membrane of pore size 0.2 μm and re-check the pH of an aliquot of the solution following filtration. Store at −20°C. Since the pH of alkaline solutions tends to fall with time (probably due to absorption of atmospheric CO_2) re-check the pH following prolonged storage and re-adjust if necessary.

   Calcium-free minimal attachment medium (CFMAM). This has the composition of 0.116 M NaCl, 5.4 mM KCl, 5.5 mM glucose, 1.0 mM $MgCl_2$, 10.0 mM Hepes. Adjust the pH to 7.55 with 10 M NaOH, sterilize by filtration through a membrane of pore size 0.2 μm and store at 4°C.
2. On day 1, trypsinize 75 cm^2 late exponential phase cultures of PSA4 and R5/3OA and prepare single cell suspensions. Set up the following cultures, each in 10 ml feeder-conditioned medium (*Table 2*) in 25 cm^2 gelatinized culture flasks:
   - 10 co-cultures of 3.3 × 10^5 PSA4 cells and 6.6 × 10^5 R5/3OA cells per flask,
   - 2 control cultures of 10^6 PSA4 cells per flask,
   - 2 control cultures of 10^6 R5/3OA cells per flask.

Table 7. continued.

3. On day 2, treat eight co-cultures and one control culture of each parent with PEG as follows, handling the flasks one at a time since rapid working is essential.
   (i) Remove the medium and wash the monolayer twice with CFMAM. Allow each flask to stand a minute or so in an upright position after the final wash to be sure of removing all the supernatant.
   (ii) Add 5 ml of the PEG solution previously prepared (step 1), initially tilting the bottle so that the solution does not contact the cells and then allowing it to run quickly across the monolayer. After 30 sec in a horizontal position stand the flask on end and allow 30 sec for the viscous solution to drain. Quickly aspirate.
   (iii) Wash the monolayer three times with CFMAM. Feed with feeder-conditioned medium.
   Feed the remaining untreated cultures with feeder-conditioned medium.
4. On day 3, trypsinize one PEG-treated and one untreated co-culture and seed 1.6×10^4 cells on to duplicate plastic coverslips (Lux Thermanox coverslips, 15 mm diameter; Flow Labs) in 17 mm wells of 24-well multiplates in feeder-conditioned medium. Feed the remaining cultures with 10 ml of feeder-conditioned selective medium. Prepare the medium as follows.
   (i) Prepare a 10 mM hypoxanthine stock solution by weighing out the solid, adding to slightly less than the required volume of water, and then adding 10 M NaOH dropwise until solution is complete. Make up to the final volume with water and sterilize by filtration through a 0.2 μm membrane.
   (ii) Prepare a stock solution of 80 μM aminopterin in water and sterilize by filtration through a 0.2 μm membrane. **(Caution — aminopterin is a suspected teratogen and it is not recommended that the solid be handled by women of childbearing age.)**
   (iii) Prepare a stock solution of 2 mM thymidine in water and sterilize by filtration through a 0.2 μm membrane.
   (iv) Prepare feeder-conditioned medium as in *Table 2* but do not supplement with further serum yet as this will facilitate the subsequent filtration step.
   (v) Dissolve ouabain octahydrate in the feeder-conditioned medium to give 2.43 mg/ml as in *Table 6*. When solution is complete, sterilize by filtration through a 0.2 μm membrane.
   (vi) Supplement with an additional 10% (v/v) foetal calf serum and add hypoxanthine, aminopterin and thymidine to final concentrations of 0.1 mM, 0.8 μM and 20 μM, respectively, from sterile stock solutions.
5. On day 4 stain the coverslips with Leishman's stain (see *Table 6*). Air dry the coverslips and mount each with the cell side uppermost on a microscope slide. Cover with a glass coverslip using DPX mountant. Examine each coverslip microscopically and score for the proportion of binucleate cells.
6. On day 6, replace the medium in the flasks with 10 ml of freshly prepared feeder-conditioned selective medium.
7. On day 9, trypsinize each flask separately and seed out the cells in 100 mm dishes at 4×10^5 cells/dish in 10 ml of feeder-conditioned selective medium.
8. Replace the medium twice weekly with fresh feeder-conditioned selective medium until colonies appear in dishes set up from PEG-treated co-cultures. Ensure that dishes set up from parental control cultures contain far fewer colonies. This should also be true of those set up from co-cultures not treated with PEG.
9. Pick colonies from PEG-treated co-culture dishes as described in Section 4 and illustrated in *Figure 1* and seed each colony into 17 mm wells in feeder-conditioned medium supplemented with hypoxanthine (0.1 mM) and thymidine (20 μM). After 2 days replace the medium by unsupplemented feeder-conditioned medium.
10. Expand the cultures in feeder-conditioned medium or on feeder layers for preparation of frozen stocks and characterization.

(*Table 7*) than in serum-free growth medium (12). pH is an important variable and it is recommended that PEG solutions are sterilized by membrane filtration rather than by autoclaving which increases their acidity. Both fusogenic capacity and toxicity increase as the molecular weight of PEG is reduced from 6000 to 600; we have successfully isolated EC cell hybrids from fusions using PEG of molecular weight 1000. PEG samples opened and stored for long periods of time develop acidic impurities which increase their toxicity. This process is accompanied by visible changes in the consistency of the solid from a hard wax or powder to a soft wax or jelly. Impure samples can be re-purified by sequential precipitation from benzene with, firstly, iso-octane, and then ethyl ether (13). The efficiency of the fusion procedure can be monitored by scoring for the frequency of binucleate cells in the population soon after fusion as detailed in *Table 7*. This is more readily achieved at lower cell density and therefore parallel fusion flasks are set up and trypsinized 24 h after fusion, while those flasks from which it is hoped to isolate viable hybrids are left without re-seeding for 7 days to minimize trauma. Using the procedure described in *Table 7* we have achieved frequencies of binucleate cells of between 10 and 15% for EC cells.

The formation of viable hybrids from binucleate heterokaryons is, however, a much rarer event and selective methods are essential for their isolation. The best known of such methods is the application of HAT (hypoxanthine/aminopterin/thymidine) selection to the fusion products of $HGPRT^-$ and thymidine kinase defective (TK^-) cells. However, TK^- variants are difficult to isolate from diploid cells because they involve an autosomal locus and are recessive. If it is necessary to introduce selective markers into cell lines for the sole purpose of hybridization, then the best strategy is to select sequentially for HGPRT deficiency and ouabain resistance in one parental cell line. This may then be fused to an unmarked cell line and hybrids isolated in HAT plus ouabain. This is the system adopted in the protocol described in *Table 7*. R5/3OA is a derivative of the feeder-independent EC cell line PC13 which is metabolic cooperation-defective, $HGPRT^-$ and ouabain-resistant. PSA4 is an unmarked, feeder-dependent EC cell line. Because the degree of feeder-dependence of the hybrids cannot be predicted, feeder-conditioned medium is used for their isolation. Feeder layers themselves cannot be used as they rescue the ouabain-sensitive PSA4 cells from ouabain toxicity by metabolic cooperation (see Sections 5 and 6). The fusogen-treated populations are allowed to recover for 7 days following treatment. This introduces an element of compromise into the isolation procedure since, while it minimizes trauma, it reduces the initial efficacy of the selective medium because of the high cell density. PSA4 cells are rescued from ouabain toxicity by metabolic cooperation with feeder cells remaining from the previous passage and with R5/3OA cells, while R5/3OA cells are rescued somewhat less efficiently from HAT toxicity by movement of purine nucleotides from PSA4 cells through gap junctions. Thus R5/3OA cells are at a selective disadvantage compared with PSA4 cells during this period, partly because of their metabolic cooperation deficiency and partly because higher concentrations of hypoxanthine (10 mM) are required in the medium for efficient rescue of $HGPRT^-$ cells from HAT toxicity even in cooperation-competent cells. For this reason more R5/3OA cells are added to the initial co-cultures than PSA4 cells. Once the cells are re-seeded at lower density, both parental cell types are efficiently eliminated.

Removal of cells directly from HAT medium into non-selective medium is sometimes accompanied by loss of viability because aminopterin remains irreversibly bound to folate reductase, resulting in a continuing requirement for hypoxanthine and thymidine. It is recommended therefore that this transition be made via medium containing hypoxanthine and thymidine only (HT medium) to allow time for the synthesis of new folate reductase enzyme, after which this requirement is no longer present. If, as is usually the case, it is desired to isolate hybrids of independent origin, only one colony should be taken from the series of dishes set up from each fusion flask.

Should appreciable numbers of colonies appear in dishes set up from parental controls this indicates the presence of variants or revertants in the starting cell population. Survivors in dishes inoculated only with R5/3OA indicate the presence of $HGPRT^+$ revertants which can be removed by passaging once or twice in 6-thioguanine immediately prior to fusion, while survivors in PSA4 dishes indicate the presence of ouabain-resistant variants which can be removed only by re-cloning.

We have successfully isolated hybrids from this fusion (14) but it is not clear to what extent the success depends upon the fact that one parent is feeder-independent and metabolic cooperation-defective. We anticipate that in the future the use of 60% BRL-conditioned medium will allow us to isolate hybrids from fusions between two feeder-dependent cell lines.

Characterization of cell lines selected from PEG-treated populations to establish their hybrid status is mandatory. Examination of the karyotype (see Chapter 4, Section 6 for methods) should always be undertaken. Ideally each parent should contain marker chromosomes in order to establish unequivocally a contribution from each parent, although of course this conflicts with the requirement to work with cell lines whose karyotype is as close as possible to diploidy. In the fusion described in *Table 7*, R5/3OA has a quasi-tetraploid karyotype with two copies of the metacentric marker chromosome, while PSA4 has a quasi-diploid karyotype with no chromosomes readily distinguishable from those of R5/3OA. The hybrids isolated possess modal chromosome numbers in the range 100–120 with one or two copies of the metacentric marker chromosome, suggesting the presence of one chromosome set from each parent with a small degree of chromosome loss. A second criterion which has been used to establish hybrid status is the use of isozyme markers characteristic of different mouse strains or different species (see Chapter 5 for methods). Hybrid cells should, unless loss of a chromosome carrying one of the structural genes has occurred, possess the isozymes contributed by both parents. To date this approach has been restricted to the products of fusion of EC cells with other cell types since no fusion studies have been reported between EC or embryonic stem cell lines isolated from different mouse strains.

8. REFERENCES

1. Hooper,M.L. (1985) *Mammalian Cell Genetics.* John Wiley and Sons, New York.
2. Patek,C.E. and Clayton,R.M. (1985) *Exp. Eye Res.*, **40**, 357.
3. van Wesel,A.L. (1973) In *Tissue Culture: Methods and Applications.* Kruse,P.F., Jr. and Patterson,M.K., Jr. (eds), Academic Press, New York. p. 372.
4. Hooper,M.L. and Subak-Sharpe,J.H. (1981) *Int. Rev. Cytol.*, **69**, 45.
5. Smith,T.A. and Hooper,M.L. (1983) *Exp. Cell Res.*, **145**, 458.
6. Smith,A.G. (1986) Ph.D. Thesis, University of Edinburgh.
7. Gillin,F.D., Roufa,D.J., Beaudet,A.L. and Caskey,C.T. (1972) *Genetics*, **72**, 239.
8. Smith,T.A., Box,K.M. and Hooper,M.L. (1986) *Exp. Cell Res.*, submitted.
9. Pitts,J.D. and Simms,J.W. (1977) *Exp. Cell Res.*, **104**, 153.
10. Pontecorvo,G. (1976) *Somat. Cell. Genet.*, **1**, 397.
11. Davidson,R.L. and Gerald,P.S. (1977) In *Methods in Cell Biology.* Prescott,D.M. (ed.), Academic Press, New York, Vol. **15**, p. 325.
12. Klebe,R.J. and Mancuso,M.G. (1981) *Somat. Cell. Genet.*, **7**, 473.
13. Klebe,R.J., Harriss,J.V., Sharp,Z.D. and Douglas,M.G. (1983) *Gene*, **25**, 333.
14. Smith,J. (1985) Ph.D. Thesis, University of Edinburgh.

CHAPTER 4

Embryo-derived stem cell lines

ELIZABETH J.ROBERTSON

1. INTRODUCTION

Tissue culture lines of pluripotential cells which are similar in phenotype to embryo cells, and whose pattern of differentiation closely resembles that of the normal embryo, provide a unique model system which has a variety of experimental applications.

Two experimental strategies have been devised for deriving cell lines which possess a stable pluripotential embryonic phenotype. Both of these approaches are based on the premise that by physically perturbing the early mouse embyro, pluripotential stem cells can be diverted from their normal pattern of growth and differentiation. The first technique to be established was the indirect route of isolating embryonal carcinoma cells (EC cells) from embryo-derived tumour masses. The experimental production of teratocarcinomas is the subject of Chapter 1. More recently, *in vitro* culture techniques have been devised whereby stem cells can be isolated directly from the normal embryo without the necessity of passing through a tumour stage. These embryo-derived stem cells have been variously termed EK cells (1) or ES cells (2) to distinguish them from EC cells.

EC cells and embryo-derived stem cells are indistinguishable in terms of the cellular morphology, cell surface antigen expression and differentiation behaviour *in vitro* and *in vivo* (1,2). However, EC cell lines have important experimental limitations; it is well documented that EC cell lines, probably as a result of the protracted procedure for their isolation, comprise a very diverse population of culture lines. This heterogeneity between lines is reflected in the range of ability for differentiation both in culture (see Chapter 2) and in chimaeric association with a host embryo (reviewed in ref. 3). Stem cells obtained directly from embryos have characteristics which give them a number of advantages over EC cell lines.

(i) The method is direct; as there is no requirement for a period of growth within a histocompatible animal there are no restrictions placed on the mouse strain from which stem cells can be derived. Within the past few years large numbers of cell lines have been generated to carry a range of genetic markers. These include isozymal variants (4), lines with marked X-chromosomes (5), translocation markers (6), coat colour markers (6) and developmental lethal mutations (7).

(ii) The isolation of the cells is immediate so that the stem cells obtained represent primary cultures which have not been subjected to deleterious selection pressures. This is reflected in a high differentiation ability of all embryo-derived cell lines both in the culture environment and during the formation of tumours.

(iii) The majority of cell lines can stably retain the euploid chromosome constitution, of the embryo from which they were derived, for many cell divisions in culture. Both euploid XY and XX cell lines have been isolated.

(iv) The most important feature which distinguishes embryo-derived stem cells from EC cells, in the experimental context, is the behaviour of the cells when returned to the embryonic environment (for methodology see Chapter 5). Embryo-derived cells differentiate totally normally generating chimaeras with a high efficiency ($>50\%$). The animals are notable in having a widespread and substantial contribution to all of the body tissues by the culture-derived cells. Test breeding of chimaeras obtained from XY cell lines has shown that the culture-derived cells commonly form functional sperm (4,8).

(v) The isolation procedure has been extended for use with parthenogenetically-derived embryos, and it is possible to derive cell lines of a homozygous diploid genetic constitution from haploid embryos (9).

This chapter describes the technique for the derivation of stem cells from normal fertilized embryos (Section 4) and from parthenogenetically-derived embryos (Section 5). The procedures used for the maintenance and differentiation of the stem cells in culture are described in Section 6.

2. TISSUE CULTURE MATERIALS AND ROUTINE PROCEDURES

2.1 General considerations

2.1.1 *Essential equipment*

A reasonably well equipped tissue culture facility is required. The following pieces of equipment are essential: a 37°C, 5% CO_2, humidified incubator, a good quality binocular dissecting microscope with transmitted and incident light sources (total magnification of up to $\times 40$), a good quality inverted microscope with a range of phase contrast objectives ($\times 10 - \times 25$ are suitable).

2.1.2 *Glassware*

If items such as pipettes and medium bottles are to be re-used it is essential that they are washed up carefully using a tissue culture grade detergent (e.g. 7×; Flow Labs) and rinsed extensively in several changes of distilled water. Both pre-implantation embryos and embryonic stem cells are intolerant of trace quantitites of either organic or inorganic residues. Where possible the use of non-toxic disposable pipettes and centrifuge tubes is advised.

2.1.3 *Water quality*

Water quality is important. We recommend a good water purification system (e.g. Milli-Q filtration system; Millipore). If this is not possible use double-distilled water. Alternatively, ultra-pure water can be bought from commercial sources (e.g. BDH Chemical Co.). If medium and tissue culture additives are to be made from scratch it is important to use only Analar chemicals.

2.1.4 *Sterile techniques*

Unless otherwise stated, all procedures described in this chapter should be carried out using sterile techniques. Because of the necessity for using a binocular dissecting microscope during many procedures, it is not easy to collect and culture embryos while working inside a standard laminar flow cabinet. If sufficient care is taken it is possible to work quite successfully on the bench. Fungal contaminations can be minimized by working inside a positive pressure filtered air facility.

2.2 **Culture media and solutions**

2.2.1 *Medium*

Unless indicated to the contrary, all tissue culture procedures use Dulbecco's modified Eagle's medium (DMEM) (no pyruvate, high glucose formulation). This can be bought from tissue culture suppliers (e.g. Flow Labs., Gibco-Biocult) in either liquid form (1× or 10×; dilute appropriately), or in powder form. If a large amount of tissue culture work is envisaged it is recommended to buy powdered medium (1 litre or 10 litre sachets). This is dissolved in distilled water (Section 2.1.3), buffered with sodium bicarbonate as directed, and adjusted to pH 7.2 using Analar HCl. The medium is sterilized by filtration through a 0.22 μm nitrocellulose filter into clean sterile bottles. The medium should be checked for sterility by incubating randomly chosen bottles at 37°C for 48 h. Prepared medium should be stored at 4°C and warmed to 37°C in a water bath immediately prior to use. DMEM can be stored for long periods. After 2 weeks it should be supplemented with 1 mM glutamine (100× stock; Flow Labs., Gibco-Biocult) as this amino acid is unstable.

2.2.2 *Trypsin EDTA*

A solution of 0.25% (w/v) trypsin in 0.04% (w/v) EDTA is used for routine subculture of both fibroblasts and stem cell lines. Trypsin is also used in the disaggregation of embryos. The composition of the trypsin solution is given in *Table 1*.

2.2.3 *Phosphate-buffered saline*

Phosphate-buffered saline (PBS) (calcium and magnesium free) is used for washing cell layers prior to trypsinization. The recipe for PBS is given in *Table 1*.

2.2.4 *Antibiotics and anti-mycotics*

DMEM is routinely supplemented with penicillin (final concentration 50 units/ml) and streptomycin (final concentration 50 μg/ml), to control bacterial contaminations (*Table 1*). Antibiotics should be included in the medium used for the recovery of embryos. They can be omitted during routine culture of established cell lines.

Fungal contaminations can be controlled by the addition of Amphotericin B (Sigma Chemical Co.) to a final concentration of 2 μg/ml (*Table 1*). Amphotericin B is toxic to cells. It is not recommended for use with embryos or embryonic stem cell lines.

2.3 **Serum**

Both foetal bovine serum (FBS) and newborn calf serum (NCS) are used as medium

Table 1. Recipes for standard tissue culture solutions.

Trypsin EDTA

To 1 litre of distilled water add the following:

2.5 g trypsin powder (Porcine; Difco)
0.4 g EDTA
7.0 g NaCl
0.3 g $Na_2HPO_4.12H_2O$
0.24 g KH_2PO_4
0.37 g KCl
1.0 g D-glucose
3.0 g Tris
1.0 ml Phenol red (Difco)

Adjust to pH 7.6

Pre-filter the solution through a glass fibre filter. Filter through a 0.22 μm nitrocellulose filter, aliquot into 10 ml plastic tubes and store at −30°C.

Phosphate-buffered saline (PBS)

To 1 litre of distilled water add the following:

10.0 g NaCl
0.25 g KCl
1.44 g $Na_2HPO_4.12H_2O$
0.25 g KH_2PO_4

Adjust to pH 7.2, dispense into glass bottles and sterilize by autoclave.

Antibiotics and anti-mycotics

(i) Antibiotics

To prepare a 100× stock solution dissolve 5000 units of penicillin (Glaxo) and 50 mg of streptomycin (Glaxo) in 100 ml of PBS. Aliquot and store at 4°C. Replace weekly.

(ii) Anti-mycotics[a]

To prepare a 100× stock, pre-dissolve 2 mg of amphotericin B[b] (Sigma Chemical Co.) in 2 ml of dimethyl sulphoxide. Make up to 100 ml in distilled water, filter sterilize, aliquot and store at −30°C.

[a]Amphotericin B can be bought commercially as 'Fungizone' (Flow Labs).
[b]Amphotericin B is highly toxic and is unsuitable for use with embryonic stem cells. Discard contaminated cultures.

supplements. Their appropriate use and relative concentrations are given in the relevant sections. For the successful growth of most tissue culture cells serum quality is critical and this is especially true for blastocysts and embryonic cells. The use of unsuitable serum is undoubtedly one of the reasons for a failure to isolate stem cells from embryos. All sera should be tested for their ability to support the growth of pluripotential stem cells. In the author's experience commercially available sera are of highly variable quality and serum samples should be requested and batch-tested prior to purchase. Serum testing is laborious and the best strategy is to simultaneously request the maximum number of available samples from all the commercial suppliers. The samples can then be tested in parallel and a bulk order placed for the most suitable sera. Serum can be stored for periods of up to 2 years at −30°C. Bottles should be thawed as required. It is not advised to re-freeze serum.

2.3.1 *Serum testing*

The most sensitive test of the ability of a given serum to support the growth of stem

Table 2. Comparative plating efficiency test for serum.

Allow 6 × 10^6 STO cels for each sample to be tested.

The following protocol is suitable for testing eight serum samples against two control sera.

1. Mitomycin C-treat 12–15 confluent 10 cm plates of STO fibroblasts (see Section 2.4). Wash the plates well with two changes of PBS and harvest the cells by trypsinization. Collect the cells in DMEM + 10% NCS and spin down (1000 r.p.m., 5 min).
2. Resuspend the cells in serum-free DMEM and count the cell density.
3. Prepare a single cell suspension of stem cells from an exponentially growing culture. (Use either an established embryonic cell line or, if not available, a feeder-dependent EC cell line, e.g. PSA-1.) Spin down and resuspend the cell pellet in serum-free DMEM. Count the suspension.
4. Place 300 ml of DMEM in a sterile 500 ml bottle. To this add 6 × 10^7 STO cells and a known number of stem cells (between 4 and 6 × 10^4 cells). Mix well.
5. Dispense uniform aliquots of 4.5 ml of the suspension into 6 cm plates. Prepare 60 plates.
6. Set plates out into an array of 10 rows of six plates and label each row with the appropriate sample number.
7. For each sample add 0.5 ml of serum (final concentration 10%) to five plates. To the remaining plate add 1.5 ml of serum (final concentration 30%) and mark appropriately.
8. Incubate the plates for 7–10 days. After this time colonies of stem cells should be clearly visible by the naked eye.
9. Aspirate all plates, wash once with PBS and flood with 2 ml of a 2% solution of methylene blue. Stain for 2–5 min, rinse in water and allow plates to air dry.
10. Invert each plate and count the total number of colonies. Calculate the average plating efficiency at 10% and 30% serum for each sample.
11. Select the serum batch which gives a plating efficiency equal to, or greater than, the appropriate control sample and which shows no toxicity at 30%.

Note: a contributory factor to toxicity may be the presence of high levels of serum complement. Complement activity may be removed by heat treating the serum (30 min at 56°C).

cells is by a comparison of the plating efficiency of single cells from either an established embryo-derived stem cell line or, alternatively, a feeder-dependent EC cell line (e.g. PSA-1). The stem cells are co-plated with feeder STO cells (Section 2.5). Control samples of FBS and NCS are included in the test. Only sera which are comparable to, or better than, the control sera should be bought.

The method for serum testing is given in *Table 2*. Basically, the technique is to set up duplicate plates containing a suspension of mitomycin C-treated STO fibroblast cells (to provide a feeder layer; Section 2.4.1) and a small number of stem cells in serum-free medium. The plates are placed in sets with each set being allocated a specific serum batch. Serum is added to a final concentration of 10%. To a single plate in each set serum is added to a final concentration of 30%. At this concentration any toxic component will readily be detected. The plates are incubated for 7–10 days and the stem cell colonies stained and counted. A good quality serum gives a plating efficiency of 20% or higher. Toxicity effects will be evidenced by a lower plating efficiency and/or the formation of smaller colonies at the 30% concentration. Serum toxicity may be due in part to high levels of complement. Heat treating serum (56°C for 30 min) to inactivate complement may dramatically decrease toxicity.

2.4 Preparation of feeder layers

Feeder cell layers are absolutely essential for both the isolation of stem cell lines from embryos and for the routine maintenance of established cell lines. Feeder layers are

Table 3. Routine subculture of STO fibroblasts.

Grow STO cells in 10 cm plates in DMEM supplemented with 10% NCS and antibiotics.	
1.	Aspirate the medium and wash the surface with 5 ml of PBS.
2.	Add 2 ml of trypsin EDTA and incubate at 37°C for 5 min.
3.	Neutralize the trypsin by adding 2 ml of DMEM + 10% NCS. Vigorously pipette the cells to give a single cell suspension.
4.	Add the suspension to 100 ml of growth medium and mix well.
5.	Dispense the suspension into 10 fresh plates and transfer to an incubator.
6.	The STO cells will grow to confluence in 5 days and should be subcultured at this time. It is advised to expand the population and freeze aliquots (Section 6.2) to act as a stock.

prepared from mitotically inactivated fibroblasts. These are plated to give a uniform monolayer of cells onto which the stem cells are seeded. Feeder layers should be prepared either from STO fibroblasts (10) or from primary embryonic fibroblasts. Other established fibroblast cell lines are largely unsuitable as, unlike STO cells, they either fail to show contact inhibition or are of an unsuitable morphology.

Cells can be rendered mitotically inactive by two means, either by exposure to gamma irradiation or by treatment with the drug mitomycin C (11).

2.4.1 *Preparation of feeder layers from STO cells*

(i) *Routine culture.* STO fibroblasts are simple to culture. They are routinely cultured in DMEM supplemented with 10% NCS on 10 cm Petri dishes. When confluent the cells are passaged, or subcultured, by trypsin treatment and diluted 1:10 into fresh plates. The method is described in *Table 3*. It is important to subculture the cells when they reach confluence. In cultures left for long periods without subculture there is a strong selection for subpopulations of cells which have lost the property of contact inhibition. These cells rapidly overgrow and the resulting cultures will make less good feeder layers. Ideally, a confluent 10 cm plate of STO cells should yield around $6-8 \times 10^6$ cells. When the cultures yield greater than 8×10^6 cells per plate the culture should be replaced by cells from an earlier freeze.

(ii) *Preparation of feeder layers.*

(1) Take confluent plates of STO cells. Aspirate the medium and replace with 6 ml of DMEM containing 10% NCS and 1 μg/ml of mitomycin C (Sigma Chemical Co.). This medium should be freshly prepared. Pre-dissolve 2 mg of mitomycin C in 1 ml of PBS, and add 1 mg of mitomcyin C to 100 ml of medium.

(2) Incubate the plates for between 2 and 3 h (minimum of 2 h).

(3) Gelatin coat tissue culture plates with a 0.1% (w/v) solution of gelatin (Swine skin, type I; Sigma Chemical Co.) in water. Sterilize the gelatin solution by autoclaving. Flood the plates to be prepared with a few ml of gelatin solution and leave for 1 h at room temperature. Aspirate.

(4) Aspirate the mitomycin C medium from the STO cells. Wash each plate three times with 5 ml of PBS. Trypsinize the cells (2 ml of trypsin EDTA/plate) and collect the cells in DMEM containing 10% NCS.

(5) Transfer the cell suspension to 20 ml centrifuge tubes and spin down the cells

(1000 r.p.m., 5 min). Resuspend the cells in DMEM plus 10% NCS. Count the cell suspension using a Coulter Counter (Coulter Electronics) or a haemocytometer.

(6) Prepare a dilution of 3×10^5 cells/ml in DMEM plus 10% NCS.

(7) Dispense the suspension into pre-gelled plates at a dilution of 5×10^4 cells/cm^2. For example, a 1 cm well should receive 0.5 ml of suspension, a 3.5 cm plate 1.0 ml and a 6 cm plate 3.0 ml. Incubate the plates.

(8) The cells will attach within 20 min and should spread to give a monolayer 12 h after plating. The cell density should be checked visually. If the cells are plated too sparsely, additional mitomcyin C-treated cells can be added. Feeder layers can be used up to 10 days after their preparation. Aspirate the medium prior to use.

2.4.2 *Preparation of feeder layers from primary embryonic fibroblasts*

As an alternative to STO fibroblasts, primary embryonic fibroblasts can be used for the preparation of feeder layers. Primary fibroblasts are used in preference by some workers for the isolation of stem cell lines (12). Although these cells can be isolated easily a major disadvantage is the limited culture lifespan of primary cells; they become senescent after about 15–20 cell divisions. Fresh isolates must be made on a regular basis to ensure that cells are continuously available. One other disadvantage is that, in comparison with STO cells, the primary fibroblasts have to be seeded at higher densities to ensure that a monolayer is formed and the culture medium becomes exhausted more quickly.

Primary embryonic fibroblasts are isolated by the following procedure.

(i) Kill a pregnant female mouse on the thirteenth or fourteenth day of pregnancy.

(ii) Swab the abdomen with 70% alcohol and dissect the abdominal cavity to expose the uterine horns. Remove the uterine horns into a 10 cm bacteriological Petri dish containing 10 ml of PBS.

(iii) Using fine scissors cut open the uterine walls and release the embryos into the saline. Remove the embryos, together with the associated placentae and foetal membranes, into a fresh dish of PBS.

(iv) Treat the embryos individually as follows. Using two pairs of watchmakers forceps dissect the embryo free from the placenta and membranes and then dissect out and discard the soft tissues (e.g. liver, heart and the other viscera). Wash the carcass in two further changes of PBS.

(v) Remove the embryo carcass into a fresh dish containing 2 ml of trypsin EDTA solution. Using curved Iris scissors finely mince the tissue. Incubate for 5 min at 37°C. Neutralize the trypsin by adding 5 ml of DMEM plus 5% NCS and transfer the contents of the dish to a 10 ml conical tube.

(vi) Allow the large pieces of cellular debris to settle out over a period of 2 min. Remove the supernatant into a fresh tube and make the total volume up to 10 ml using standard growth medium. Plate out the suspension into a 10 cm tissue culture plate and incubate.

(vii) Leave the plates undisturbed for 24 h. After this time change the medium to remove the cellular debris.

(viii) When the plates are confluent expand the cultures using the procedure described in *Table 3*.

(ix) Feeder layers are prepared using the method described for STO cells (Section 2.4.1). The cell density should be increased by at least a factor of two to ensure a confluent layer. It is important, however, not to plate cells at too high a density as the layer will detach from the substrate and 'roll up' after 5–6 days or on re-feeding.

3. COLLECTION OF EMBRYOS

The starting material of choice for the isolation of stem cells is the blastocyst stage embryo. As the time course of mouse embryogenesis is well documented, and the time of fertilization can readily be determined, the recommended method for obtaining the maximum number of fully viable embryos is to leave the embryos to develop within the reproductive tract until they have reached the appropriate blastocyst stage. Blastocysts are amenable to culture in standard tissue culture media and do not pose special requirements. This is not the case for embryos at pre-blastocyst stages which have to be grown in a simpler serum-free medium (see Section 5).

The mouse strain from which a cell line is to be made is another important consideration. The majority of inbred strains will, for example, yield relatively few embryos (normally between five and eight). If approaching the techniques outlined in this chapter for the first time it is recommended that more fecund outbred stocks of mice be used to generate embryos. Outbred mouse stocks are readily available from commercial breeders and matings provide large numbers of embryos on which to practice the various techniques before turning to the particular strain, or mutations, of interest. The rate of recovery of stem cell lines is rarely higher than 30%, and may be lower than 10%, and therefore a large percentage of the embryos are wasted. If the mouse strain of interest is available in restricted numbers, or if the desired genotype is carried by only a fraction of the embryos from a given mating, attempts to isolate a stem cell line can be frustrating.

3.1 Timed matings

Mouse stocks should be kept in a cycle of 12 h light (normally 6.00 hrs – 18.00 hrs) to 12 h dark. Under these conditions the females will pass through the reproductive cycle every 4–5 days. Female mice are only receptive to males if they are in oestrus and mating normally occurs at the mid-point of the dark period (i.e. midnight). Mating can readily be ascertained by inspecting the females the following morning for the presence in the vagina of a white copulation plug (see Chapter 1).

In the late afternoon take females from store cages and place with individual stud males (two females per stud). At this time it should be possible to guarantee mating by selecting females which display external signs of being in oestrus (13). If females are chosen at random it is necessary to set up more mating cages to generate plugs. Remove the females the following morning and check for the presence of a copulation plug. Separate out plugged females and place in a clean cage. The day of finding the copulation plug is termed day 1/2 post-coitum (p.c.). It should be noted that different systems are adopted in the literature to denote the day of finding the plug (day 0 or day 1).

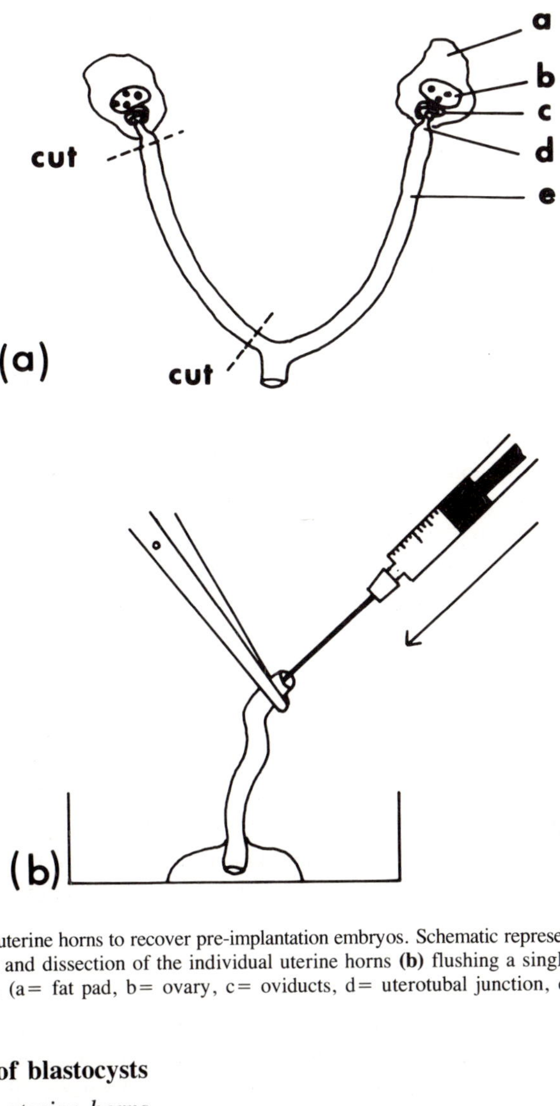

Figure 1. Flushing uterine horns to recover pre-implantation embryos. Schematic representations of **(a)** isclated reproductive tgract and dissection of the individual uterine horns **(b)** flushing a single uterine horn with a stream of medium. (a= fat pad, b= ovary, c= oviducts, d= uterotubal junction, e= uterus).

3.2 Recovery of blastocysts

3.2.1 *Flushing uterine horns*

On day 3.5 p.c. the females are killed and the blastocysts recovered from the uterine horns by the following procedure.

(i) Kill the mouse by cervical dislocation.

(ii) Place the animal ventral side uppermost on several layers of paper towels. Sterilize the abdominal skin by swabbing with 70% alcohol.

(ii) Grasp the skin with blunt forceps, lift, and using a pair of dressing scissors, open the abdominal cavity to display the contents. Displace the guts to one side and locate the two uterine horns. Remove the individual horns by cutting below the uterotubal junction and above the cervical bifurcation (for guidance see *Figure 1a*).

(iv) Remove the horns onto a fresh layer of paper towels. Trim away the associated adipose tissue and mesenteries using a pair of fine scissors (e.g. Iris scissors).
(v) Fill a 1 ml disposable syringe with DMEM plus 5% NCS and attach a 25-guage hypodermic needle. The function of the serum is to provide protein to prevent the embryos from sticking to glass and plastic surfaces. Carefully introduce the needle into one end of the horn, for a distance of 2–3 mm, and grasp the needle through the uterine wall using a pair of blunt forceps.
(vi) Lift the horn off the paper towel and lower the free end into a 6 cm bacteriological Petri dish until it touches the bottom surface. Recover the contents of the horns by flushing a stream of medium through the uterine lumen (see *Figure 1b*). Discard the uterus and repeat with remaining horn.

3.2.2 *Transferring embryos into culture*

Pre-implantation embryos are most conveniently handled using a drawn-out Pasteur pipette held in a mouth-controlled tube. The pipette is held at both ends and the narrow part is melted in a Bunsen burner flame. When the glass has softened the pipette is withdrawn from the flame and the end is pulled out sharply to create a fine capillary (approximate diameter of 100 μm). The capillary is snapped off to give a clean right-angle break. The pipette is held in the end of a piece of tubing to which a mouth holding piece has been fitted.

The most convenient way to culture blastocysts is in small drops of medium under inert liquid paraffin oil (14). This not only enables embryos to be located easily but also the paraffin oil prevents rapid alterations to temperature and pH. Prepare a drop culture in the following way.

Take a 6 cm tissue culture plate and place on to it an array of small drops (drop size of ~50 μl) of medium (DMEM + 5% NCS). Flood the plate with sufficient light weight paraffin oil to cover the drops and place the plate in the incubator. Note that it is important to check the oil for toxicity effects. This may be done by assessing the viability of embryos incubated in drop cultures for several days, or layering the oil on top of a culture of stem cells and observing for cell death.

Recover the embryos by the following procedure.

(i) Place the plate containing the contents of the uterine horns on the stage of a stereo dissecting microscope. Using transmitted illumination and a magnification of ×20 locate the blastocysts. This is most easily done by focusing first on the bottom surface of the plate using a scratch mark for reference.
(ii) Draw out a Pasteur pipette and introduce the end of the pipette into the plate. Medium will be drawn into the pipette by capillary action and the embryos will enter the end of the pipette without the necessity to suck on the mouth holder. Collect two or three embryos at one time.
(iii) Remove the plate and replace with the drop culture. Gently blow the embryos into a drop of medium taking care not to introduce bubbles of air.
(iv) Repeat until all the embryos have been transferred.
(v) Return the drop culture to the incubator.

The embryos may not form a synchronous population, even from the same female, and not all the embryos may have cavitated but may still be at the compacted morula

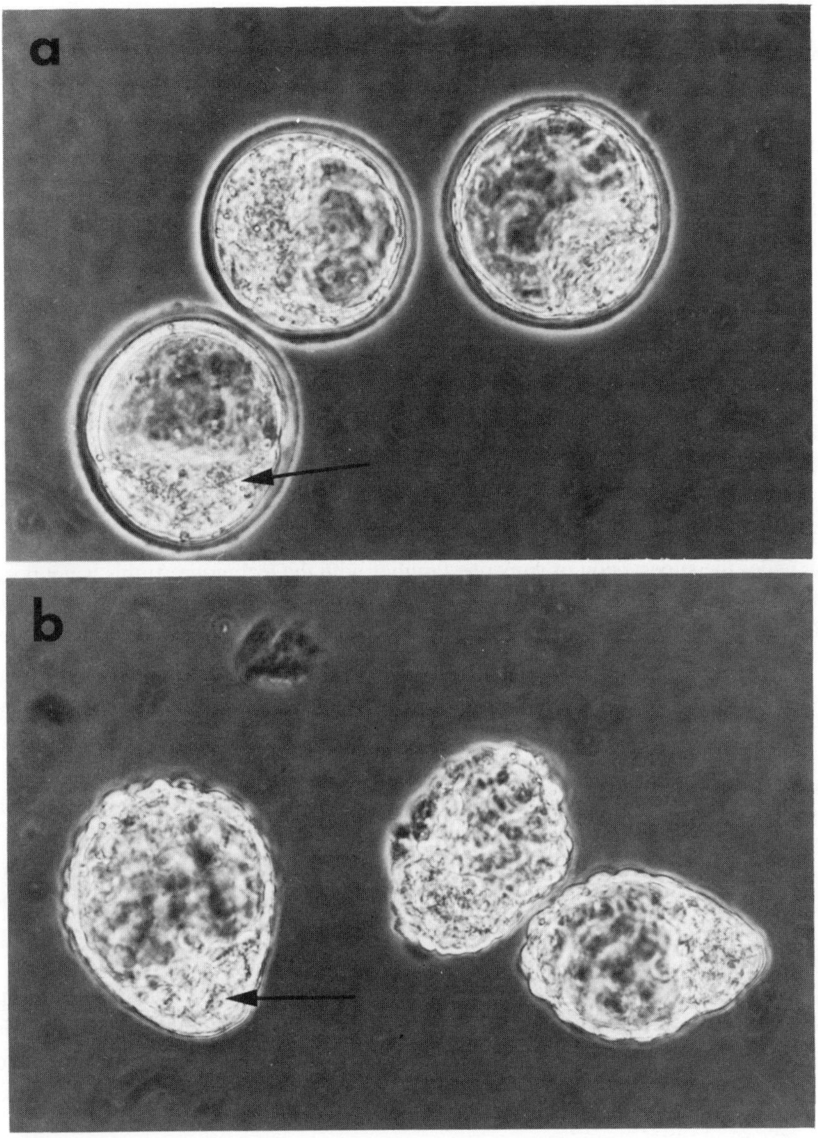

Figure 2. Appearance of embryos recovered from the uterus. **(a)** fully expanded blastocysts with an outer layer of trophoblast cells surrounding the inner cell mass (arrowed) **(b)** delayed blastocysts. Hatched from the zona pellucida these embryos lose their spherical form. Trophoblast cells and inner cell mass (arrowed) are clearly visible.

stage. Morulae should be retrieved for culture as the majority will progress normally and form blastocysts within a few hours. Additionally, in many instances a proportion of the embryos are abnormal having arrested at an early cleavage stage. As a guide, and to aid identification, representative embryos are shown in *Figure 2*.

3.3 Production of delayed blastocysts

Blastocysts which have undergone a period of implantational delay prior to flushing may be more suitable material from which to recover stem cells. In the author's experience the success rate can be significantly improved by the use of delayed blastocysts.

Implantational delay is brought about by altering the hormonal status of the pregnant female (15). The source of cestrogen production is removed by surgical ovariectomy 2.5 days p.c. Lowering the level of oestrogen, together with the post-operative administration of a synthetic progesterone, prevents the embryos from implanting. The embryos develop normally to the blastocyst stage but as the tissues of the uterine wall have been artificially rendered non-receptive, the embryos arrest and remain free floating within the uterine lumen. In the delayed state the embryos remain viable for up to 10 days. For the purposes of isolating stem cells a 4–5 day delaying procedure is normally used.

3.3.1 *Ovariectomy*

Ovariectomy is a simple and quick procedure. As with all surgical procedures it is best learnt under the supervision of someone familiar with the technique. If this is not possible it is advised that practice operations be carried out on a dead animal in order to become familiar with the anatomy. The technique of ovariectomy is performed as follows.

(i) Take pregnant females at 2.5 days p.c. At this stage of pregnancy the embryos have developed to the morula stage and are located in the oviducts. Weigh the female and anaesthetize with Avertin (tribromoethanol; for details of the Avertin anaesthetic see Chapter 5, Section 8).

(ii) When the animal is fully unconscious (no eye reflex) place it on layers of paper towels and swab the lower back with 70% alcohol. Using blunt forceps lift the skin on the lower back and holding a pair of sharp scissors in a vertical position make a small incision (~0.5 cm) across the midline. Wipe the cut with a tissue in a head to tail direction. This serves to both open the wound while removing the cut hairs.

(iii) Move the skin laterally passing over the lateral back muscles and locate the ovarian fat pad and attached ovary, which are readily visible through the body wall. Grasp the body wall over the region of the ovary using a pair of fine watchmakers forceps (e.g., Inox, No. 5) and make a small incision (~0.5 cm) in the body wall using a pair of sharp fine pointed scissors.

(iv) Push down gently on the flank at the sides of the cut to encourage the ovarian fat pad to emerge from the abdominal cavity. Grasp the fat pad using blunt forceps and with a gentle teasing action pull clear of the incision. The ovary, which is embedded in the fat tissue, will follow. The oviducts and the top of the uterine horn will also be seen.

(v) Attach a small serriphine clip to the fat pad and use the weight of the clip to hold the ovary in position and prevent it from slipping back inside the body cavity.

(vi) Place the mouse under a binocular microscope with a good incident lighting source. With reference to *Figure 3* identify the parts of the reproductive tract. Using two pairs of watchmakers forceps locate and gently tear open the transparent

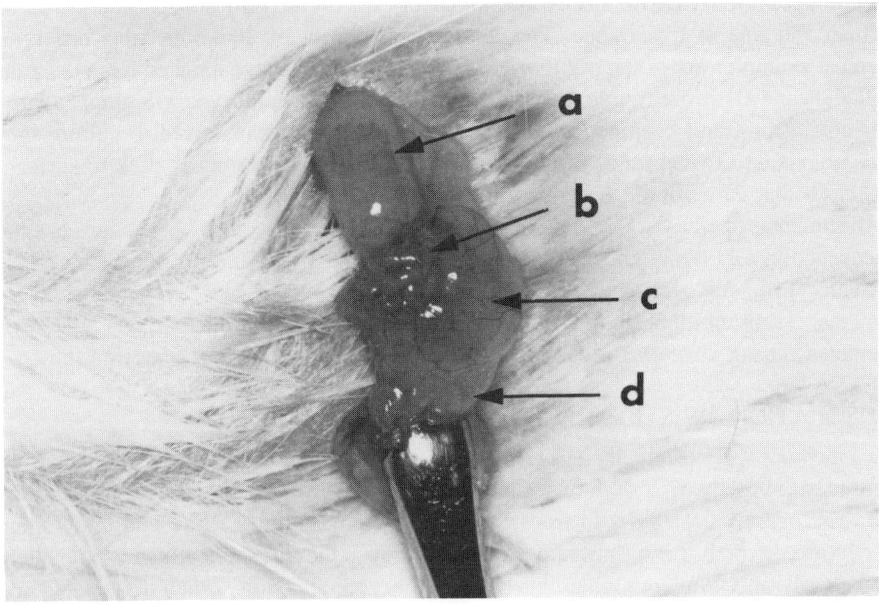

Figure 3. Anatomy of the ovarian region of the reproductive tract. Clearly visible are the top of the uterine horn **(a)** the coiled tubes of the oviducts **(b)** closely opposed to the ovary **(c)**. The ovary and oviducts are held in close proximity by a transparent membrane. The ovary is partly embedded in the peri-ovarian fat pad **(d)**.

membrane which encapsulates the ovary and oviducts. Take a small length of fine suture cotton and tie a small loop (diameter of ~0.5 cm). Using the forceps manoeuvre the cotton loop between the ovary and the oviducts taking care not to damage, or trap, the oviducts. Slowly tighten the loop while lifting the cotton at both ends. When the cotton is correctly positioned so that it completely encircles the ovary, pull the loop firmly closed to ligature the blood vessels. Using the sharp scissors cut out the ovarian tissue taking care not to cut the cotton knot. It is important to remove all the ovarian tissue.

(vii) With a pair of watchmakers forceps grasp one cut edge of the body wall and lift slightly. This opens up the incision. Using blunt forceps return the fat pad and oviducts to the abdominal cavity.

(viii) Repeat the procedure on the other side.

(ix) Close the incisions in the body wall with single sutures. Clip the cut edges of the skin together using one or two metal Michel clips (7.5 mm). Do *not* suture the skin as the mouse will quickly bite through the cotton.

(x) While the animal is still under the influence of the anaesthetic, administer 0.1 ml of a 10 mg/ml solution of Depo-Provera (UpJohn), diluted in saline, subcutaneously in the belly region.

(xi) Place the mouse in a cage and leave in a warm location until the animal has regained consciousness.

3.3.2 *Recovery of delayed blastocysts*

The delayed blastocysts are recovered by the same method as for fertilized blastocysts

(Section 3.2). The uterine horns are noticeably different from normal and are often thinner and pale in appearance. The delayed blastocysts are also noticeably different. Typical examples are given in *Figure 2*. Delayed embryos have hatched from the zona pellucida and are more fragile than normal blastocysts. The embryos will often collapse during flushing and become more difficult to identify. If transferred to culture they will re-expand. The absence of a zona pellucida makes the embryos stick to surfaces easily so that care should be taken during handling.

If implantation sites are evident in the uteri then the ovariectomy procedure has not succeeded in removing all traces of the ovarian tissue. Even small fragments of tissue provide sufficient oestrogen to allow implantation to take place. The most likely explanation for an inability to find embryos on flushing is that the oviducts have been damaged during surgery preventing the embryos from entering the uterine lumen.

4. ISOLATION OF STEM CELL LINES

Two methods were originally described to recover stem cells (1,2). The technqiue given here is based on the first of these which was described initially by Evans and Kaufman (1). The overall strategy is to transfer intact blastocysts into culture and allow them to continue growth, for a limited period, until they achieve a stage which is equivalent to the early post-implantation embryo. The embryonic portion is then dissociated and culture of the cells is continued under conditions designed to be amenable to cell growth while encouraging stem cells to maintain the undifferentiated phenotype. Stem cell lines can be retrieved from a proportion of the embryos treated in this way.

The technique described by Martin (2) is slightly more complicated in two respects. Firstly the inner cell mass (ICM) of the embryo has to be selectively isolated by immunosurgical removal of the trophectoderm cells prior to culture. Secondly, culture takes place in medium which has been conditioned by exposure to a culture of EC cells. However, if attempts are to be made using this technique it is relevant to note that a comparative study (16) has shown that conditioned medium has no signfiicant effect on the overall efficiency of isolation of stem cells.

Of the two techniques the method of Evans and Kaufman (1) is the simpler and in the author's experience has been found reliably to give rise to stem cells at an acceptably high frequency (10–30% of embryos).

The procedure for the derivation of stem cells from blastocysts is divided into two stages. During the first stage the embryos are transferred into tissue culture and left undisturbed to attach and grow. The second stage involves the mechanical disruption of the embryos and the culture of the embryo-derived cells. In a percentage of cases it is possible to identify cells of a stem cell phenotype in the embryo cultures. Colonies of these cells are selected and expanded to give permanent cell lines.

4.1 **Culture medium**

The composition of the medium devised for the culture of blastocysts and the subsequent stem cell derivatives is given in *Table 4*. Briefly, the medium is DMEM supplemented with both foetal calf serum and NCS. To this basic medium is added 2-mercaptoethanol at low concentration. 2-Mercaptoethanol, together with other sulphydryl-containing compounds, has been shown to increase the viability and plating efficiency of primary

Table 4. Composition of culture medium used for the isolation of stem cells.

To prepare approximately 100 ml of medium

80 ml DMEM
10 ml foetal bovine serum
10 ml newborn calf serum
1 ml antibiotic stock (*Table 2*)
1 ml mercaptoethanol stock
1 ml non-essential amino acids (basal stock; Gibco-Biocult, Flow Labs)
1 ml nucleosides stock

Warm to 37°C before use. Store at 4°C.

Mercaptoethanol stock

To prepare 10 ml of stock add 7 μl of 2-mercaptoethanol (Analar) to 10 ml of PBS. Filter sterilize and store at 4°C. Replace at weekly intervals.

Nucleosides stock

To prepare 100 ml

Adenosine 80 mg
Guanosine 85 mg
Cytidine 73 mg
Uridine 73 mg
Thymidine 24 mg

Add to 100 ml of distilled water and dissolve by warming to 37°C. Filter sterilize and aliquot while warm. Store at 4°C. The nucleosides will come out of solution. Warm to 37°C before use to re-solubilize.

cell cultures (17) and EC cell lines (18). The levels of non-essential amino acids are raised by a factor of two. Additionally, the levels of free nucleosides are raised. A 100× stock of nucleosides is made up from scratch using Analar ingredients. Medium which has been stored for longer than 2 weeks should be supplemented with 1 mM glutamine.

4.2 **Blastocyst culture**

During this first stage embryos are transferred into tissue culture. The embryos hatch from the zona pellucida and attach within 24−36 h by outgrowth of the trophoblast cells. The spreading of the trophoblast onto the substrate results in the ICM component, which during normal development is sequestered within the trophoblast cells, becoming exposed to the tissue culture environment. The blastocyst outgrowths are left undisturbed for a period of between 4 and 6 days continuous culture. During this time the ICM cells proliferate giving rise to a small clump of cells.

There are a variety of technical considerations to be made regarding the best strategy for the culture of blastocysts during this initial stage. These are as follows.

(i) Embryos collected on a single day can be kept together, and cultured as one group, or split up into individual embryo cultures. For most purposes it is recommended that embryos be cultured singly. This has the advantage of avoiding the necessity for recloning the stem cell cultures obtained, as a cell line derived from a single embryo is effectively a clonal population in terms of genotype and sex chromosome complement. An additional advantage is that the monitoring and recording of embryos on an individual basis provides information about the

Table 5. Method for blastocyst culture.

1.	Prepare feeder layers in 1 cm wells[a] (Section 2.4). Ideally, these should be prepared the previous day to allow the feeder cells to attach and spread.
2.	Aspirate the wells and replace with 1 ml of embryo culture medium (*Table 4*).
3.	Remove the drop culture containing blastocysts from the incubator.
4.	Under a binocular microscope transfer the individual blastocysts into the wells using a drawn out Pasteur pipette. Take care to place the embryos in the centre of the well[b].
5.	Return the well plates to the incubator.
6.	Inspect daily over a period of 4–6 days paying particular attention to the appearance and relative size of the ICM-derived component.

[a]It is convenient to use plates containing four wells (Nunc). Feeder layers should be prepared by adding 5×10^4 STO cells, or 10^5 primary fibroblasts, in 1.0 ml of medium. The cells should be left for 24 h and inspected to ensure that the cells have formed a confluent monolayer.
[b]Embryos which settle at the edge of the well are difficult to examine under the microscope.

overall viability of cultures and the efficiency of recovering stem cells. The major disadvantage to this approach is that it takes considerably more time and materials to maintain large numbers of individual cultures.

(ii) Blastocysts can either be grown in drop cultures (Section 3.2) under a layer of liquid paraffin oil or in larger tissue culture dishes. The advantage of drop cultures is that embyros are easily located for observation as they are physically confined to a small area. In our experience, however, embryos grown in drop cultures for the necessary 5 or 6 days fare less well. This is probably due to depletion of an essential growth requirement in the medium or, more likely, results from the gradual accumulation into the medium of traces of toxic residues from the overlying oil. The most satisfactory method is to use small tissue culture wells (10 mm) which hold about 1 ml of medium.

(iii) Embryos may either be cultured directly on tissue culture plastic or on feeder layers. The use of feeder layers during initial culture apears to maximize the viability of blastocyst cultures.

The method for culturing blastocysts is summarized in *Table 5*. The progressive change in the appearance of cultured blastocysts over a 4-day period is shown in *Figure 4*. Note that it is normal for a small percentage of the embryos to fail. This may be seen as a failure to hatch, attach or develop an ICM outgrowth.

4.3 Disaggregation of the inner cell mass

The stage at which an individual ICM-derived component is selected for disaggregation is fairly critical. *Figure 5* shows examples of the morphologies of blastocyst outgrowths. The most suitable cultures are indicated as well as examples of less promising cultures. Under the conditions of culture described in Section 4.2 embryos will normally attain a suitable morphology following a 5-day culture period. A degree of variability in the size of the ICM-derived cell clump between embryos of the same batch is a common observation. As a result it may be necessary to disaggregate ICM clumps over a 2-day period according to the morphology of individual cultures. It is important to note that there is by no means an absolute correlation between the phenotype of the colony and the successful isolation of a stem cell line; stem cells can be obtained from both vigorously growing ICMs and from small outgrowths. In our experience, however,

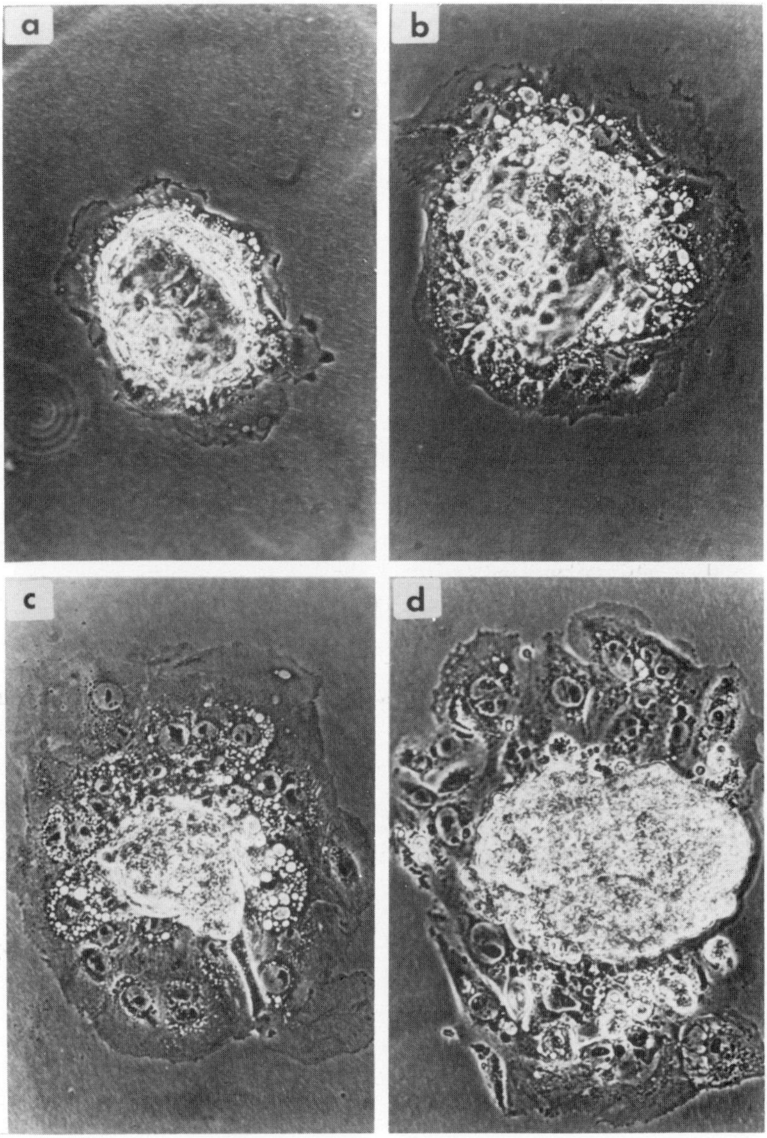

Figure 4. Attachment and growth of a single blastocyst on a gelatinised tissue culture surface. (**a**) 1.5-day culture, blastocyst attaching by trophoblast cell outgrowth; (**b**) 2.5-day culture, ICM component is revealed by the continued spreading of the trophoblast cells; (**c**) 4-day culture, cells of the ICM proliferating rapidly; (**d**) 5-day culture, ICM has formed a mass of cells.

ICM clumps in which there is extensive endoderm formation, or a relatively rapid progression to an overtly multilayered egg cylinder-like structure, tend to have a reduced chance of retaining pluripotential cells.

ICM-derived clumps are disaggregated as follows.

(i) Aspirate the medium from the well and replace it with 1 ml of PBS.

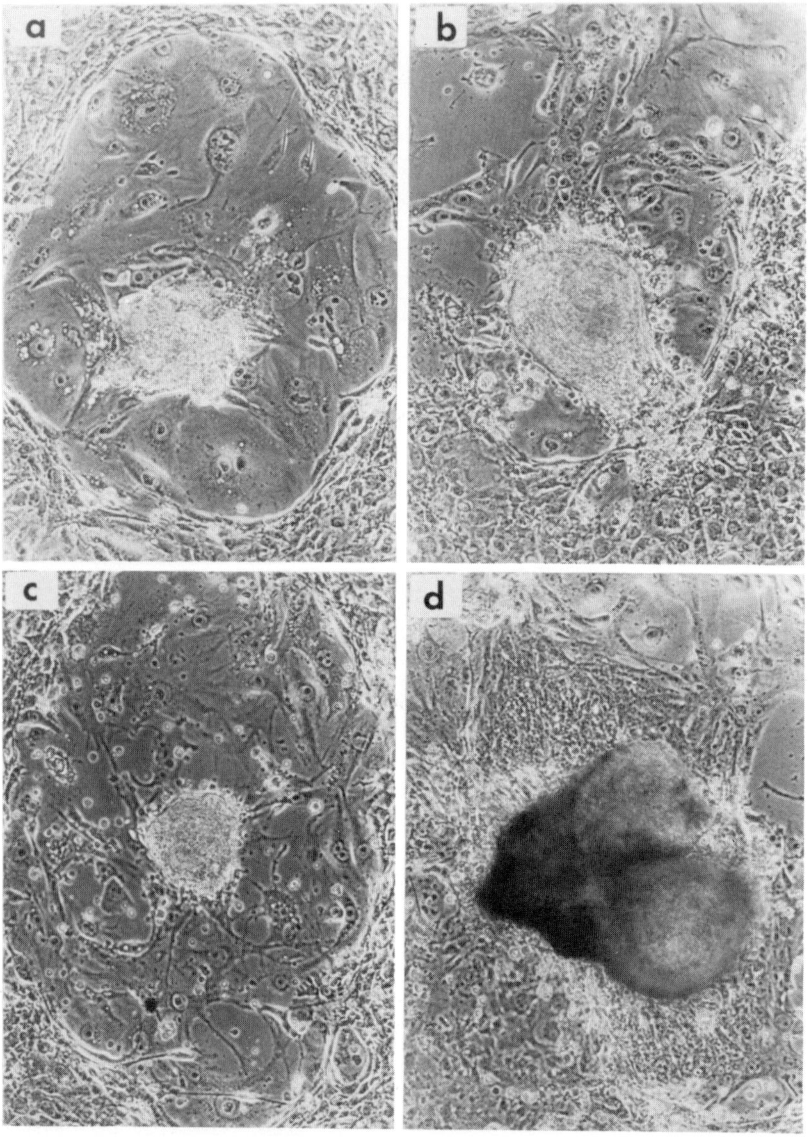

Figure 5. A series of blastocyst outgrowths on feeder layers of STO cells. **(a)** and **(b)** appearance of blastocysts after a 5-day period in culture; **(c)** 4 days in culture; **(d)** 6 days in culture. The outgrowths in **(a)** and **(b)** are suitable for disaggregation. The outgrowth seen in **(c)** has not yet attained the size viewed as being most suitable for disaggregation. Outgrowths similar to that shown in **(d)** are not suitable; the ICM component has proliferated and differentiated extensively. Large quantitites of endoderm-like cells can be seen covering the surface of the outgrowth and there is evidence for extensive internal differentiation.

(ii) Place the well under a binocular microscope. Pull out a Pasteur pipette to give a fine capillary and heat-seal the tip in a Bunsen burner flame. Using this blunt end gently push the ICM clump free from the trophoblast cell layer taking care not to fragment the ICM clump.

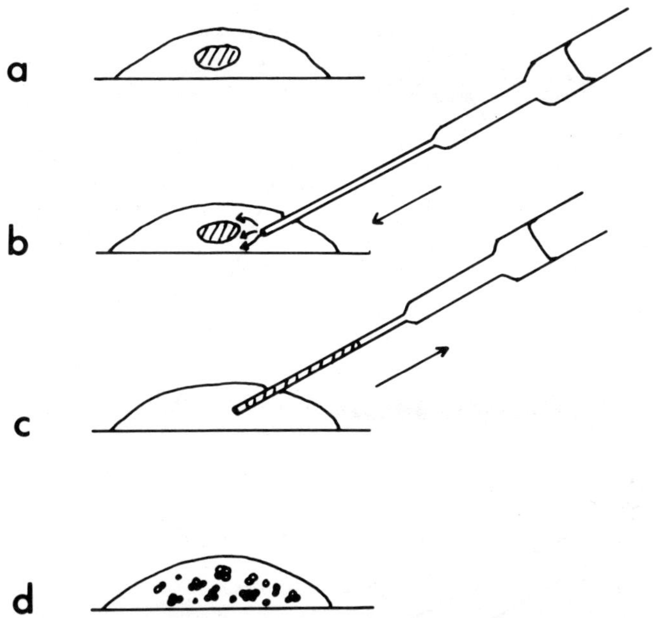

Figure 6. Schematic representation of the procedure used for disaggregation. **(a)** the ICM clump is placed in a drop of trypsin EDTA and allowed to digest; **(b)** the end of a drawn-out Pasteur pipette, filled with serum-containing medium, is introduced into the drop and a stream of medium is used to neutralise the trypsin; **(c)** using mouth control the ICM is drawn into and out of the capillary tube. The procedure is repeated until the ICM is broken up **(d)**.

(iii) Using a Pasteur pipette and bulb pick up the cell clump in a minimal volume of PBS and transfer into a small drop of trypsin EDTA ($\sim 50\ \mu l$) under oil.

(v) Transfer the plate containing the drop into the incubator and leave to digest for 5 min.

(v) Pull out a Pasteur pipette to give an extremely fine capillary. The diameter of the end should be approximately a quarter of the diameter of the ICM clump to be disaggregated. Place in a mouth-controlled holder and by suction pick up a small volume ($\sim 50\ \mu l$) of embryo culture medium.

(vi) Return the plate to the binocular microscope and using the Pasteur pipette break up the ICM clump. First blow a small quantity of medium into the drop to neutralize the trypsin and then, using mouth control, draw the ICM clump in and out of the end of the pipette. This should reduce the ICM clump into smaller cellular fragments and a few single cells. It may be necessary to repeat this action several times. Draw the contents of the drop into the pipette and transfer them into the centre of a fresh feeder well containing 1 ml of embryo culture medium.

(vii) Transfer the feeder well plate to the incubator.

(viii) Repeat the procedure with the remaining embryos.

The procedure is summarized diagrammatically in *Figure 6*. Photographs of the different stages are given in *Figure 7*.

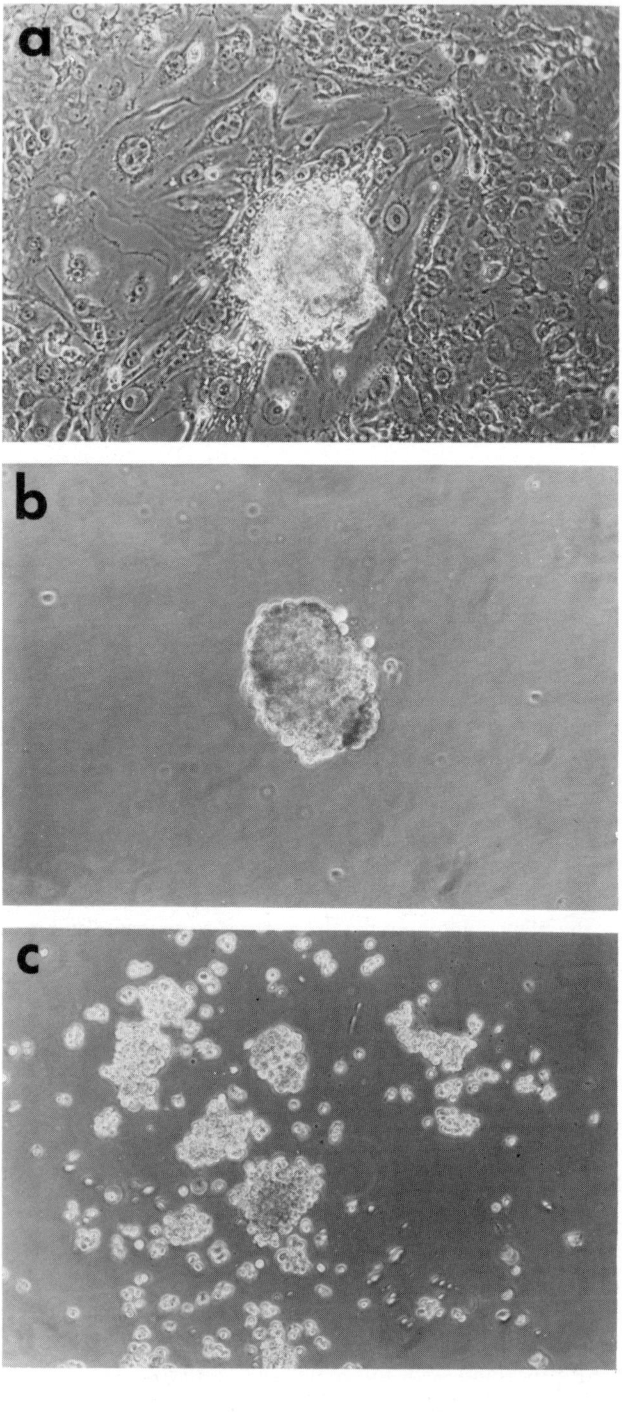

Figure 7. Appearance of the ICM-derived cell clump during the disaggregation procedure. **(a)** intact outgrowth on a feeder layer; **(b)** isolated ICM in a drop of trypsin; **(c)** fragmented ICM after disaggregation.

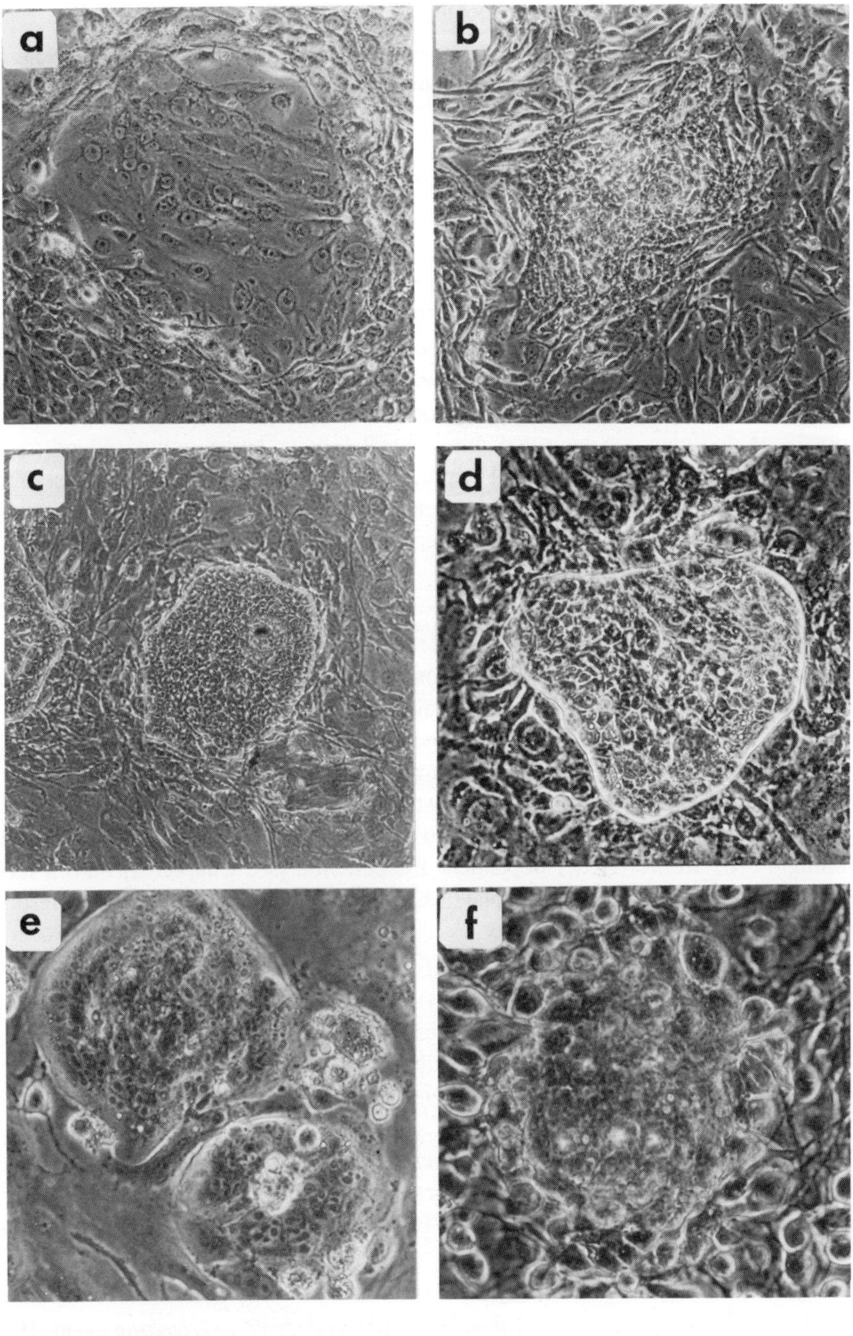

Figure 8. Morphologies of primary colonies which result from the disaggregated ICM. **(a)** patch of 'spread' trophoblast cells; **(b)** fibroblast-like cells; **(c)** epithelioid-like cells; **(d)** epithelioid cells, single colony of cells at high power; **(e)** appearance at high power of trophoblast-like cells before cell flattening has occurred; **(f)** colony of stem cells.

4.4 **Identification of ICM-derived colonies**

The majority of the ICM-derived cellular aggregates, generated as described in Section 4.3, attach to the feeder layer and continue to divide to give rise to small colonies of cells. Initially the feeder cells may obscure these colonies but after about 2 days of culture they should become apparent by scanning the well under low power magnification (×100). On average between five and ten discrete colonies of cells appear in each well.

4.4.1 *Colony morphologies*

Cultures should be left for an initial 4-day period to allow the colonies to become established. Inspect the colonies individually under high power phase contrast (×200). Generally speaking, colonies are composed of cells of the same cellular phenotype although 'mixed' colonies do sometimes arise. Colonies of cells can be classified into several broad types on the basis of morphology. These are illustrated in *Figure 8*. The majority of colonies can be ruled out immediately as being composed of overtly differentiated cell types *(Figure 8)*. The predominant colony type is trophoblast. These appear as areas of giant cells which push aside the feeder cells or as colonies which develop giant cells at their periphery after a few days in culture.

4.4.2 *Identification of stem cells*

It is important to be able to accurately identify colonies of pluripotential stem cells. Positive identification is difficult unless one is familiar with the appropriate cellular morphology. Pluripotential cells are typically small, have a large nucleus and minimal cytoplasm, and the nuclei contain one or more prominent dark nucleoli structures. The cells pack tightly together in small nests in which it is difficult to discern the individual component cells. Individual cells are most easily seen at the edges of the colony.

A helpful exercise, prior to attempting to isolate stem cells from embryos, is to plate out single cells from a feeder-dependent EC cell line (see Chapter 2) onto a feeder plate (say 5×10^2 cells on a 10 cm plate). Carefully observe the growth of the resulting colonies over a period of 5–8 days.

The best method for the identification of stem cell colonies is to mark potential colonies after 4 days of culture. This is easily done using a fine-tipped indelible marking pen on the under surface of the well. These colonies should be inspected daily over the following 3 days. Colonies which exhibit cellular heterogeneity, i.e. spreading of the cells at the periphery and general flattening, can be progressively eliminated. Only retain colonies of cells which maintain a stable phenotype.

Figure 9 illustrates the progressive differentiation of a putative stem cell colony. Such colonies are disregarded. *Figure 10* illustrates the progressive growth of a colony in the absence of overt differentiation. This is characteristic of a true stem cell colony.

4.5 **Expansion of embryonic stem cells**

Subculture colonies of stem cells 7 days after the primary disaggregation event. If colonies are selected before this time there is a risk, firstly, of selecting non-stem cell colonies and, secondly, of loosing the cells as the constituent cells of a small colony will replate less efficiently. If the colonies are cultured in excess of 7 days they reach

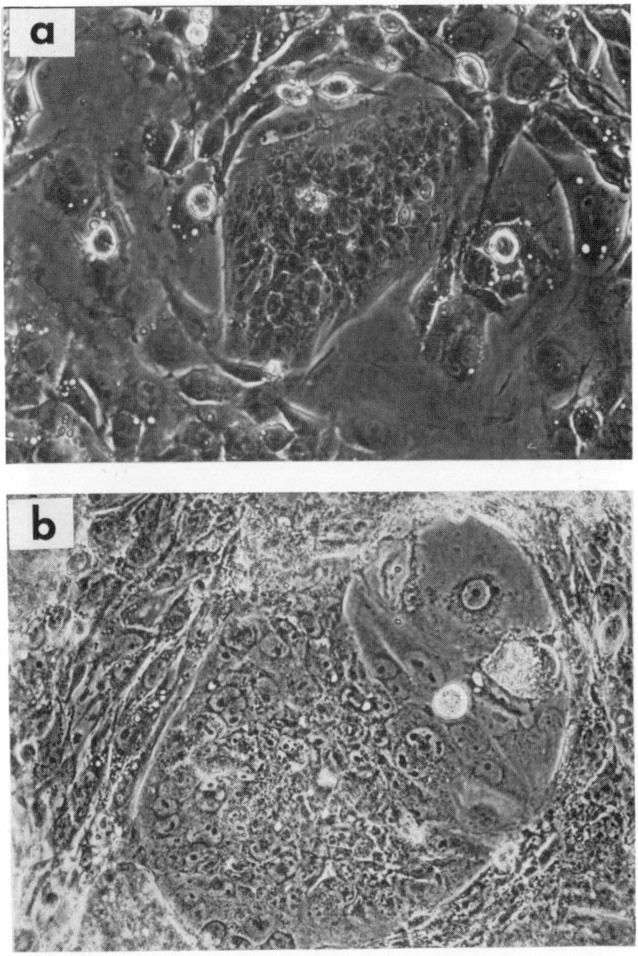

Figure 9. Progressive alteration to the morphology of a putative stem cell colony over a 3-day culture period. **(a)** when first located 2 days after disaggregation of the ICM the constituent cells superficially resemble stem cells; **(b)** following a further 3 days' culture the colony has spread and flattened to form a patch of giant trophoblast-like cells.

a size at which cellular differentiation is initiated. This is normally seen as endoderm formation on the outside free surface of the colony. The size and morphology of a colony appropriate for dissociation is shown in *Figure 10c*.

Stem cell colonies should be disaggregated individually. This eliminates the possibility of cross-contamination with other cell types. Colonies are disaggregated as follows;

(i) Aspirate the well and replace with 1 ml of PBS.
(ii) Using a drawn out Pasteur pipette dislodge the colony and transfer it into a drop of trypsin EDTA (see Section 4.3).
(iii) Incubate the drop for 2–3 min at 37°C.
(iv) Using a finely drawn-out Pasteur pipette disaggregate the colony as described

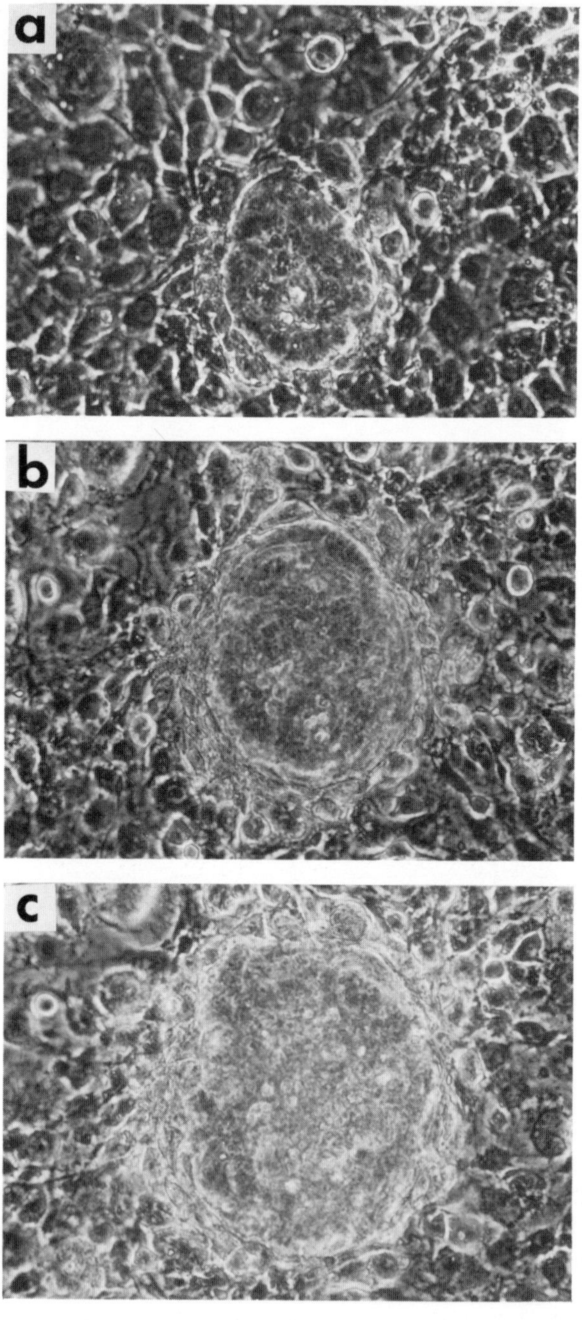

Figure 10. Progressive growth of a colony of stem cells. **(a)** day 3, **(b)** day 4, **(c)** day 5. Note that the colony increases in size in the absence of overt differentiation. The cells pile up and maintain a discrete colony on top of the feeder cells. The constituent cells adhere tightly to one another making it difficult to see individual cells. This growth pattern is characteristic of stem cells although the colonies may appear slightly flatter and less regularly shaped than the colony illustrated here.

in Section 4.3. In this case it is important to reduce the colony into a single cell suspension. This should be achieved easily as stem cells are more readily dissociated than the original ICM-derived clump.

(v) Transfer the cells into a fresh feeder well.

Small colonies of stem cells should be apparent within 2 days. The wells should contain only colonies of a stem cell phenotype. Some of the potential stem cell colonies will fail to passage and no progressively growing cells will appear. This is a result of a mis-identification of the primary colony. If stem cells are not apparent within 2 days the well should be discarded.

For a given embryo all possible stem cell colonies should be picked and dissociated as described. Positive wells containing stem cells should be identified and assigned a code number. The cultures should be incubated for 4–5 days and re-fed during this period if necessary. After this time most cultures should be passaged by the following procedure.

(i) Aspirate the well and wash with 1 ml of PBS.

(ii) Add 100 μl of trypsin EDTA and incubate for 3–4 min at 37°C.

(iii) Using a Pasteur pipette and bulb pick up about 0.5 ml of embryo culture medium and add to the well. Vigorously pipette the suspension to break up the cellular aggregates.

(iv) Transfer the suspension to a 3 cm feeder plate containing 2 ml of embryo culture medium. Label the plate 'passage 1' and return the plate to the incubator.

(v) Inspect the plate daily and re-feed when necessary.

Treating the cells in this way encourages the cells to divide rapidly. The cultures should be re-fed at 1–2 day intervals and the plates passaged onto larger dishes when confluent. This is normally 4 days after subculturing. *Figure 11* shows the morphology of a sub-confluent established cell line at 'passage 1'.

When sufficient cells are available freeze samples of the culture. For the freezing procedure see Section 6.2.

4.6 Delayed blastocysts

In the authors's experience blastocysts which have undergone a period of implantational delay (see Section 3.3) are more successful in producing stem cell lines. Treat delayed embryos exactly as for normal blastocysts described in the previous sections. The one major difference is in the timing of the disaggregation of the ICM-derived clump. Delayed embryos behave in a similar manner to normal embryos and cannot be distinguished from them in terms of morphological criteria. They are more slow growing during the initial stages of culture due to the cessation of cell division imposed by the delaying period. A period of 6 days in culture prior to disaggregation has been found to give the best results. The primary colonies of cells which form are identical to those which form from normal embryos (see *Figure 8*).

The reasons why it is easier to recover pluripotential cells from delayed blastocysts are unclear. It has been suggested that one effect of delay is to selectively increase the number of cells in the ICM (1). Alternatively, the delaying procedure artificially prolongs the time the embryo spends in the blastocyst stage, during which time the ICM is sequestered within the layer of trophectoderm cells. Additionally the embryo

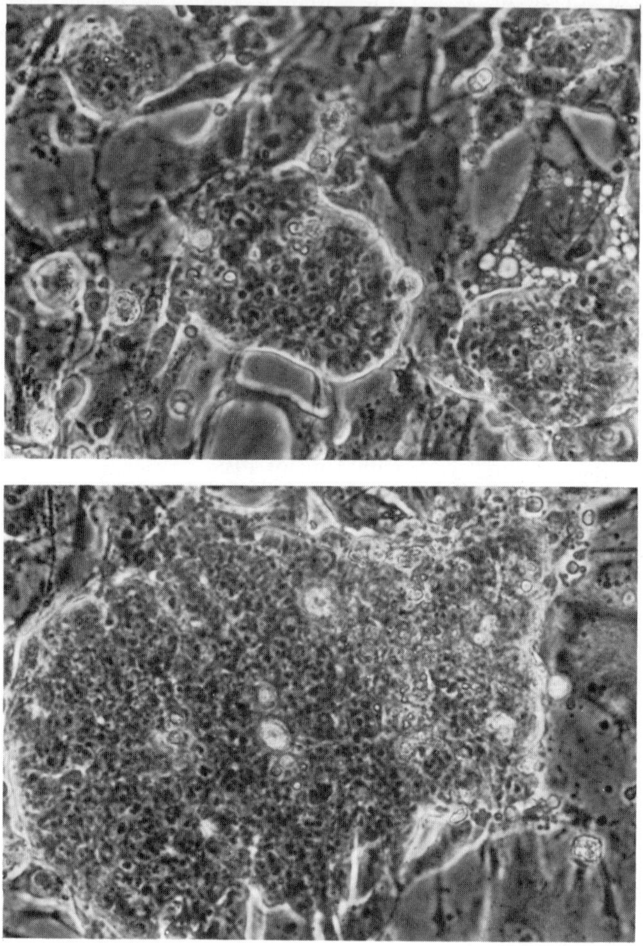

Figure 11. Examples of typical areas of cells found within a sub-confluent culture of established stem cells, at 'passage 1', to illustrate the cellular morphology exhibited by pluripotential cells. The culture was seeded initially as single cells. These divide and the sister cells stay together to form small nests. With time the nests increase in size and merge to form a confluent layer.

ceases growth. This imposed reduction in cell division and differentiation may be the most important factor.

Through the use of delayed blastocysts the efficiency of isolation of permanent cell lines is normally in excess of 30%.

5. STEM CELL LINES FROM PARTHENOGENETICALLY-DERIVED BLASTOCYSTS

Mouse oocytes can be induced to initiate development in the absence of fertilization. A variety of methods can be used for the experimental production of parthenogenetic mouse embryos (reviewed in ref. 19). Activation of oocytes results in the production of various classes of parthenogenones; these are either haploid or diploid in genetic constitution (see Section 5.4). The majority of parthenogenetically-derived embryos will develop normally through the pre-implantation stages of development. In a small number of cases embryogenesis may proceed well beyond implantation before the embryos fail in development (reviewed in ref. 19).

The method described here is the ethanol activation technique whereby oocytes are stimulated to develop by a brief exposure to a dilute solution of ethanol. The success of the method varies according to the mouse strain used. F_1 hybrids are generally the most successful and high levels of *in vitro* development to the blastocyst stage have reliably been achieved with (C57BL × CBA)F_1 hybrid eggs (19). For most other mouse stocks activated oocytes maintained in culture normally arrest in development after the first cleavage division – the so-termed '2-cell block'. For these strains successful development is only achieved if the eggs are transferred back to the oviducts of pseudopregnant recipient females. This is not only a technically very difficult procedure but relatively high losses of embryos occur as a result of the failure of a proportion of the recipients to maintain the pregnancy.

This section describes the experimental production of activated F_1 hybrid oocytes and the *in vitro* culture techniques necessary to produce the blastocyst stage embryos from which stem cells can be isolated.

5.1 **Super-ovulation**

In order to collect large numbers of oocytes at a given time, female mice are superovulated by the following hormone regime. Use 8–10 week old (C57BL × CBA)F_1 hybrid mice.

(i) Day 1. Select 10 females which show no visible signs of oestrus (13). Inject mice intraperitoneally with 5 i.u. of pregnant mares serum gonadotrophin (PMSG: Intervet) at 1600 hrs. The PMSG should be diluted at 50 i.u./ml in PBS, aliquoted and stored at –30°C. Thaw immediately before use.

(ii) Day 3. Inject mice intraperitoneally with 5 i.u. of human chorionic gonadotrophin (HCG; Intervet) at 16.00 hrs (i.e. 48 h following PMSG). Dilute and store the HCG as for PMSG.

Ovulation occurs approximately 12 h following administration of the HCG.

5.2 **Collection of cumulus masses**

At 9.00 hrs on day 4 (i.e. 17 h post-HCG) the ovulated oocytes, shed from the ovaries within the cumulus mass, are collected. The oocytes are most conveniently recovered and handled within the cumulus mass structures.

(i) Kill the female by cervical dislocation and sterilize the skin with 70% alcohol. Open the abdominal cavity and locate the reproductive tract. Using a pair of fine forceps cut through each uterine horn just below the uterotubal junction. Holding the cut end of the uterus with a pair of blunt forceps pull the oviducts away from the ovary and carefully cut between the oviducts and ovary. This procedure should free the oviducts from the reproductive tract.

(ii) Dissect and transfer the oviducts to a dish containing PBS. Place the dish under a dissecting microscope. Using two pairs of watchmakers forceps manipulate the tissue and locate the ampullary region. This structure is located slightly below

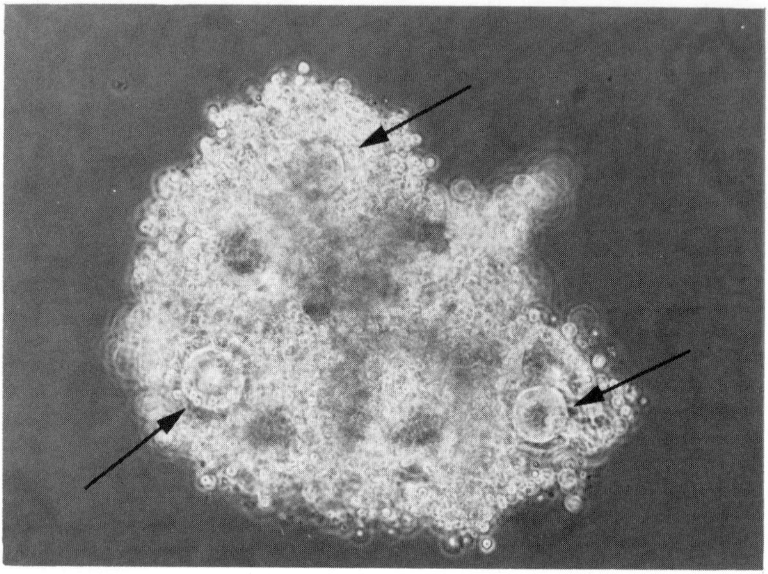

Figure 12. An individual Cumulus Mass recovered from the ampullary region of the oviducts. Oocytes are shed from the ovary within a 'cloud' of adhering follicle cells. The individual oocytes (arrowed) can be discerned within the mass. The follicle cells are normally dispersed by the sperm by about 12 h after mating has taken place.

the end of the oviduct opening (the infundibulum), is normally the largest part of the oviduct and is typically longitudinally striated. The cumulus mass is contained within the ampulla and in most cases, using reflected lighting, can clearly be seen within the structure. Using the forceps gently tear the ampulla open. This releases the cumulus mass into the saline. If the cumulus mass is not seen gently squeeze the tissue to encourage release. To enable easy identification an example of the appearance of a single cumulus mass is given in *Figure 12*.

(iii) Using a wide bored Pasteur pipette and bulb transfer and pool the cumulus masses in a fresh dish of PBS.

5.3 **Ethanol activation**

It is convenient to treat 8–10 cumulus masses as a single group.

(i) Using a wide bored Pasteur pipette and bulb transfer the cumulus masses into a small dish containing a freshly prepared 7% (v/v) solution of Analar quality ethanol in PBS containing both Ca^{2+} and Mg^{2+} (*Table 6*). Leave for 4.5–5 min. Transfer the cumulus masses through three changes of ethanol-free PBS and two changes of medium 16 (20). The recipe for medium 16 is given in *Table 6*.

(ii) Transfer the individual cumulus masses into separate drops of medium 16 under oil and incubate for 4–5 h in a 37°C, 5% CO_2 incubator.

(iii) Remove the adhering follicle cells by digestion with a solution of hyaluronidase (*Table 6*). Transfer the cumulus masses into drops of hyaluronidase and wait for the cells to disperse (3–4 min).

Table 6. Solutions for the ethanol activation technique.

PBS (calcium and magnesium containing)	
For 100 ml	
NaCl	0.8 g
KCl	0.02 g
KH_2PO_4	0.02 g
$Na_2HPO_4.12H_2O$	0.289 g
$CaCl_2.2H_2O$	0.014 g
$MgCl_2.6H_2O$	0.01 g
Na pyruvate	0.0036 g
D-glucose	0.1 g
Adjust to pH 7.4 and filter sterilize.	
Medium 16[a] (0.240 osmolarity)	
For 100 ml	
NaCl	0.517 g
KCl	0.036 g
KH_2PO_4	0.016 g
$MgSO_4.7H_2O$	0.029 g
$CaCl_2.2H_2O$	0.025
Na pyruvate	0.006 g
D-glucose	0.1 g
DL-lactic acid syrup	0.32 ml
Phenol red	0.001 g
$NaHCO_3$	0.211 g

Add to about 30 ml of distilled water. Dissolve and make up the volume to 100 ml. Add the bicarbonate last. Filter sterilize and store at 4°C. Before use add 4 mg/ml of bovine serum albumin (BSA, Fraction V; Sigma Chemical Co.).

Hyaluronidase

For a 1× stock dissolve hyaluronidase (Sigma Chemical Co.) in PBS at a concentration of 1 mg/ml. Aliquot and store at −20°C.

[a]Modified according to ref. 19.

(iv) Using a finely drawn-out Pasteur pipette transfer the oocytes into a drop of medium 16 under oil.

From 10 females approximately 100 oocytes should be obtained. In excess of 90% of the oocytes will be activated by this procedure.

5.4 Classification of oocytes

Mouse oocytes are ovulated at metaphase of meiosis II. Activation, which is normally brought about by the fertilizing sperm, causes the oocyte to resume and complete the second meiotic division. One haploid set of maternal chromosomes is extruded into the second polar body. The remaining haploid set forms the female pronucleus. Parthenogenetic activation may disturb this normal pattern and several distinct classes of activated oocytes, of different genetic constitution, result. Possible routes of development are summarized in *Figure 13*.

Four to six hours following activation the oocytes are classified and segregated. Inspect the oocytes individually under low-power phase contrast to ascertain whether ac-

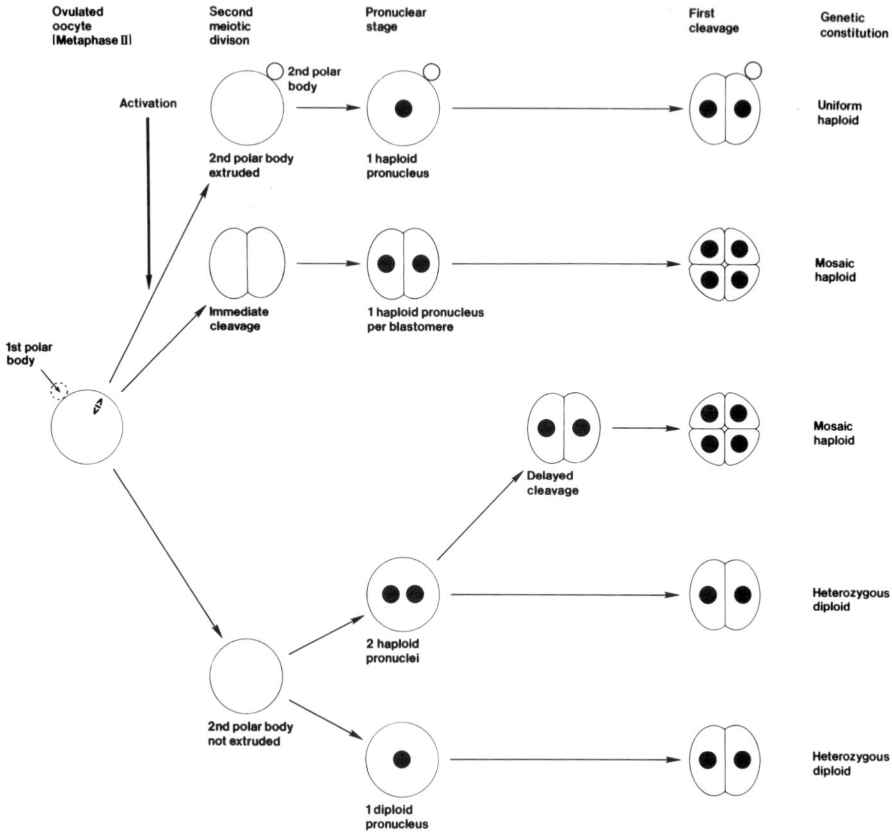

Figure 13. Possible routes of development and the genetic constitution of parthenogenetically-activated oocytes. Figure reproduced from ref. 19.

tivation has taken place. The presence of one or more pronuclei indicates that the oocyte has initiated development. Classify each oocyte on the basis of the presence or absence of a second polar body and number of pronuclei present. At this pronuclear stage the activated oocytes are distinguishable into four distinct categories (see *Figure 13*). Using a finely drawn-out Pasteur pipette segregate the different classes of oocytes into marked drops of medium 16 under oil.

The rate of activation and the incidence of the different types of parthenogenones induced from an experimental series of ethanol activations is given in *Figure 14*.

5.5 Culture of parthenogenetic embryos

The two largest classes of oocyte, namely the 1-pronuclear haploid class (60% of the population) and the 2-pronuclear diploid class (15% of the population) should be selected for further culture. The oocytes are left, in the original drops, for a 4-day incubation period. During this time the embryos will be observed to cleave and progress to the blastocyst stage. Not all embryos initiate cleavage and those that do not should be discarded. Pre-implantation embryos progress more slowly in culture than *in vivo* and par-

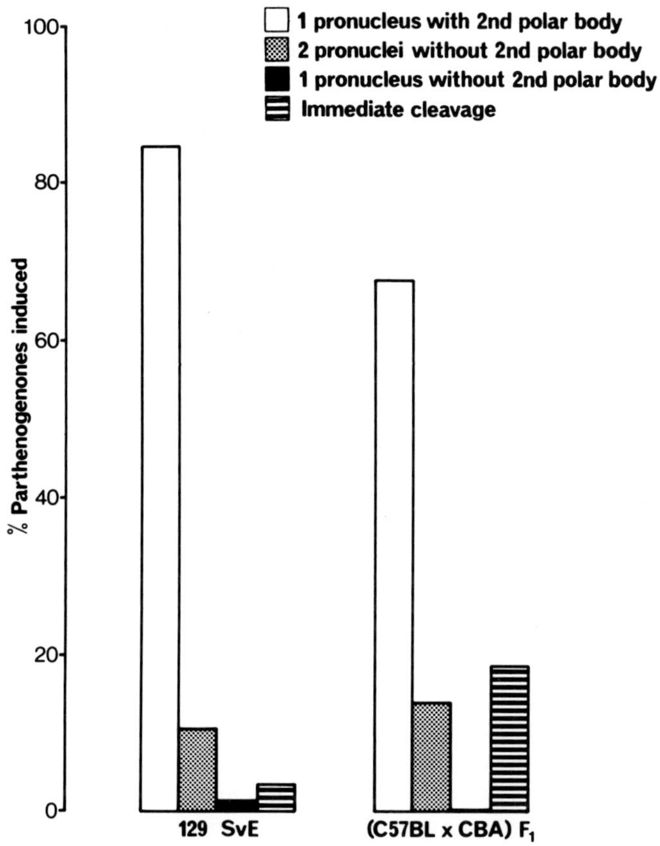

Figure 14. Incidence of the different types of parthenogenone induced when eggs from **(a)** 129 SvE inbred strain mice and **(b)** (C57BL × CBA)F_1 hybrid mice were activated by the ethanol activation technique. The oocytes were classified 4–5 h following activation. The total number of eggs scored for the 129 SvE series was 1351 (activation frequency 80.9%). In the F_1 series 1211 eggs were scored (activtion frequency 95.7%). Figure reproduced from ref. 9.

thenogenones develop more slowly than their fertilized counterparts. If culture conditions are optimal approximately 50% of the oocytes will form blastocysts after a total of 4–5 days culture.

5.6 Derivation of stem cell lines

Transfer the individual blastocysts to feeder wells in complete embryo medium (*Table 4*). Treat the blastocysts using the method described for fertilized blastocysts (Section 4). One major difference is in the timing of the first disaggregation. Parthenogenetically-derived ICM cells should be disaggregated after a total of 6 days culture. In the author's experience stem cell lines can readily be obtained from parthenogenetic blastocysts and numbers of pluripotential cell lines have been obtained from the 1-pronuclear haploid and 2-pronuclear diploid classes of parthenogenone (9).

It should be noted that the 1-pronuclear haploid class of embryo give rise to diploid cell lines. The time at which diploidization occurs is unclear although it is known that haploid cells persist until the late morula stage (9). Additionally, the X-chromosome constitution has been found to be unstable in established stem cell lines (21).

6. CULTURE OF EMBRYO-DERIVED STEM CELLS

Embryo-derived stem cell lines should be grown exclusively on feeder layers. The use of feeder layers, together with careful culture in high quality medium, acts to prolong the embryonic phenotype of the cell line specifically in terms of maintaining a high differentiation ability and euploid chromosome complement. As with all permanently established cell lines, long-term tissue culture will, however, select for abnormal cells within the population. For this reason it is advised that many samples are frozen as soon as possible after founding a specific cell line. Cultures can be replaced as necessary or recloned to re-establish euploid cultures. Recloning can be achieved by picking single cells into feeder wells (see Chapter 2). The resulting colonies of cells are re-expanded by the method described in Section 4.5.

For experimental procedures in which it is desirable to minimize the number of contaminating fibroblasts from the feeder layer it is possible to passage the stem cells once or twice on gelatinized plates. If cultures are kept under these conditions for longer periods of time extensive cell death occurs.

6.1 **Routine maintenance**

Once established, stem cells grow rapidly, dividing every 18–24 h. The cells should be kept at relatively high densities to ensure that a high rate of cell division is maintained as this minimizes the level of spontaneous differentiation. The cultures should be re-fed daily, or according to the acidity of the medium, and will need subculturing at 3–4 day intervals. Cells are routinely grown in the same medium as for the original embryos from which they were derived (*Table 4*) although the foetal bovine serum content can be reduced to 5%. When passaging it is important to trypsinize the cells well to ensure that a single cell suspension is generated. Carrying over of cellular aggregates will stimulate the cells to differentiate producing quantitites of endoderm-like cells.

6.1.1 *Sub-culturing*

Passage the cells when the plates are fully confluent. Cell viability is improved if the culture is re-fed 2–3 h prior to passaging. Passage a 6 cm plate as follows.

(i) Aspirate the medium and wash the surface with 5 ml of PBS.
(ii) Flood the plate with 0.5 ml of trypsin EDTA and return the dish to the incubator for 4–5 min.
(iii) Remove the plate and swirl it to dislodge the cells from the bottom of the plate. Using a sterile plugged Pasteur pipette and bulb add 1 ml of complete medium and pipette the suspension vigorously.
(iv) Check the cell suspension visually under low-power magnification to ensure that it is relatively free of cellular aggregates.
(v) Collect the cell suspension into a plastic centrifuge tube and pellet the cells by low-speed centrifugation (1000 r.p.m., 5 min).

(vi) Aspirate the supernatant and resuspend the cells in 10 ml of complete medium. Determine the cell density and re-plate the cell suspension into feeder plates containing 4 ml of complete medium. Add approximately 1×10^6 cells per 6 cm plate.

(vii) When confluent (3–4 days) a 6 cm plate will yield up to 2×10^7 cells.

If a cell line fails to grow according to the schedule outlined above the problem may lie in either passaging the cells too sparsely or through the use of poor quality feeder plates. It is also important not to let the cells overgrow and become highly acidic.

6.2 **Freezing and thawing**

It is advisable to freeze large numbers of aliquots of cells as soon as possible after establishing a cell line. Keep a careful record of the relative age of the cultures. This is most conveniently done by recording on the freezing ampoule details of the number of passages the cells have undergone prior to freezing.

A general rule, which holds true for all cultured cells, is that cells should be frozen slowly and thawed rapidly. The cryoprotectant is dimethyl sulphoxide (DMSO) used at a concentration of 10%. Cells will survive almost indefinitely if stored under liquid nitrogen and, in our experience, are viable for 2–3 years in a −70°C freezer. The cells should be thawed rapidly and removed from the DMSO-containing medium as soon as possible as they deteriorate rapidly if left in the freezing medium for any period of time. Stem cells freeze well and on thawing there is a high recovery of cells.

6.2.1 *Freezing cells*

(i) Harvest cells by trypsinizing (steps i–iv, Section 6.1).

(ii) Count the cell suspension and calculate the total number of cells in the suspension. Pellet the cells by low-speed centrifugation (1000 r.p.m., 5 min).

(iii) Ideally the cells should be frozen at a density of between 5×10^6 and 10^7 cells per 1 ml capacity freezing ampoule (e.g. Nunc). For smaller cell numbers use either a smaller total volume or add additional feeder cells. On this basis calculate the volume of supension required. Resuspend the cell pellet in complete medium to give half the final required volume. For example, a culture plate yields 2×10^7 cells. These should be frozen in three 1 ml apoules. The cell pellet should be resuspended in 1.5 ml of complete medium.

(iv) Make the suspension up to the final volume by the addition of freezing medium (20% DMSO, 20% FBS, 60% DMEM). The freezing medium should be added gradually while shaking the tube.

(v) Working quickly, dispense the suspension into freezing ampoules and place in the air phase compartment of a liquid nitrogen storage vessel. (Alternatively, place inside a polystyrene box and cool in a −70°C freezer.) Leave the ampoules for 24 h before transferring into liquid nitrogen.

6.2.2 *Thawing cells*

(i) Retrieve the ampoule from storage and place directly in a 37°C water bath. Wait until all the ice crystals have disappeared, remove the ampoule and sterilize with 70% alcohol.

(ii) Using a sterile Pasteur pipette transfer the contents of the ampoule to a 10 ml centrifuge tube. Add 5 ml of complete medium slowly while agitating the tube. Pellet the cells by low-speed centrifugation. Aspirate the supernatant and resuspend the cells in complete medium.
(iii) Transfer the suspension to a fresh 6 cm feeder plate and incubate.
(iv) Change the medium the following day and wash away any dead cells. If the correct procedures have been followed the majority of the cells (up to 90%) in the freeze will be viable.
(v) Passage the culture when the plate reaches confluence.

6.3 **Induction of differentiation**

One of the most important features of embryo-derived stem cell lines is that they universally exhibit a high differentiation ability in comparison with EC cell lines. EC cell lines are very heterogeneous with respect to differentiation ability (see Chapter 2).

Differentiation of embryo-derived stem cells can be brought about by three experimental means, all of which alter the growth environment of the cell.

(i) Cells will grow and differentiate to form a teratocarcinoma tumour if inoculated into syngeneic host animals (see Chapter 2, Section 4 for methods). These tumours are typically composed of a large diversity of cell and tissue types (see Chapter 1).
(ii) A more controlled pattern of differentiation can be achieved when the stem cells are returned to the pre-implantation stage embryo (see Chapter 5 for methods). The cells participate in normal development forming a wide range of normal cell and tissue types in th adult chimaeras. For some lines tested this has included formation of functional germ cells (4,8).
(iii) Large populations of undifferentiated cells can be synchronously induced to enter a programme of differentiation in culture by growing the cells in suspension.

6.3.1 *Differentiation in culture*

Embryo-derived cells will differentiate readily in the absence of chemical inducers such as retinoic acid (RA) and hexabisacetamide (see Chapter 2, Section 3). Indeed the addition of RA results in extensive cell death.

A small proportion of the stem cells in a feeder layer culture will differentiate spontaneously to form endoderm-like cells. A synchronous pattern of differentiation can be obtained by the technique of suspension culture to induce the formation of 'embryoid body' structures. This method was first described for certain feeder-dependent EC cell lines (22–24).

The procedure is divided into two phases. In the first stage the cells are stimulated to form endoderm cells by their removal from the substrate into suspension culture. The cellular aggregates formed by this procedure will quickly delineate a layer of endoderm cells on their outer surface forming structures termed 'simple embryoid bodies' within 2–4 days. The formation of endoderm cells seems to be a necessary prerequisite for the initiation of a more extensive pattern of differentiation. In this second phase two strategies can be adopted. If suspension culture is continued after day four the 'simple embryoid bodies' differentiate further to form 'cystic embryoid bodies'. These typically contain an inner layer of columnar ectoderm-like cells and accumulate fluid in the in-

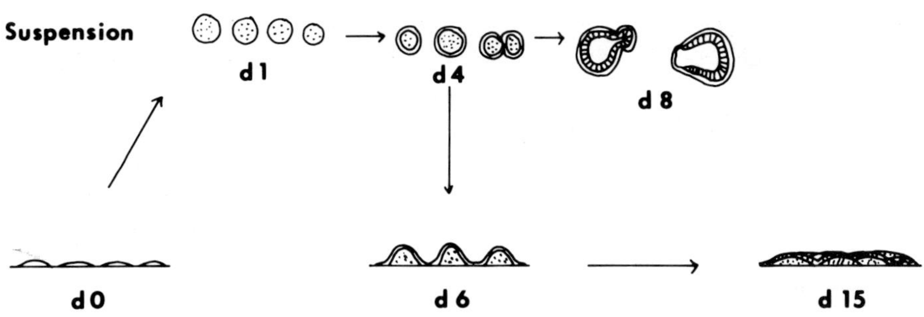

Figure 15. Schematic representation of the methods for the induction of differentiation of embryo-derived stem cells in culture. Cells growing on a substrate as a monolayer **(d0)** are transferred to suspension culture and form small aggregates **(d1)**. After 4 days suspension culture they form simple embryoid bodies **(d4)**. If maintained for longer periods in suspension they form cystic embryoid bodies **(d8)**. If simple embryoid bodies are replated onto a tissue culture surface they attach via outgrowth of the endoderm cells **(d6)**. Long term culture results in extensive chaotic differentiation **(d15)**.

terior of the structure. Alternatively, simple embryoid bodies can be re-plated onto tissue culture surfaces and allowed to re-attach. If the cultures are maintained for a period of 10 days or longer a wide variety of differentiated derivatives will form in the culture. This cellular differentiation is chaotic.

The pattern and sequence of *in vitro* differentiation is summarized in *Figure 15*.

(i) *Formation of simple embryoid bodies*. Cell populations are induced to form simple embryoid bodies by the following method. The medium used throughout is DMEM supplemented with 10% NCS. Start with a 10 cm plate.

(1) Free the culture of contaminating fibroblast cells by passaging the cells once on gelatinized plates. To achieve this, passage the cells at high density as a single cell suspension. For example, seed 6×10^6 cells per 10 cm plate, 2×10^6 cells per 6 cm plate.
(2) Incubate for 3 days and inspect the cultures. The cells should be approaching confluency and few feeder cells should be evident. Wash the surface with PBS and add 2 ml of trypsin EDTA (or enough to just cover the surface). Incubate the plates for ~2 min. Gently rock the plate until flakes of cells are seen to detach. Add an excess of complete medium to neutralize the trypsin action.
(3) Using a wide bore pipette, and taking care not to break up the cellular aggregates, dispense the suspension into bacteriological Petri dishes. The suspension should be split into three 10 cm plates each containing 10 ml of complete medium.
(v) The cellular aggregates formed in this way are unable to attach to the bacteriological surface. The cells adhere together and form small three-dimensional structures within 24 h. Culture in suspension for a total of 4–5 days.

After this time approaching 100% of the aggregates will have formed a clearly discernable outer layer of large endoderm cells. Both visceral and parietal endoderm cell types form. The parietal endoderm cells secrete a thick layer of basement membrane material termed Reichert's membrane. This is seen as a dark layer separating the endoderm cell from the undifferentiated core cells (*Figure 16*). Following endoderm formation the aggregates are termed simple embryoid bodies.

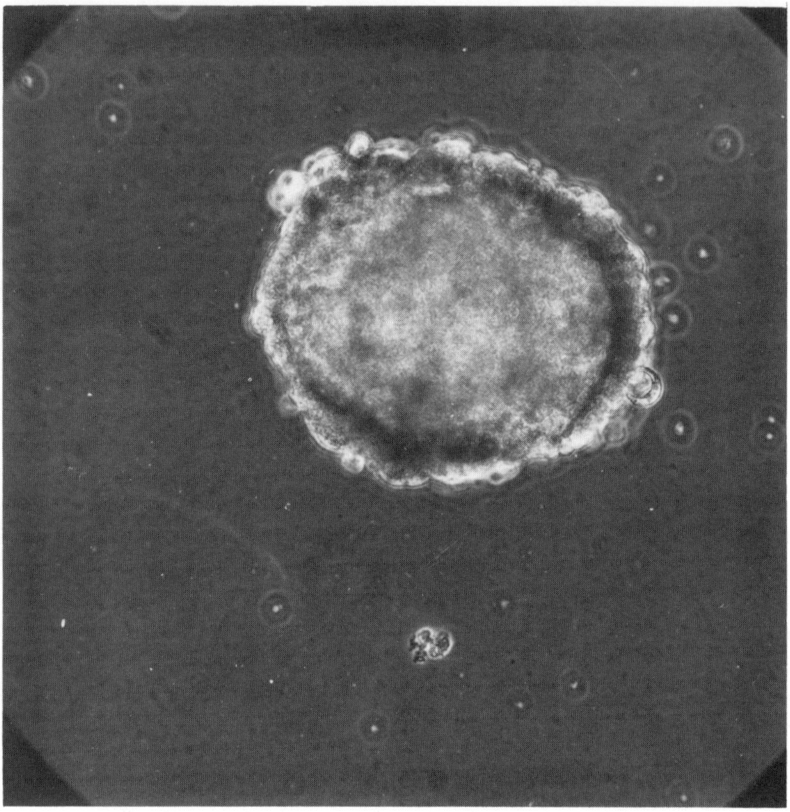

Figure 16. The appearance of simple embryoid bodies following a period of 4 days in suspension culture. Note the rind of endoderm cells (endo) and underlying Reichert's membrane (RM).

If the initial suspension of aggregates is not diluted sufficiently in step (3) they will tend to adhere to one another forming large clumps, or strings. Once aggregates have coalesced in this way they are difficult to separate without damage. If this occurs discard the cultures and start again. If the cells are over trypsinized [step (2)] a large number of single cells will be released. Single cells are not viable in suspension and will die quickly leading to a build up of cellular debris in the medium. If this occurs clean up these cultures as follows. Transfer the contents of the affected plate(s) to a conical-bottomed tube (using a wide bored pipette). Allow the aggregates to settle out by gravity sedimentation. Aspirate the supernatant containing the debris and resuspend the aggregates in 10 ml of fresh medium. Swirl the tube and tip the contents into a fresh plate.

(iii) *Formation of cystic embryoid bodies.*

(1) Collect 4 day old simple embryoid bodies in a conical-bottomed tube and allow them to settle. Aspirate the medium and replace. Dilute the embryoid bodies into 3–4 fresh bacteriological plates in a total volume of 10–12 ml per plate.

(2) Incubate the cultures for a further 4–5 days re-feeding every other day. This can be most simply done by tilting the plate so that the embryoid bodies ac-

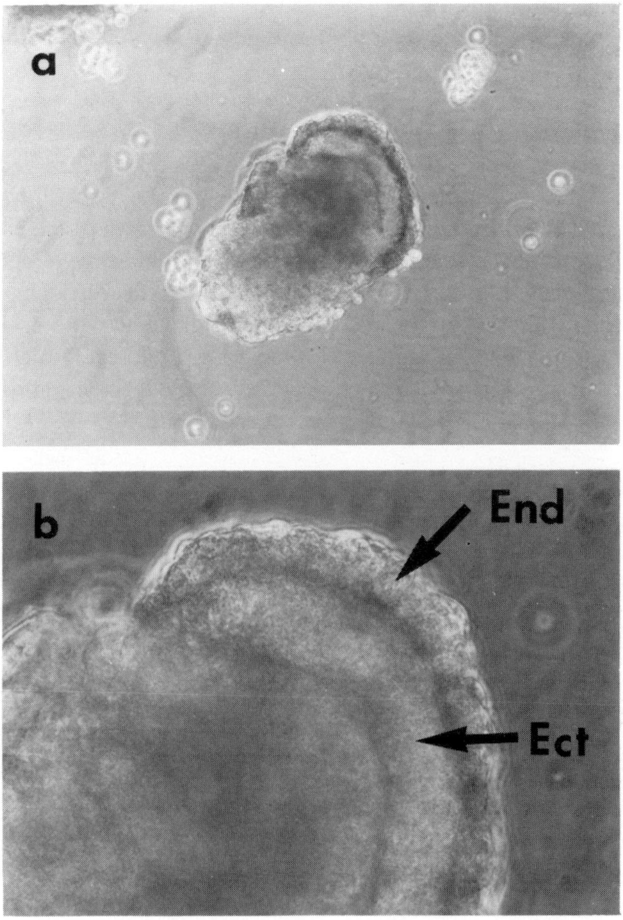

Figure 17. Cystic embryoid bodies viewed under phase contrast to show the delineation of an internal layer of ectodermal cells. These are often difficult to make out because of the dense overlying layer of endoderm cells and the accompanying basement membrane layer (embryoid bodies cultured for a total of 8 days in suspension). **(a)** low power showing intact cystic embryoid body **(b)** higher power, focusing on equatorial region, to show the ectoderm-like cells (ect) interior to the endoderm (end) cells.

cumulate and settle in one area. A high proportion of the medium can be removed without disturbing the aggregates. As the population of embryoid bodies becomes cystic this technique becomes more difficult as removal of medium causes them to ground on the bottom of the plate. If the cultures become too dense and cause the medium to become acidic quickly, split into further plates.

(3) Culture for a period of 8–10 days in total.

After this time a high percentage of the embryoid bodies will have developed fluid-filled cavities. This may be accompanied by a dramatic increase in diameter. Under phase contrast it is often possible to distinguish an internal structure within the embryoid bodies. The formation of an inner layer of 'ectodermal-like' cells is common (*Figure 17*). With longer term culture further differentiation will occur. This should include the formation of mesodermal derivatives such as beating muscle. Cystic embryoid bodies remain viable for long periods (up to 20 days) if the cultures are re-fed regularly.

(iii) *Differentiation on a substrate.* Take a sample of embryoid bodies (4 days culture), re-feed with fresh medium and plate out at a suitable density on gelatinized tissue culture plates. Return to the incubator and leave undisturbed for 2 days to allow the embryoid bodies to attach. The embryoid bodies attach by the outgrowth of the endoderm cells. Continued culture of these aggregates will give rise to an array of cell phenotypes, for example, nerve, muscle, cartilage, etc. (2). These cells are difficult to see by visual inspection as the differentiation occurs within a multilayered culture. The plates can be fixed and prepared for conventional histological procedures. If such an analysis is to be carried out it is recommended to re-plate the embryoid bodies on solvent-resistant tissue culture plates (e.g. 'Permanox'; Lux Scientific Corp.).

6.4 **Karyotype analysis**

Karyotype analysis procedures should be used to ascertain the chromosome complement and sex chromosome constitution of a cell line as soon as possible after it has been established. Such an analysis performed on a large number (>35) of embryo-derived stem cell lines derived in the author's laboratory has shown the majority (>70%) to have a euploid chromosome complement and a strong modal number of 40. The euploid chromosome complement has been shown to be retained by at least three independently derived stem cell lines held in culture for in excess of 15 passage generations (~70 cell generations) by the demonstration that cells can form functional germ cells in chimaeras (4,8). It is inevitable that cell populations will drift away from the normal genotype and subsets of aneuploid cells will be selected over a period of continuous culture. For this reason it is important to routinely check the chromosome complement of stem cells in culture. Simple chromosome counts are used to determine the modal number and estimate the range of variability of the population. Ideally, this should be complemented by G-banding analysis which enables the exact chromosomal constitution to be determined.

It is a relatively simple exercise to carry out a chromosome analysis on a tissue culture cell line. The procedure consists of collecting metaphase spreads by treatment of the cells with a spindle acting agent such as colcemid. The cells caught in mitosis are fixed and the cell contents released onto glass slides to visualize the chromosome spreads. The chromosomes can then be stained immediately for counting purposes or G-banded to allow identification of the individual chromosomes and the construction of standard karyograms.

6.4.1 *Collection of mitotic cells*

Cells can be arrested in metaphase by the addition of colcemid to the culture medium. For best results it is best to use an exponentially dividing population of cells, approximately 2 days after passage, which have been re-fed 3 h previously. If a high percentage of the cells are dividing actively then relatively few cells are needed to collect sufficient spreads. The following method is suitable for a 6 cm plate of cells, although smaller plates (e.g. 3.5 cm) can be used successfully.

(i) To an exponentially growing plate of cells add colecmid to a final concentration of 0.02 μg/ml. The colcemid stock is 2 μg/ml dissolved in distilled water and stored at 4°C. Return the plate to the incubator and leave for 1 h.

(ii) Collect the cells by the standard procedure (Section 6.1), transfer the suspension to a 10 ml conical centrifuge tube and pellet the cells.

(iii) Aspirate the supernatant as completely as possible and tap the tube to disrupt the pellet. Using a Pasteur pipette add approximately 1 ml of hypotonic (0.56% w/v KCl solution) and resuspend the cells. Add an excess of hypotonic (6 ml) and invert the tube to ensure thorough mixing.

(iv) Leave for 10 min at room temperature to allow the cells to swell. Pellet the cells by gentle centrifugation (500 r.p.m., 5 min). Aspirate the hypotonic as completely as possible. Flick the tube vigorously. Using a Pasteur pipette add ice-cold fixative (3:1 volumes of absolute methanol to glacial acetic acid, freshly prepared). This should be done dropwise while flicking the tube to ensure that the cells do not form large clumps.

(v) Leave the tube for 5 min at room temperature. Change the fixative a further three times by spinning out the cells. The volume of the final suspension should be approximately 1 ml.

It is important to ensure that the cells are maintained as a single cell suspension throughout. The pellet of swollen cells should be thoroughly disrupted before addition of the fixative in order to prevent the formation of large clumps, or strings, of cells. The fixative is changed at least three times to ensure that the preparation is completely dehydrated.

In the author's experience best results are obtained if spreads are prepared on the same day. If this is not possible the preparation should be stored at −20°C and the fixative changed immediately before use.

6.4.2 *Preparation of mitotic spreads*

The fixed swollen cells are air dried on glass microscope slides. The spreading of the fixative layer and subsequent evaporation causes the cell membranes to rupture and release the chromosomes.

A variety of techniques have been described for spreading chromosomes. The following procedure gives reproducibly good results.

(i) Pre-clean the slides by wiping with 70% alcohol or fixative.

(ii) Take a small quantity of suspension in a Pasteur pipette. Hold the pipette vertically, position the end approximately 10 cm above the centre of the slide and release a single drop of suspension.

(iii) Wait until the edges of the drop have spread to reach the sides of the slide. Pick up the slide and blow gently across the surface until all traces of the fixative have evaporated. Take care not to inhale the fixative vapour.

(iv) Prepare 8−10 slides in this way. Examine the individual slides carefully under phase contrast (×200) to check the number and quality of the chromosome spreads.

Table 7. Solutions for G-banding.

Saline (standard saline citrate)	
2 × SSC	0.03 M NaCl 0.03 M trisodium citrate
Trypsin	
	0.25% (w/v) trypsin powder (Porcine, Difco) dissolved in Gurr's pH 6.8 phosphate buffer[a] Cool to a working temperature of 4°C on ice.
Buffer	
	Gurr's pH 6.8 phosphate buffer. Commercially available in tablet form for the preparation of 100 ml or 1 litre quantitites (e.g. BDH Chemical Co.)
Staining solution	
	5% (v/v) solution of Giemsa Gurr's R-66 stain (e.g. BDH Chemical Co.) in pH 6.8 buffer. Prepare immediately before use and replace frequently.

[a]The solution will be noticeably turbid. This is normal.

Several factors control the number and quality of mitotic spreads. The drying procedure is critical for good spreading of the chromosomes. The relative length of the chromosomes is determined by the colcemid treatment. If the colcemid is omitted, or is too dilute, the chromosomes fail to contract and form tangled spreads. Overexposure to colcemid has the opposite effect causing severe contraction of the chromosomes which are then totally unsuitable for banding. Although there is always a wide range in variability between spreads the general aim is to obtain a preparation in which a high proportion of the spreads appear as straight rods.

For the purposes of chromosome counting the slides can be stained immediately by immersing the slides in a 3% (v/v) solution of Gurr's Giemsa stain in PBS for 13–15 min. Rinse the slides in two changes of distilled water and allow to air dry before examining.

6.4.3 *G-banding analysis*

A variety of banding techniques are available to enable a detailed chromosome analysis to be made. One of the most reliable methods is G-banding (Giemsa banding) in which mitotic spreads are subjected to various denaturation treatments before staining (25). The procedure described here is a modification of the original methods and has been devised to give best results with stem cells. The method is a combination treatment of denaturation in hot saline followed by partial digestion with trypsin. The solutions required for G-banding are given in *Table 7*.

(i) Incubate the slides at 60°C in a bath of 2× standard saline citrate (SSC) for 1 h.
(ii) Rinse the slides extensively in several changes of distilled water and store the slides in a rack under water.
(iii) Treat the slides individually. Immerse in trypsin (4°C) for 3–4 min. Rinse in two changes of buffer.
(iv) Immerse in staining solution for between 5 and 8 min. Rinse in two changes of buffer followed by two changes of distilled water. Allow the slides to air dry.
(v) Examine the slides under a high-power objective (e.g. ×65 oil immersion). Photograph suitable metaphase spreads.

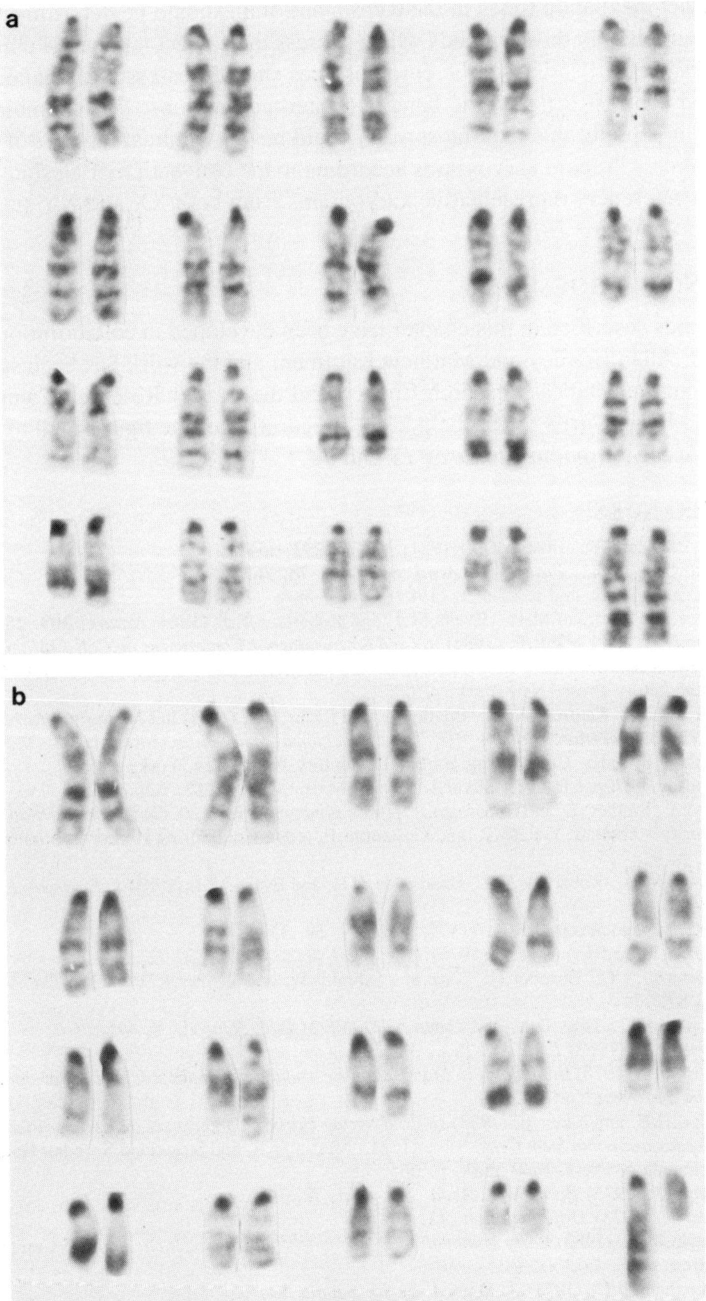

Figure 18. Karyograms constructed from photographs of G-banded metaphase spreads. **(a)** Female cell line showing two X chromosomes. **(b)** Male cell line showing XY sex chromosome complement.

The exact incubation times in the trypsin and stain should be determined empirically. Photograph only those spreads which possess the correct number of chromosomes. Ignore fragmented spreads. The chromosomes should possess clear banding patterns with good contrast between the band and interband regions. For unambiguous identification it is preferable that the spread should have a minimal number of overlapped chromosomes. Prepare karyograms according to the convention of Nesbitt and Franke (26). For reference, representative karyograms from both XX and XY stem cell lines are given in *Figure 18*.

7. ACKNOWLEDGEMENTS

The methods described in this chapter have been developed in collaboration with Martin Evans, Allan Bradley and Matthew Kaufman, and the work has been supported by grants from the Medical Research Council and the Cancer Research Campaign. I am grateful to Brian Curtis for his help in the preparation of the figures and to Matt Kaufman for kindly providing *Figures 13* and *14*.

8. REFERENCES

1. Evans,M.J. and Kaufman,M.H. (1981) *Nature,* **292**, 154.
2. Martin,G.R. (1981) *Proc. Natl. Acad. Sci. USA,* **78**, 7634.
3. Papaioannou,V.E. and Rossant,J. (1983) *Cancer Surv.,* **2**, 165.
4. Bradley,A., Kaufman,M.H., Evans,M.J. and Robertson,E.J. (1984) *Nature,* **309**, 255.
5. Martin,G.R. and Lock,L.F. (1983) In *Cold Spring Harbor Conferences on Cell Proliferation. Volume 10. Teratocarcinoma Stem Cells.* Silver,L.M., Martin,G.R. and Strickland,S. (eds), Cold Spring Harbor Laboratory Press, New York, p. 635.
6. Robertson,E.J., Kaufman,M.H., Bradley,A. and Evans,M.J. (1983) In *Cold Spring Harbor Conferences on Cell Proliferation. Volume 10. Teratocarcinoma Stem Cells.* Silver,L.M., Martin,G.R. and Strickland,S. (eds), Cold Spring Harbor Laboratory Press, New York, p. 647.
7. Magnuson,T., Epstein,C.J., Silver,L.M. and Martin,G.R. (1982) *Nature,* **298**, 750.
8. Evans,M., Bradley,A. and Robertson,E. (1985) *Banbury Report. 20. Genetic Manipulation of the Early Mammalian Embryo.* Jaenish,R. and Costantini,F. (eds), Cold Spring Harbor Laboratory Press, New York, p. 93.
9. Kaufman,M.H., Robertson,E.J., Handyside,A.H. and Evans,M.J. (1983) *J. Embryol. Exp. Morphol.,* **73**, 249.
10. Ware,L.M. and Axelrad,A.A. (1972) *Virology,* **50**, 339.
11. Martin,G.R. and Evans,M.J. (1975) *Proc. Natl. Acad. Sci. USA,* **72**, 1441.
12. Doetschman,T.C., Eistetter,H., Katz,M., Schmidt,W. and Kemler,R. (1985) *J. Embryol. Exp. Morphol.,* **87**, 27.
13. Champlin,A.K., Dorr,D.L. and Gates,A.H. (1973) *Biol. Reprod.,* **8**, 491.
14. Brinster,R.L. (1963) *Exp. Cell Res.,* **32**, 205.
15. Bergstrom,S. (1978) In *Methods in Mammalian Reproduction.* Daniel,J.C. (ed.), Academic Press Inc., London and New York, p. 419.
16. Axelrod,H.R. and Lader,E. (1983) In *Cold Spring Harbor Conferences on Cell Proliferation. Volume 10. Teratocarcinoma Stem Cells.* Silver,L.M., Martin,G.R. and Strickland,S. (eds), Cold Spring Harbor Laboratory Press, New York, p. 665.
17. Toohey,J.I. (1975) *Proc. Natl. Acad. Sci. USA,* **72**, 75.
18. Oshima,R. (1978) *Differentiation,* **11**, 149.
19. Kaufman,M.H. (1983) *Early Mammalian Development: Parthenogenetic Studies.* Cambridge University Press.
20. Whittingham,D.C. (1971) *J. Reprod. Fertil. Suppl.,* **14**, 7.
21. Robertson,E.J., Evans,M.J. and Kaufman,M.H. (1983) *J. Embryol. Exp. Morphol.,* **74**, 297.
22. Martin,G.R. and Evans,M.J. (1975) *Cell,* **6**, 467.
23. Martin,G.R. and Evans,M.J. (1975) In *Roche Symposium on Teratomas and Differentiation.* Sherman,M. and Solter,D. (eds), Academic Press Inc., London and New York, p. 169.
24. Martin,G.R., Wiley,L.M. and Damjanov,I. (1977) *Dev. Biol.,* **61**, 230.
25. Gallimore,P.H. and Richardson,C.R. (1973) *Chromosoma, Berl.,* **41**, 259.
26. Nesbitt,M.N. and Franke,U. (1973) *Chromosoma, Berl,* **41**, 145.

CHAPTER 5

Production and analysis of chimaeric mice

ALLAN BRADLEY

1. INTRODUCTION

The formation of chimaeras by the re-introduction of cultured stem cells into early embryos originated from the development of microsurgical techniques for manipulation of the early mouse embryo (1) and the realization that the differentiation of embryonal carcinoma (EC) cells recapitulates the early events of murine embryogenesis (2). The differentiation of cultured stem cells in association with a normal mouse embryo provides an extremely potent experimental system for the developmental biologist. The technique can utilize the full potential offered by the tissue culture system to modify and select stem cells. Clonally-derived and characterized cells may then be transferred to embryos where their developmental potential can be assayed following differentiation to any adult or embryonic tissue type.

In the first experiments, carried out in the 1970s, chimaeras were constructed with tumour-derived teratocarcinoma stem cells (EC cells). Using these cells the efficiency of chimaera formation was, however, generally relatively low and EC-derived contributions to specific tissues were shown to be largely unpredictable (reviewed ref. 3). More recently embryo-derived stem cells (see Chapter 4), which exhibit important similarities (as well as crucial differences) to EC cells, have proved to be more suitable for the formation of chimaeras; the efficiency of formation is high, the levels of contribution are substantial and generally the cultured cells are included in all of the tissues in a chimaeric mouse (4). The demonstration of a high efficiency of functional germ line contributions by these cell lines has opened up the avenue to introduce 'directed' mutations induced in tissue culture into the mouse genome (5). For those experiments which require cells to participate in embryogenesis it seems that embryo-derived stem cells will completely supercede tumour-derived EC cells. However, the techniques considered in this chapter are equally applicable to both cell types.

The success with which chimaeric mice can be constructed depends on three criteria.

(i) The availability of an adequate mouse facility. This is required to generate sufficient numbers of embryos for manipulation and to provide a suitable environment for post-operative foster mothers to produce and rear viable healthy offspring.

(ii) *In vitro* culture of stem cells. Tissue culture conditions must be such that stem cells retain both their normal embryonic phenotype and a normal euploid karyotype (see Chapter 4). These factors are important pre-requisites if the cells are to eventually re-populate the embryo and form functional germ cells.

(iii) Embryo manipulation and culture. This starts with the removal of embryos from the reproductive tract and continues until the embryo is successfully re-implanted in a foster mother. The experimentor must be familiar with embryo retrieval, manipulation and transfer. The physical association of stem cells with the host embryo is most commonly achieved by their injection into a 3.5 day post-fertilization host blastocyst. This necessitates the use of sophisticated micromanipulatory equipment. More recently, a simple morula aggregation technique has been developed which can be carried out in any tissue culture laboratory without the necessity for expensive or elaborate equipment (refs 6–8).

2. MOUSE STOCKS

2.1 General facilities

The availability of a suitable mouse facility is an essential pre-requisite for the techniques described in this chapter. Ideally the mouse rooms should be very well ventilated as excessive odour has an important restrictive effect on the fertility of both male and female mice; poor ventilation diminishes embryo yield to unacceptable levels.

2.2 Mouse strains

2.2.1 *Generation of embryos*

Outbred strains of mice are normally used to generate embryos, because they yield relatively large numbers of embryos (12–15 from each natural mating). However it should be noted that outbred strains have different characteristics. The embryos obtained from some strains such as CFLP (Interfauna) tend to yield very synchronous populations of embryos, while embryos obtained from others, for example MFI (Olac), are somewhat asynchronous, and the embryos may differ by up to 4–5 h in their developmental stages at 3.5 days post-coitum (p.c.). This asynchrony is not necessarily a problem because the embryos can be cultured to the appropriate stage. Inbred strains of mice can also be used to produce embryos but typically the embryo yield is reduced, averaging only 6–8 from individual females.

An important consideration in the choice of strain is the characteristics of the cell line being used to construct the chimaeras. It is useful if the two cell populations can be distinguished visually in a chimaera without the necessity to use a biochemical assay. Coat and eye pigmentation markers are very convenient for this purpose. Generally, pigmented (*CC*) cell lines are combined with albino (*cc*) host embryos. Any pigmentation observed in live born and pre-natal embryos clearly demonstrates a contribution derived from the introduced cell line. If required this combination of markers can of course be used in reverse.

In addition to pigmentation markers it is convenient to have biochemical markers to enable contributions to internal body organs to be assessed. One of the most widely used enzymes for this purpose is glucose phosphate isomerase (GPI) which exists in three isozymal forms 1^a, 1^b and 1^c in laboratory mice (see Section 9 for details). This enzyme is very convenient because it is ubiquitously expressed, it is very stable and the electrophoretic assay can be accomplished in under 1 h (see Section 9).

2.2.2 *Foster mothers*

Foster mothers can be of any genetic strain, but F_1 hybrid strains have proved to be more satisfactory than either inbred or outbred strains. The reasons for this are 2-fold. Firstly, that 80–90% of operated F_1 mothers maintain the pregnancy (9). In comparison other strains may rarely achieve a 50% pregnancy rate. Secondly, F_1 females generally make better mothers after delivery of the pups and the majority of operated embryos that come to term survive to weaning age and give healthy, fertile animals. The F_1 strain we use is C57BL × CBA.

Although it is often convenient to purchase large numbers of females for generating embryos and for recipient foster mothers, this often presents problems. This is because mice from commercial breeders are often bred under specific pathogen-free (SPF) conditions. Although such animals remain healthy until mating, the majority of laboratory animal houses are not SPF, and the pups born from these embryos do not generally have the immunity levels required to survive common infections. This results in a high post-natal loss of young animals. Breeding females 'in house' will effectively eliminate post-natal losses caused by endemic disease.

2.3 **Setting up matings**

2.3.1 *Timings*

Although superovulation (see Chapter 4, Section 5) can be used to obtain increased numbers of fertilized embryos for manipulation, it is better if this can be avoided. In superovulated females, by 3.5 days p.c., many of the ovulated embryos are developmentally retarded or morphologically abnormal. This is especially noticeable when immature females are used. Additionally, superovulation may not always successfully override the normal ovulation cycle of the female so the yield is often variable. For these reasons natural matings are preferred. We have, however, successfully superovulated mature females to produce embryos from specific inbred strains of mice which have limited fertility.

The time of natural mating is controlled by the light cycle of the mouse room. Usually under a controlled regime of 12 h light to 12 h dark, females pass through the oestrus cycle every 4–5 days. In oestrus females ovulation occurs 3–5 h after the onset of the dark period and the males will copulate with oestrus females about 6 h after the onset of the dark period. Embryonic development is initiated at this time. The light cycle-dependent ovulation of females allows the experimentor to subject the mice to a regime which is calculated to provide embryos of the correct developmental stage at a convenient time. In our laboratory, for example, for blastocyst injection experiments, the mice are maintained on a cycle of 13.00 h–01.00 h dark to 01.00 h–13.00 h light. This regime usually ensures a supply of expanded blastocysts if the embryos are recovered at 09.00–10.00 h 3 days after mating.

2.3.2 *Selecting oestrus females*

Although mice can be set up to mate on a random basis, the synchronous cycling of female mice kept together in a single storage cage results either in the generation of few matings or, occasionally, an excessive number. Fortunately it is possible to preferentially select females that are about to ovulate (in oestrus) by examining the

Table 1. Appearance of the vagina during the oestrus cycle (adapted from ref. 10).

Stage of cycle	*Opening*	*Colour*	*Moistness*	*Other features*
Diestrus	Small	Bluish	Wet	None
Pro-oestrus	Gaping	Pink/red	Moist	Swelling folds and striations on both lips
Oestrus	Gaping	Pink	Less moist	Swelling in pronounced folds and striations
Menstrous 1	Gaping	Pale	Dry	Pronounced swelling and folds
Menstrous 2	Gaping	Pale	Moist	White debris in vagina

colour, moistness and degree of swelling of the vagina (10). *Table 1* details the appearance of the vagina in the various stages of normal cycle.

Females used for mating should be over 6 weeks old. Store the animals in groups of 15–25 in large cages as this is most convenient for the rapid selection of oestrus individuals. Select females shortly before the onset of the dark period and cage them singly, or in pairs, with one stud male. Males mature slightly later than females (9–14 weeks) and are often infertile for the first few matings. Ideally, stud males should not be used more frequently than once every 2 days. Once separated, males must be maintained in separate cages, otherwise they will fight. Keep mating records from individual studs so that non-productive males can be eliminated.

Natural matings are also required to generate pseudo-pregnant foster mothers to act as recipients for the operated embryos. Oestrus females should be selected using the criteria listed in *Table 1*. The pigmented skin of the F_1 mice used as recipient foster mothers can make oestrus selection by inexperienced workers a little more difficult, but with practice successful matings can be guaranteed. Pseudo-pregnant foster mothers are generated by mating to sterile males. It is possible to use males from specific infertile strains for this purpose but, in practice, these animals are not always readily available and their mating behaviour can be unpredictable. The best source of sterile male mice with a good libido is vasectomized (C57BL × CBA) F_1 hybrid males. The vasectomy procedure is described in Section 8.2.

2.3.3 *Detecting successful matings*

The following morning after setting up matings check each female for evidence of mating. Successful mating is indicated by the presence of a solid seminal copulation plug in the vagina (including vasectomized matings). The plug is usually of white/yellow colour. The copulation plug will vary in size and position. Often the plug is very visible, however if it is deep in the vagina it is necessary to insert a blunt probe to detect if it is present. The presence of a hard plug will prevent the probe entering the lower cervical part of the vagina. The copulation plug slowly dissolves in the vagina, so the longer after mating the female is checked, the more likely it is that the plug will not be detected. To maximize the detection of mating check the females for plugs as early as possible (although matings can be detected as late as 10 h after the onset of the light period).

Place the plugged mice in groups in small cages and return unplugged females to the large stock cages. It is important to note that maintenance of pregnancy in females can be affected by the odour of strange males. Therefore always put the plugged females into a clean cage and place the cage in a well ventilated room.

3. COLLECTION OF EMBRYOS

For the manipulations described in this chapter embryos are retrieved from pregnant females either at 2.5 days p.c. (for morulae) or 3.5 days p.c. (for blastocysts). Embryos are removed from females by dissection and flushing of the reproductive tract. The general technique is the same for both embryonic stages but the embryos are removed from different parts of the reproductive tract. The following procedure is used to dissect out the rcproductive tract. The technique is illustrated in *Figure 1*.

(i) Kill the mouse by cervical dislocation. Place it on absorbent paper with the ventral side uppermost and apply a liberal quantity of 70% ethanol to the abdomen and thorax to sterilize the skin.

(ii) Using a large pair of dressing scissors make a small incision (~2.0 cm) in the skin (see *Figure 1a*).

(iii) Hold the skin on each side of the incision between the fingers and thumb of each hand and pull the skin back towards the head and tail, respectively. The skin should peel back around the body and be pulled over the head and rear legs (see *Figure 1b*).

(iv) The abdomen is now exposed. Open the body wall using a pair of fine iris scissors. The position of the cut is indicated in *Figure 1b*.

(v) The contents of the abdomen are now visible. The guts obscure the reproductive tract. Expose the reproductive tract by pushing the coils of the gut up towards the head (see *Figure 1d*).

Subsequent dissection depends on whether blastocysts or morulae are being collected.

3.1 Collection of blastocysts

At 3.5 days p.c. the blastocysts are present in the uterine horns. They are removed by flushing 1 ml of medium [Dulbecco's modified Eagle's medium (DMEM) plus 10% newborn calf serum (NCS)] through each uterine horn using the following method. The procedure is illustrated in *Figure 2*.

(i) Cut through the two horns at the distal cervix end. Grasp the horn with blunt forceps and trace the uterine horn up to the ovary (located just below the kidney). Remove the tract by cutting the fat pad away from the kidney and cutting through the mesenteries.

(ii) Remove the mesentery tissue to prevent contamination by blood which may subsequently obscure the embryos. Place the uterine horn on absorbent paper and spread the mesentery away from the uterus (see *Figure 2a*). Hold one end of the uterus, stretch it away from the mesenteries and cut close to the uterus using a pair of fine iris scissors (see *Figure 2b*).

(iii) Remove the blastocysts from the uterus by inserting a 25-gauge needle into the lumen of the uterus at the oviduct end. Grasp the needle through the uterine wall using fine forceps and flush 1 ml of flushing medium (DMEM plus 10% NCS)

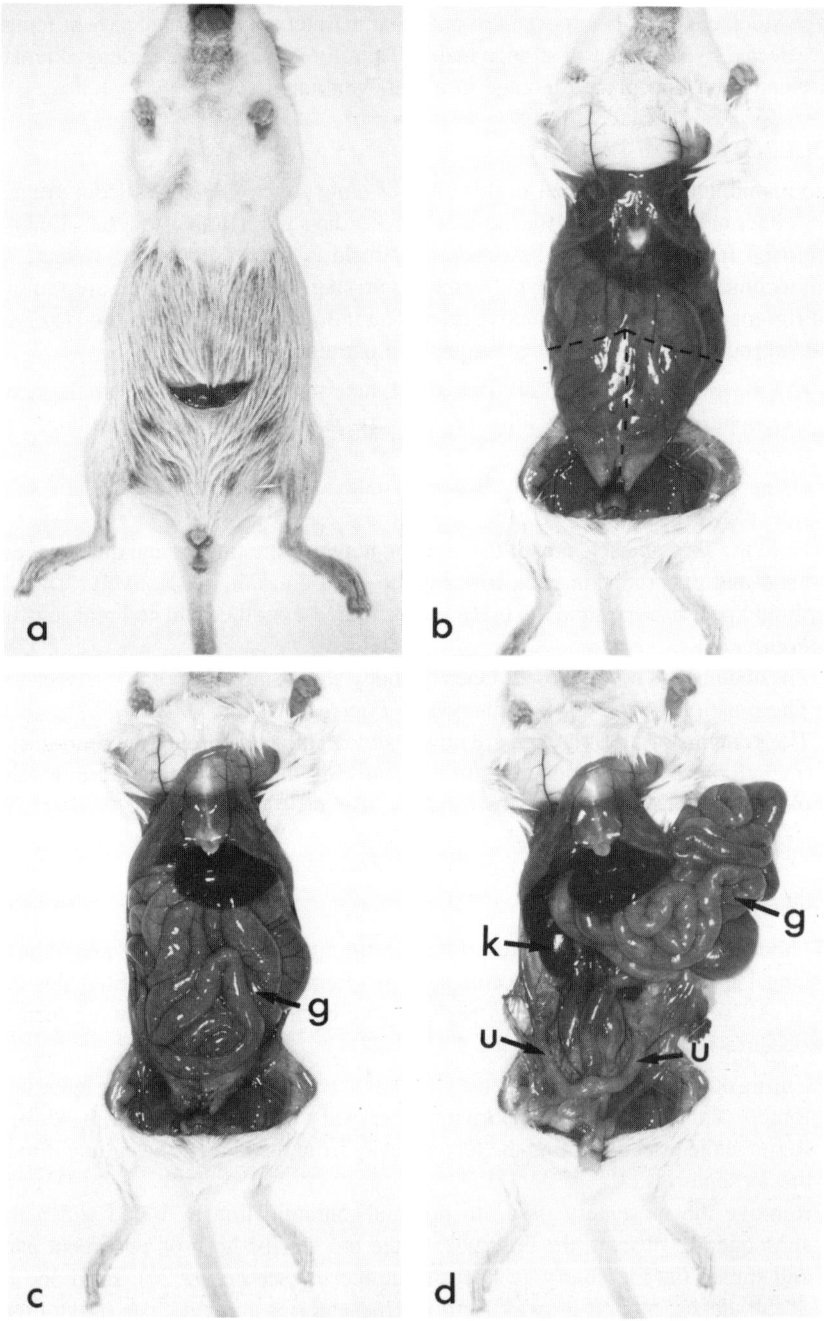

Figure 1. Procedure for exposing the reproductive tract. (**a**) Killed female, ventral side uppermost showing the position of the incision in the skin. (**b**) Exposu:e of the body wall and abdomen after pulling the skin towards the head and tail. The position of the subsequent incision is illustrated by the dotted line. (**c**) Contents of the abdomen displayed. (**d**) Coils of gut pushed up and to one side to expose the reproductive tract. (K = kidney, G = gut, O = ovary.)

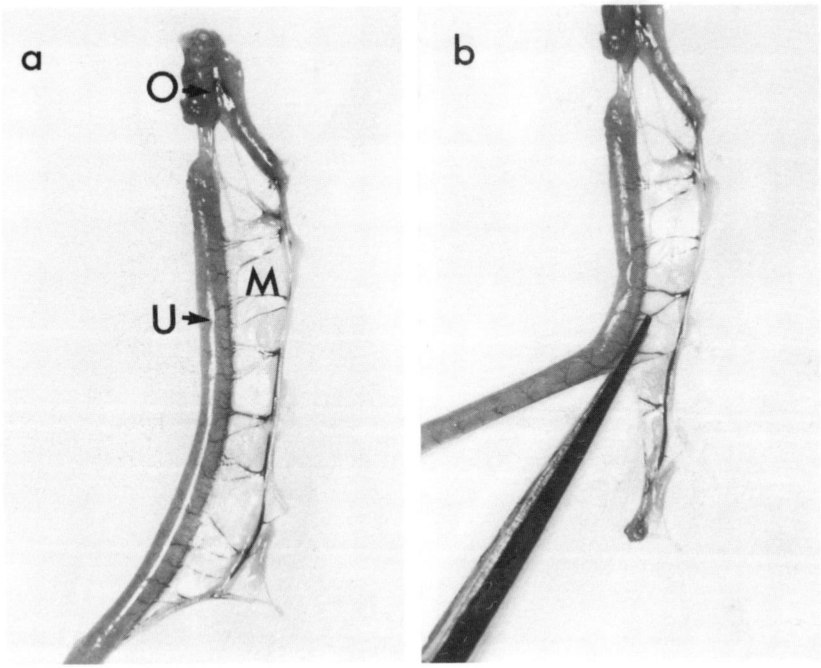

Figure 2. Isolation of uterine horn for blastocyst collection. (**a**) Uterus displayed on absorbent paper. (**b**) Trimming away the mesometrium. (O = ovary and oviducts, U = uterine horn, M = mesometrium.)

through the uterus into a sterile watch glass or plastic Petri dish. If the uterus has not been removed cleanly the medium will not flow through the lumen. If this occurs remove the blocked end by cutting off a small section of the uterus at the cervix end.

3.2 **Collection of morulae**

At 2.5 days p.c. the morulae are present in the oviducts. It is sufficient therefore to only remove the oviducts and leave the remainder of the reproductive tract in the mouse. The procedure is given below and illustrated in *Figure 3*.

(i) Grasp the uterus with fine forceps (e.g. Watchmakers No. 5) just below the oviducts and pull it away from the ovary. Insert the points of a pair of fine scissors through the mesometrium and tear the membranes free from the oviducts (see *Figure 3a*).
(ii) Cut through the uterus just below the uterotubal junction and below the forceps (see *Figure 3b*).
(iii) Grasp the cut end below the oviducts with blunt forceps and pull the tissue taught (*Figure 3c*). Cut the oviduct free from the ovary with fine scissors (*Figure 3d*).
(iv) Place the oviducts in a drop of Medium 16 (see Section 4 for details) supplemented with 4 mg/ml bovine serum albumin (BSA) in a Petri dish.
(v) Transfer the dish to the stage of a stereo dissecting microscope. Using watchmakers forceps manipulate the oviducts and insert a 32-gauge hypodermic

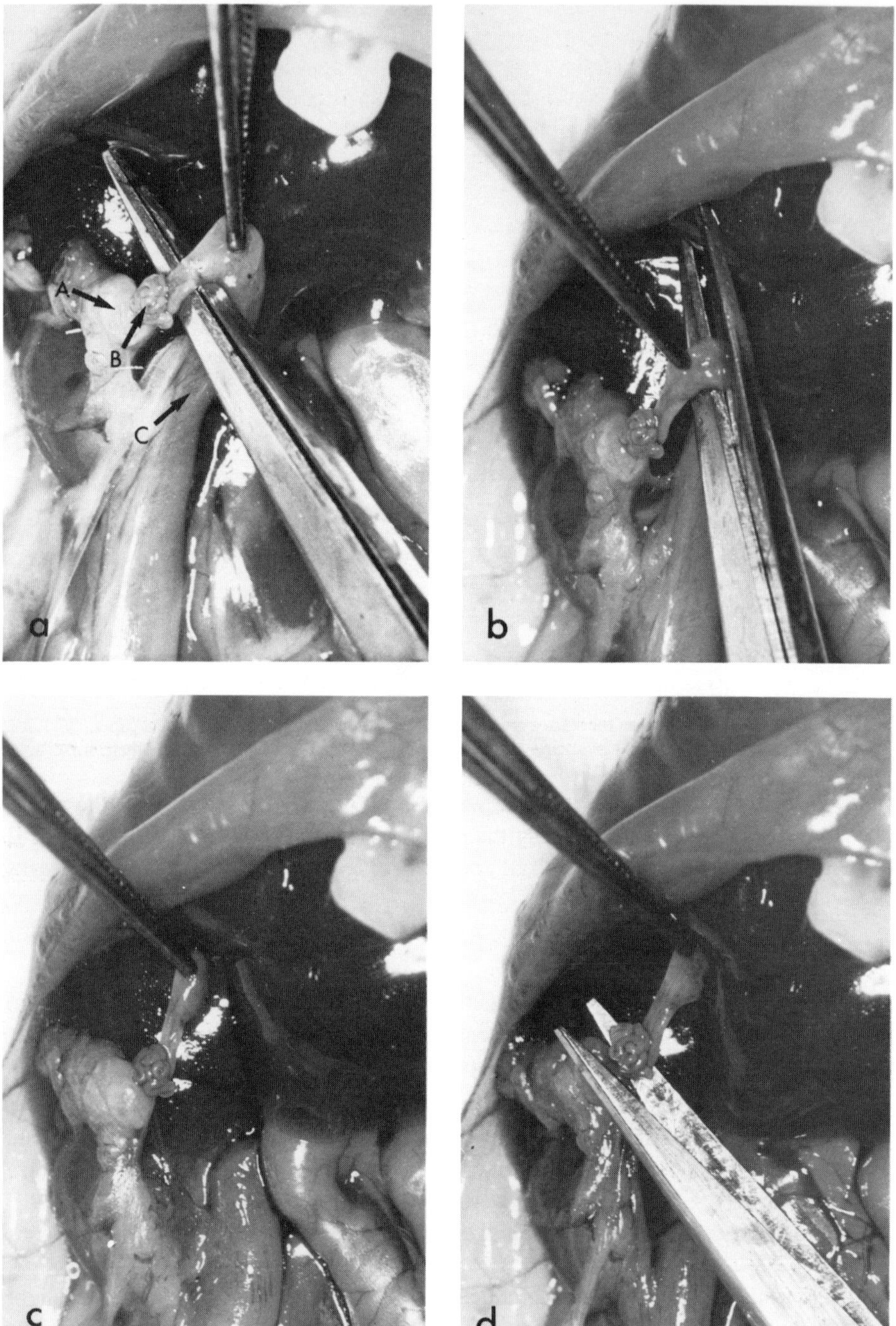

Figure 3. Dissection of oviducts for collection of morulae. (**a**) Tearing the mesometrium away from the ovary, oviducts and uterus. (**b**) Cutting the oviducts away from the uterus. (**c**) Holding the cut end of uterus in preparation for oviduct removal. (**d**) Cutting the oviducts away from the ovary. (A = ovary, B = oviducts, C = uterus.)

needle with a polished tip (connected to a 1 ml syringe filled with Medium 16) into the end of the oviduct. The end of the oviduct is very small, often difficult to find among the coils of the oviduct and is easily damaged. Using fine forceps gently move the coils until the end is seen. Once the end has been located, insert the needle, gently hold it in place with the fine forceps and slowly flush the oviducts with a stream of medium. Usually 0.1 ml of fluid is sufficient to flush the embryos from the oviducts.

3.3 Recovering embryos into culture

Using a stereo dissecting microscope locate the blastocysts or morulae in the Petri dish or watch glass (a total magnification ×20 is adequate). Remove the embryos with a mouth controlled drawn-out Pasteur pipette and place in micro-drop cultures of the appropriate culture medium (see Section 4).

4. CULTURE OF EMBRYOS

4.1 Culture media

The choice of culture medium depends on the particular embryonic stage being cultured. Cleavage stage embryos do not have a glucose-based metabolism and they are therefore cultured in a simple bicarbonate-buffered medium supplemented with pyruvate as the carbon source (11). The medium of choice is Medium 16. The composition of Medium 16 is given in *Table 2*. It is essential to use only Analar chemical grades in the preparation of medium (including the water). Toxicity tested BSA [pentax fraction V (Sigma)] is added to a concentration of 4 mg/ml before use and the medium is sterilized by filtration through a 0.22 μm nitrocellulose filter. Complete Medium 16 (without BSA) may be stored at 4°C for up to 3 weeks, however if acidified with CO_2 it may be stored indefinitely at −20°C. Blastocysts can be cultured in standard tissue culture medium (e.g. DMEM) supplemented with selected batches of foetal bovine serum (FBS).

Table 2. Medium 16 (from ref. 11).

To prepare 100 ml of medium	
NaCl	0.553 g
KCl	0.036 g
$KHPO_4$	0.016 g
$MgSO_47H_2O$	0.029 g
$CaCl_2.2H_2O$	0.025 g
Na lactate (60% syrup)	0.32 ml
Na pyruvate	0.006 g
D-glucose	0.100 g
Phenol red	0.001 g
$NaHCO_3$	0.211 g
Water to 100 ml	

Add ingredients, in the order listed, to 20−30 ml of warmed water (25−30°C), making sure they have dissolved before adding the next ingredient. Make the volume up to 100 ml and filter sterilize.

4.2 Micro-drop cultures

Embryos are generally cultured in tissue culture dishes in micro-drops of medium under light weight paraffin oil. This technique allows embryos to be readily located within a small area in the dish and facilitates the separation of different groups of embryos within the same dish. The micro-drop cultures can be varied in volume depending on the particular requirement of the experiment. When large numbers of embryos are being cultured in a single drop the volume of the drops should be about 100 μl. The micro-drops used for the morulae aggregation technique (Section 7) should be only 1–5 μl. If it is necessary to distinguish the medium drops it is convenient to label the underside of the dish before the culture has been set up. The procedure for setting up micro-drop cultures is illustrated in *Figure 4*.

Add embryos to the micro-drops using a mouth controlled drawn-out Pasteur pipette. Make these pipettes by the following procedure.

(i) Rotate the narrow portion of a 150 mm Pasteur pipette in a bunsen burner flame (approximately half way along the length of the narrow portion) until the glass is molten and flexible.

(ii) Remove the pipette from the flame and pull the two ends apart rapidly. The pipette should form a thin walled capillary tube. The diameter and length of the capillary

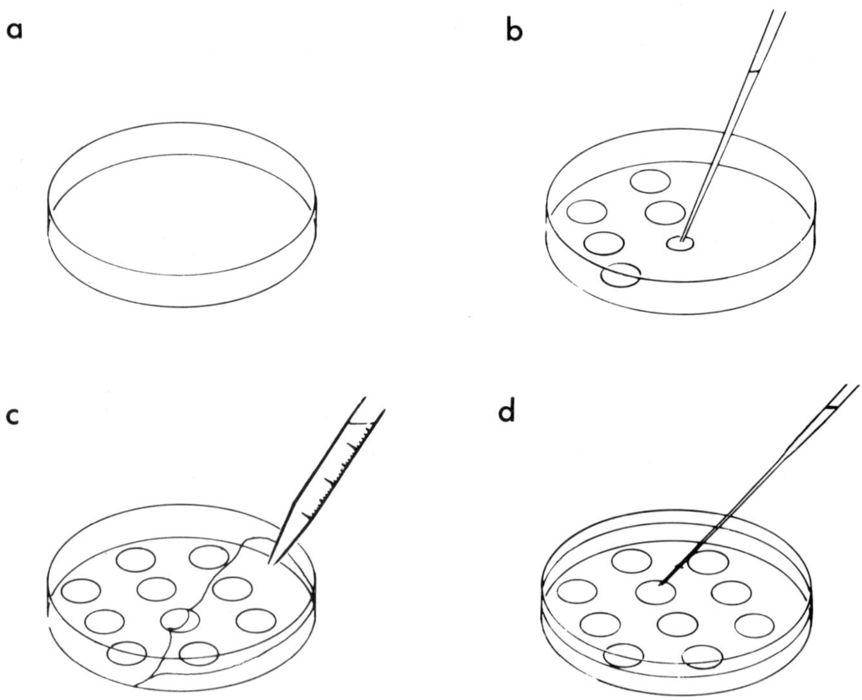

Figure 4. Setting up micro-drops. (**a**) Label the tissue culture dish on the underside. (**b**) Set up an array of micro-drops using a drawn-out Pasteur pipette, syringe or micro-pipetter. (**c**) Flood the dish with light weight paraffin oil (e.g. BDH). Equilibrate the dish in an incubator at 37°C and 5% CO_2 in air. (**d**) Add embryos to the micro-drops using a drawn-out Pasteur pipette.

section created depend on the length of the molten section, the speed of the pull and the degree to which the glass has been allowed to melt. By varying the degree of melting and the strength of pulling, it is possible, with practise, to control the length and diameter of the capillary formed.

(iii) The capillary section is very flexible and may be bent through 180° or more without snapping. Bend the glass capillary section over and scratch one piece against the other at a position which will give the appropriate diameter (approximately 130 μm). Gently flex the glass to cause the capillary to snap at this point with a clean square break.

(iv) Connect the Pasteur pipette to a length (40–50 cm) of flexible plastic tubing. At the free end insert a plastic mouth piece.

5. PREPARATION OF STEM CELLS

5.1 Routine culture

Stem cell lines are routinely cultured according to the protocols described in Chapter 2, Section 2 and Chapter 4, Section 6.

It is important to maximize the viability of cells used for constuction of chimaeras because this improves the efficiency of their incorporation into the embryonic cell pool. There are a number of important considerations concerning cell viability. Clearly culture conditions are very important and, given sufficient attention to other factors (see Section 5.3), the cloning efficiency of the cell preparation has the single most important effect on the rate of chimaera formation. Cloning efficiency can be improved if the following points are noted:

(i) Cultures should be re-fed regularly or when the medium becomes acidic (usually daily).

(ii) The stem cells should be maintained on good feeder layers (see Chapter 4, Section 2).

(iii) The cells should be re-fed 2–3 h before passaging and seeded at relatively high density (0.5×10^6 per 6 cm plate). This will ensure confluence after 2–3 days.

Cells for injection are prepared by trypsinization as for routine passaging. Inspect all available cultures and prepare a cell suspension from a culture which displays a healthy cellular morphology. When the cell lines are grown on a 2–3 day passaging regime it is acceptable to use cells on day 1 and day 2, but by day 3 they are normally somewhat overgrown and the viability of the constituent cells can be a little more variable.

5.2 Preparation of cells for the construction of chimaeras

5.2.1 *Preparation of cells for blastocyst injection*

Make a single cell suspension by trypsinizing a culture (see Chapter 4, Section 6) and introduce a small volume of the suspension directly into the injection chamber (Section 6). The cells remain in the chamber until they are taken up by the injection needle and introduced into the blastocyst cavity.

5.2.2 *Preparation of cells for morula aggregation*

Prepare the cells for morula aggregation experiments by very lightly trypsinizing a culture of cells and preparing a suspension composed mainly of small clumps containing 5–8 cells. Under the microscope select clumps of this size using a drawn-out Pasteur pipette. Place single clumps into individual micro-drop cultures which have been previously seeded with one or two zona-free morulae (see Section 7).

5.3 **General considerations**

(i) The chemicals and water used to make up basic tissue culture supplies and additives should be of the highest quality available.

(ii) While most batches of serum will support the growth of fibroblasts, in general only a limited sub-set of these are suitable for stem cells. The majority of sera allow only restricted growth and some serum batches are positively toxic for stem cells. It is therefore necessary to test sera before purchase (see Chapter 4, Section 2 for the serum testing method).

(iii) Careful washing of glassware and pipettes is essential. Stem cells are generally considerably more sensitive than most other culture lines to organic and inorganic residues. The use of liberal quantities of distilled water and a large number of distilled water rinses in the washing up process is normally sufficient to ensure clean glassware. If necessary, and where the budget permits, the use of sterile disposable plastic or glass pipettes is recommended.

(iv) In many experimental applications of the chimaera-forming technique the stem cells will have been modified to carry the *neo* gene (see Chapter 6) and as such are resistant to the tissue culture additive G418. In practice although levels of resistance can be achieved which are apparently sufficient to maintain a normal growth rate in culture, cell viability upon passaging is often visibly diminished in the presence of this drug. This problem is well illustrated by experimental results using cells grown continuously in G418 until just prior to blastocyst injection. Example results are given in *Table 3*. This problem can be minimized by growing the cells in the absence of G418 for 5–6 days prior to incorporation into host embryos.

(v) It is important to check cultures for mycoplasma contamination. The problems associated with mycoplasma infection are somewhat alarming in terms of pre-natal loss of embryos. It appears that the infection is transmitted from the stem

Table 3. Examples of how various factors can affect the rate of chimaera formation from an XY cell line.

		% Embryos born	*% Chimaeric animals*	*% Male*	*Average contribution to tissues*
Normal stem cell line		65	55	75	40–60%
Mycoplasma-infected cells		40	20–30	50	5–10%
G418	a	65	55	70	40–60%
Cells	b	65	0	–	

[a]Cells grown in the absence of G418.
[b]Cells grown in G418.

cells to the embryo where it multiplies in an uncontrolled fashion until the embryo dies and/or the maternal response kills the embryo. The level of contamination in the cells will affect the numbers and quality of the chimaeras born. The major source of mycoplasmal infection is usually from other contaminated cell lines. This can be a particular problem when tissue culture facilities are shared, and simple rules of tissue culture hygiene are not adhered to. Once a stem cell line has become contaminated we have found no satisfactory way of de-contaminating it. Such cultures should be discarded. To avoid mycoplasmal contamination there is no substitute for meticulous aseptic techniques. Culturing cells without antibiotics will quickly affirm how effective those techniques have been in practice. Clearly where stem cells have by necessity been exposed to culture supernatants from other cells (for example viral producer cells, see Chapter 6) the risks are greatly magnified. Some of the signs of infection are rapid production of acidic medium and poor growth in the absence of antibiotics but reasonable growth when they are present. The other sign of infection is the dramatic effect on pre-natal embryo loss, with an associated poor quality of chimaeras which do survive to term. The experimental results given in *Table 3* illustrate the effect of using contaminated cells.

6. BLASTOCYST INJECTION

The technique described here was developed as a technically simple method designed specifically for the injection into blastocysts of cells prepared as a single cell suspension. Clearly the method does not have the flexibility of the method described by Gardner (1,12,13) which can be used to inject large aggregates of cells.

The injection of embryos is achieved by placing the expanded blastocysts into a cooled injection chamber held on the microscope stage. A freshly prepared cell suspension is placed in the same chamber. The cells (10−15) are selected individually and are collected into the injection pipette. This is then pushed through the zona pellucida and trophoblast layers into the blastocoel cavity of a blastocyst immobilized by suction on the end of the holding pipette. The cells are expelled into the blastocoel, the pipette is withdrawn and the blastocyst released from the holding pipette. This procedure is repeated for each expanded blastocyst.

6.1 **Schedule**

The following schedule is found to be convenient.

(i)	Day 1	p.m.	Set up mice for fertile matings (see Section 2).
(ii)	Day 2	a.m.	Plug check mice for fertile matings (day 0.5 p.c.).
		p.m.	Set up the appropriate number of F_1 female mice with vasectomized males to generate recipient females.
(iii)	Day 3	a.m.	Plug check F_1 females for vasectomized matings.
(iv)	Day 5	a.m.	Collect fertilized embryos (day 3.5 p.c.) (see Section 3.1).
		a.m./p.m.	Inject the blastocysts.
		p.m.	Transfer the injected embryos to pseudo-pregnant recipients (day 2.5 p.c.).

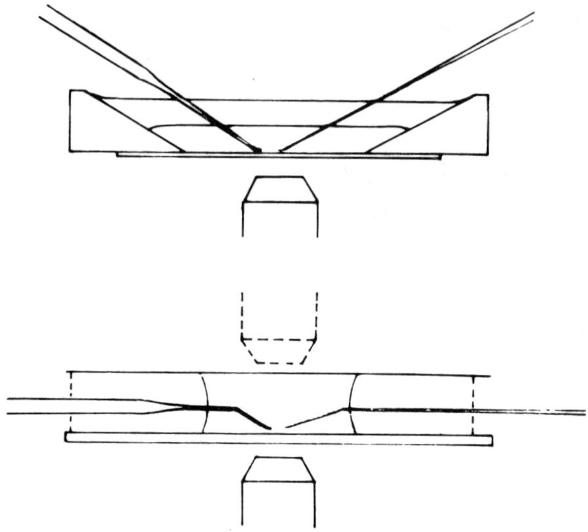

Figure 5. Sectional views of injection chambers and positions of the micro-instruments. (**a**) Typical set up for injection chambers designed for inverted microscopes. (**b**) Typical set up for hanging-drop cultures.

6.2 Manipulators and optics

The Leitz double manipulator set up (Leitz Instruments) is used by most workers who perform microsurgery on pre-implantation mouse embryos. These manipulators are mechanical inertia-free devices which have a joystick movement in two planes (horizontal plane) and the third plane adjustment (vertical) controlled by a screw thread positioned close to the joystick. The direct mechanical coupling and joystick operation are important parameters for any manipulator used for blastocyst injection. The newer design electronic joystick manipulators, where the movement is adjusted by servos, do not have the flexibility of the old style mechanical devices and are not satisfactory for blastocyst injection. The Leitz manipulator also has built in coarse adjustment screws and a very flexible instrument head, which allows an almost infinite variation in the preliminary positioning of the micro-instruments.

In general, there are more options available for the optical set up than with the manipulator equipment. The most important feature of the microscope is a mechanical consideration as, ideally, the focus should be provided by movement of the objectives, not the microscope stage. This requirement is not essential but is one of exceptional convenience. If the stage moves while focusing on an embryo, the relative positions of the embryos and the micro-instruments are altered. The Leitz Diavert with a fixed stage adaption has proved to be an ideal microscope for this type of work, because it is an inverted microscope which gives a large working distance above the injection chamber (unless an Interference Contrast condenser is fitted). Normal upright microscopes can be used though this necessitates the use of hanging drop injection chambers (Section 6.3). Typical sectional views of the set up for upright and hanging drop chambers are illustrated in *Figure 5*. The microscope should be fitted with a range of phase contrast or interference contrast objectives to give a magnification range of ×40 to ×200.

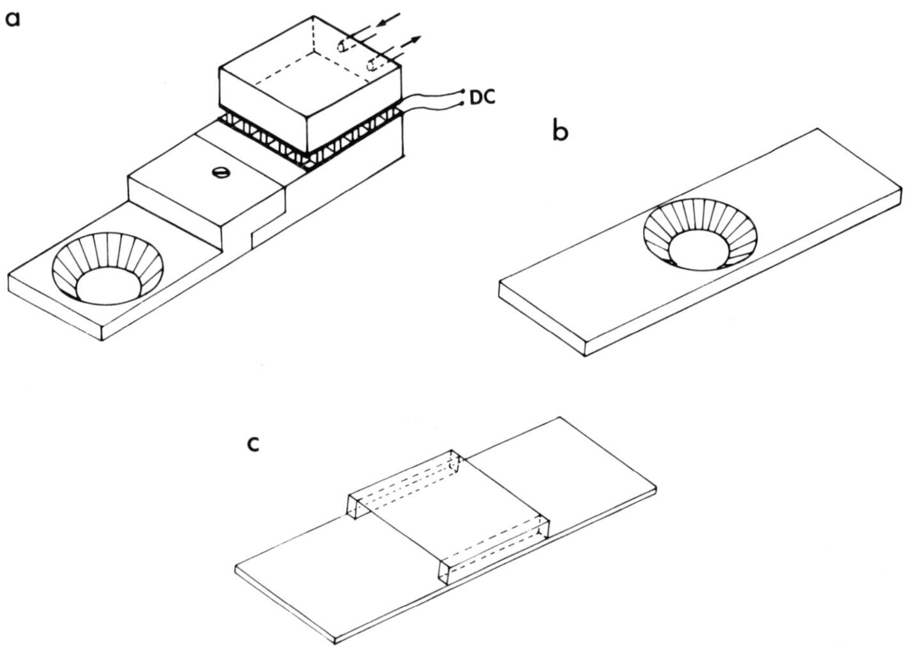

Figure 6. Cooling devices and injection chambers. (**a**) Solid state cooler with separate injection and cooling modules. (**b**) and (**c**) Typical injection chambers that may be fitted to microscopes that employ sub-stage or complete stage cooling. (**b**) is perhaps more convenient to use but may only be fitted to inverted microscopes. The hanging-drop chamber (**c**) is required with normal upright microscopes.

6.3 Cooling devices and injection chambers

Injections should be carried out at 10°C. Cooling is important because it gives the cell membranes a degree of rigidity. It also helps to prevent the 'stickiness' associated with cell lysis which otherwise necessitates the frequent changing of the injection instruments. There are a number of options with regard to the method used for cooling and chamber design. The system employed depends on whether the injection chamber, the movable sub-stage or the whole microscope stage is cooled.

6.3.1 *Solid state cooler (see Figure 6a)*

This cooling device provides localized cooling of the injection chamber. By attaching this device to a plastic sub-stage very little heat gain occurs by conduction. The nature of the solid state device requires that excess heat is carried away by a slow trickle of tap water. The device is constructed from milled aluminium with a glass microscope slide glued to the bottom of the chamber. The device is constructed in two detachable parts so that the injection chamber may be simply removed for cleaning purposes without the necessity for disconnecting the electrical circuits and re-fitting the water tubes. The injection chamber is 15 mm in diameter at the lower glass-covered surface, is 5 mm deep and is chamfered to an angle of 30°. The ceramic solid state cooling chip (Hanovia Ltd) is driven by a low voltage, high current powerpack.

6.3.2 *Sub-stage cooling*

Sub-stage cooling is achieved using a chilled water bath, or chilled flow cooler, connected to a heat sink attached to the sub-stage plate. This set up has been particularly useful where the working distances are very small (on normal upright microscopes and where an interference contrast condenser has been fitted to an inverted microscope). In practice the heat sink needs to be very slim and of a size and shape which will not interfere with the normal movement of the sub-stage. In addition it should allow relatively easy access to the injection chamber so that embryos can be introduced and withdrawn without having to dis-assemble the whole set up. Once a suitable design has been fitted water circulating at a temperature of 6–8°C provides effective cooling. It should be noted that water circulating at high rates can cause vibration.

Sub-stage cooling can be operated with different injection chamber designs (see *Figure 6b* and *c*). The author uses the one illustrated in *Figure 6b*, constructed from milled aluminium and a glass plate. The dimensions of the actual chamber are essentially the same as the solid state device (Section 6.3.1), though the chamber is accommodated in a piece of aluminium essentially no bigger than a microscope slide (though it is deeper). This design allows the chamber to be simply fitted into the slide holder on the microscope stage. Cavity slides or hanging drop chambers (*Figure 6c*) are also suitable injection chambers for use with sub-stage cooling.

6.3.3 *Microscope stage cooling*

If a suitable slimline heat sink cannot be manufactured it is possible to cool the whole microscope stage. Direct clip-on devices are sold commercially, though their suitability should be investigated bearing in mind the physical constraints described above. Direct stage cooling increases the range of injection chambers that can be used. This can include the use of disposable plastic Petri dishes.

6.4 **Micro-syringes**

Micro-syringes are used to control the suction applied through the micro-instruments. The micro-syringe controlling unit should be placed on the opposite side of the manipulator to the side which controls the movement of the instrument. This allows suction to be applied with one hand while operating the joystick with the other hand. The design of the syringe is relatively simple and is illustrated in *Figure 7*. Both the

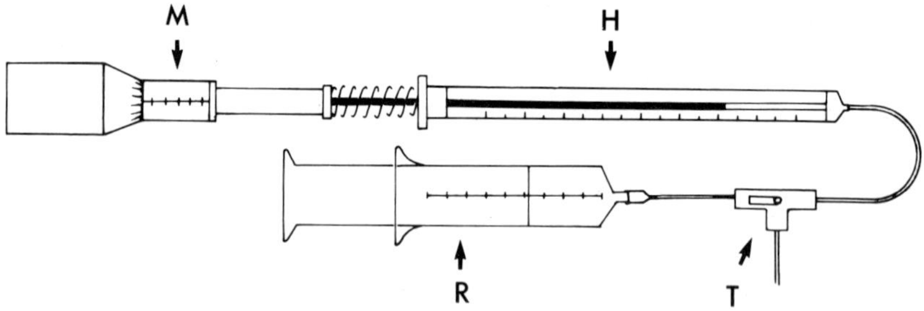

Figure 7. The design of a micro-syringe. H = Hamilton syringe, M = micrometer screw, R = 10 ml reservoir syringe, T = 3-way tap.

injection pipette and the holding pipette are controlled by a micrometer-driven spring-loaded syringe (a 1 ml ground glass syringe for the holding pipette and a 10–50 μl Hamilton syringe for the injection pipette). A large reservoir syringe (10 ml) is connected into the system via a 3-way tap. This reservoir syringe is necessary to remove trapped air bubbles from the inter-connecting tubing prior to loading the micro-instruments. The system is filled either with water or with light weight paraffin oil and is connected to the Leitz instrument tubes by a length of 1.0 mm diameter flexible polythene tubing.

6.5 **Preparation of pipettes**

Pipettes should be pulled on an automatic electro-magnetic electrode puller, although in the absence of a puller it is possible to improvise with a variety of gravity devices. We have used both flying coil (Campden Instruments) and flying magnet (Narishige Ltd) types with equal success. The parameters of the pipettes do not need to be very exact. Any electrode puller will probably suffice if a little time is spent adjusting the controls to pull capillaries of the correct shape and size. The ideal dimensions of the pulled capillary are illustrated in *Figure 8a*.

The glass capillary tubing used to construct the pipettes should be thin walled (1.0 mm external diameter, 0.8 mm internal diameter) borosilicate glass without a fibre (e.g. Campden Instruments No. B110815N; Clarke Electromedical). Once a number of pipettes have been pulled, then they should be fashioned into either injection or holding pipettes.

6.5.1 *Injection pipettes*

Injection pipettes are basically micro-hypodermic needles, which should have an internal diameter only marginally larger than the stem cells they are designed to inject and a sharp point which allows the needle end to penetrate cleanly into the blastocoel cavity. Although it is possible to fashion a sharp point by grinding on fine abrasive paper, in practice this is time consuming and the finished needles are not always as sharp as they appear. The best method we have found is to snap the glass capillary by hand under a stereo dissecting microscope, at a magnification of ×40, using a sharp scalpel blade and a spongy transparent silicon rubber sheet. Providing thin walled borosilicate glass is used the pulled glass will usually break with a sharp bevelled end when touched with the edge of the scalpel blade. The design of the pulled pipette is such that a relatively long section of the capillary is at the appropriate diameter (18–20 μm; see *Figure 8a*) and it is possible to repeatedly snap the same pulled piece of glass to achieve a suitable end of the correct size. With practice about half the needles have the appropriate diameters (see *Figure 8c*). The appropriate secondary structure can then be introduced into the pipette (Seciton 6.5.3).

6.5.2 *Holding pipettes*

Holding pipettes are blunt flame-polished pipettes through which suction is applied to immobilize the blastocyst. They are constructed using a Defonbrune microforge with a platinum wire heating element. Construct holding pipettes by the following procedure. The stages in the construction are illustrated in *Figure 9*.

(i) Melt a spare piece of capillary onto the apex of the platinum element of the microforge to form a glass bead 0.2–0.5 mm in diameter.

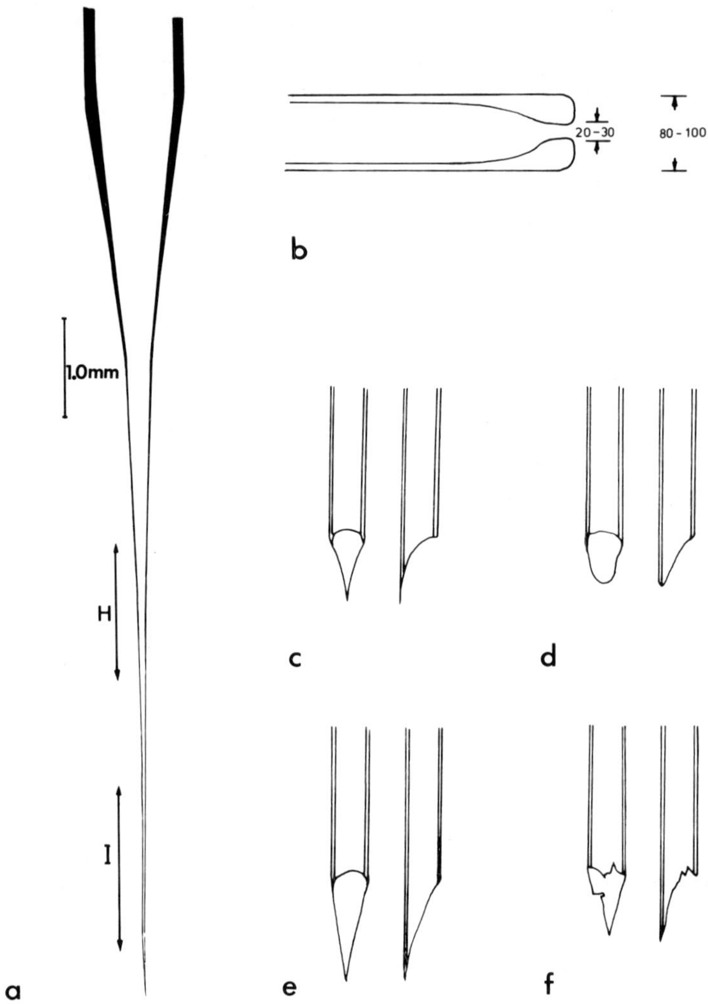

Figure 8. Construction of micro-instruments. (**a**) Preliminary shape of pulled capillary. H = region of snapping to construct a holding pipette, I = region of snapping to construct an injection pipette. (**b**) Ideal holding pipette, external diameter 80–90 μm, internal hole diameter, 20–30 μm. (**c**) Ideal injection pipette. Diameter, 18–25 μm, length of taper, 25–40 μm. (**d**) Pipette tip too blunt. (**e**) Pipette is correct shape, but end is too long (in excess of 50 μm). (**f**) Pipette end too jagged, unlikely to enter blastocoel cleanly.

(ii) Bring the pipette into contact with a glass bead at the appropriate diameter (80–100 μm). It is important that this is in an equatorial position, preferably at the North pole (*Figure 9a*).

(iii) Switch on the heat, and adjust until it is just sufficient to cause the glass bead and pipette to fuse (the filament should glow a dull red). The filament will expand across the field, but the pipette should not bend significantly (*Figure 9b*).

(iv) Switch off the heat. As the filament cools it contracts and the pipette will snap with a square end *(Figure 9c*).

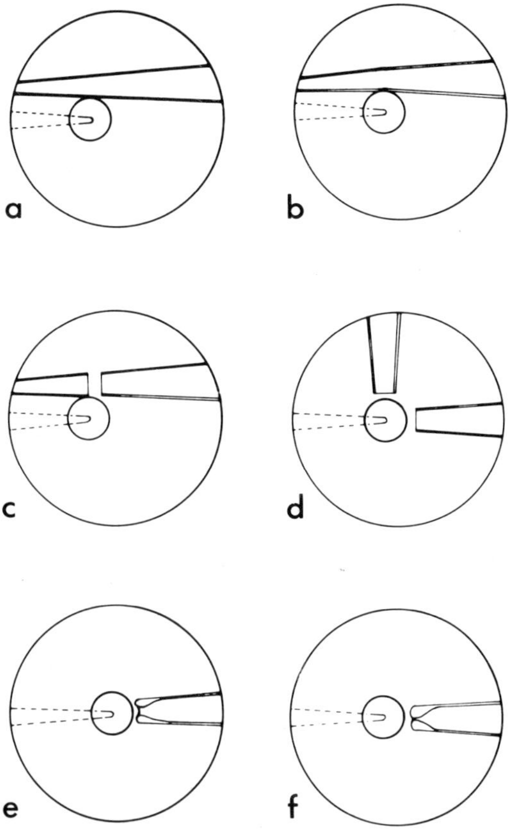

Figure 9. Fashioning a holding pipette. (**a**) Pipette in contact with the glass bead at an approprirate diameter. (**b**) Element switched on causes the glass to fuse. (**c**) Element switched off, pipette snaps. (**d**) and (**e**) Flame polishing the tip. (**f**) Finished holding pipette.

(v) Heat polishing the end may take place in any number of orientations (*Figure 9d*). Obtain an equatorial positioning of the pipette relative to the glass bead ensuring that both are in the same focal plane. Bring the pipette into close proximity to the bead, ensuring that it does not touch and that a sufficient distance remains between the two glass surfaces to allow for expansion of the filament. Switch on the heat to high (the glass head should glow bright orange). The rate of polishing is controlled by the temperature of the bead and the proximity of the pipette to the bead (*Figure 9e*).

(vi) Switch off the heat when the pipette has the parameters of the pipette illustrated in *Figure 8a* and *Figure 9f.*

6.5.3 *Secondary structure*

Secondary structure is necessary to enable the micro-instruments to be moved freely in the injection chamber (using the joystick manipulator arm) in the plane of focus without the necessity of re-adjusting their position. The degree of secondary structure required on each pipette depends to a great extent on the set up being used by the experimenter.

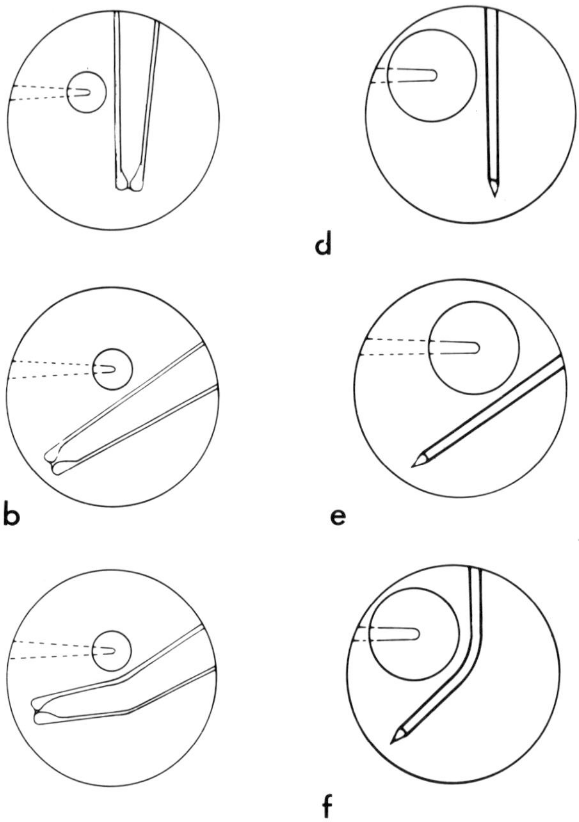

Figure 10. Fashioning secondary structure. (**a**) and (**b**) Alternative orientations of the holding pipette with respect to the glass bead. (**c**) Switching on the heat causes the pipette to bend towards the bead. (**d**) and (**e**) Alternative orientations of the injection pipette. Note the higher magnification and the orientation of the bevel. (**f**) Switching on the heat causes the pipette to bend.

For example, if using an inverted microscope and a milled depression chamber (see *Figure 5a*) to achieve the required position, a 30° bend is introduced into both the holding and injection pipettes (see *Figure 10*). If the microscope is of the normal upright type, then 'hanging-drop' chambers are usually required. This type of chamber often necessitates the construction of pipettes with shallow double bends (see *Figure 5b*).

The appropriate bends are placed in the micro-instruments using the microforge. The technique for bending holding and injection pipettes is identical. Holding pipettes are radially symmetrical so the orientation of the bend is unimportant, however the orientation of the bend with respect to the bevel on an injection pipette is important. The method of bending pipettes is described below and illustrated in *Figure 10*.

(i) Bring the pipette towards the heating element, ensuring that an equatorial position is achieved. The position should be such that the bend occurs 0.3–0.7 mm from the end. For injection pipettes make sure the bevelled end is in the correct orientation (see *Figure 10*).

(ii) Switch on the filament and adjust the heat until the glass bead glows a red/orange. This causes the pipette to bend towards the glass bead. Switch off the heat when the appropriate angle has been achieved (approximately 30°).

6.6 Blastocyst injection

6.6.1 *Summary of the setting up procedure*

(i) Switch on the cooling apparatus.

(ii) Wash out the injection chamber using 70% ethanol and wipe dry using absorbent tissue.

(iii) Set up a hanging-drop or depression chamber using blastocyst injection medium (DMEM supplemented with 20 mM Hepes and 10% FBS). Add the medium to a depth of 2–3 mm. Overlay the surface of the medium with a layer of paraffin oil and place the chamber onto the microscope stage.

(iv) Backfill the micro-injection pipettes with silicone oil.

(v) Purge air bubbles (using a reservoir syringe) from both instrument tubes.

(vi) Insert the holding and injection pipettes into the instrument tubes ensuring that no air bubbles are trapped.

(vii) Position the pipettes using the coarse adjustment controls and bring the ends into focus. For this manoeuvre it is helpful if a few blastocysts have been placed into the chamber and the focus adjusted. Without this information it is easy to focus on the wrong focal plane and break the pipettes on the bottom of the chamber.

(viii) Turn the 3-way taps to fine control and adjust the micro-syringes until the oil has stopped flowing out of the ends of the pipettes and the medium/oil meniscus is held stationary near the end of the pipettes.

(ix) Place the cells from a freshly prepared cell suspension into the injection chamber using a drawn-out Pasteur pipette and allow them to settle and cool for a few minutes before proceeding.

6.6.2 *Injecting the blastocyst*

The diameter of the cells in the suspension may vary between experiments and additionally the blastocysts are often different sizes due to the relative size of the blastocoel cavity (some are even partially or completely hatched). For each occasion the instruments must be tailored to suit the experimental material. If the first attempt at injection fails normally either the injection needle or the experimenter is at fault. If operator error is suspected a second embryo might be attempted. If this also fails change the injection pipette. The need to change the pipette at this early stage cannot be over stressed. There is nothing more frustrating than repeatedly failing to inject a dwindling population of embryos. The detailed injection procedure, illustrated in *Figure 11*, is described below.

(i) Using a magnification of ×200 carefully *select* the cells for injection on an individual basis from the population seeded on the floor of the injection chamber. Collect the cells into the injection pipette by suction. From a healthy population select 10–12 cells. If cell viability is suspected to be low, larger numbers of cells (15–18) should be chosen. Position the cells near the tip of the injection pipette.

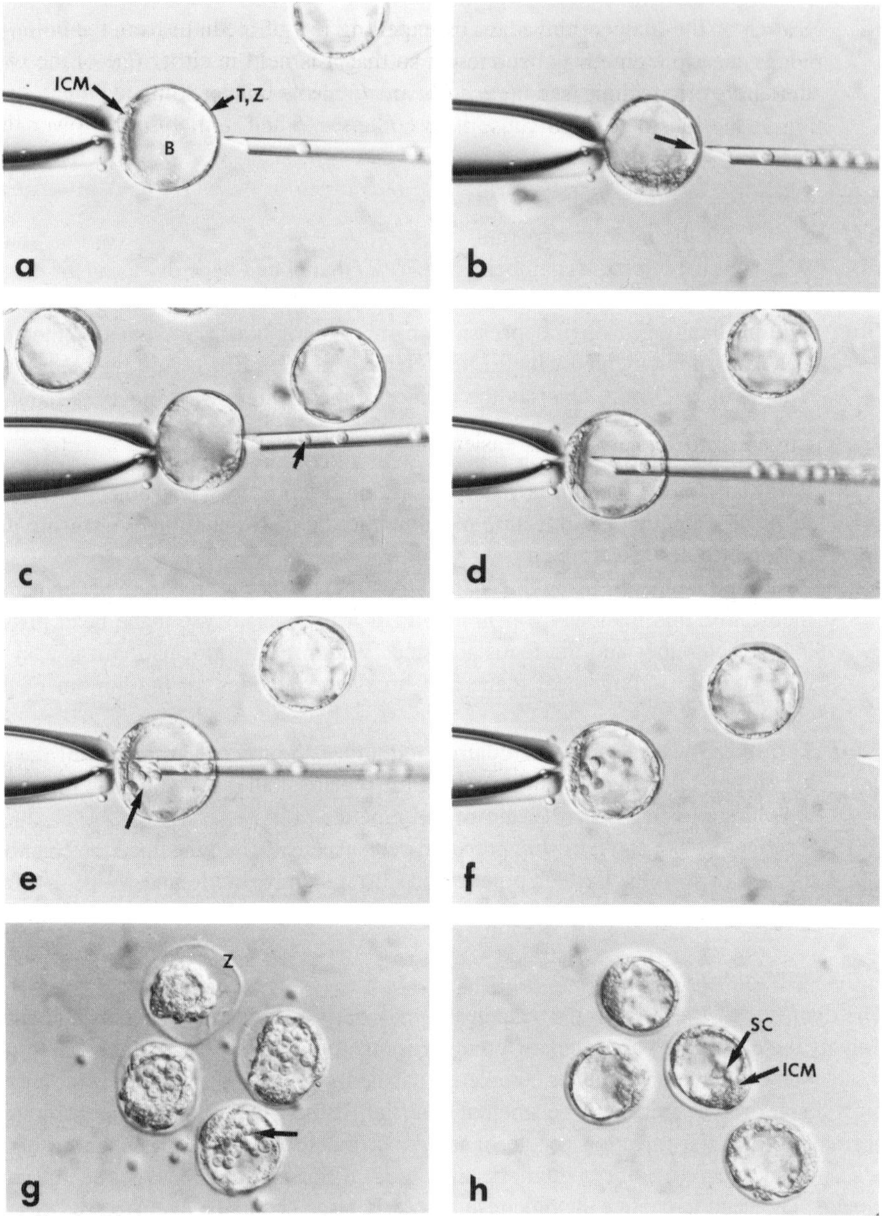

Figure 11. Stages in blastocyst injection. (**a**) Pre-injection position of blastocyst, holding and injection pipette. Blastocyst is held by suction in region of ICM. (**b**) Alternative blastocyst orientation. 'Gap' between trophoblast cells arrowed. (**c**) 3D positioning of injection needle is obtained by using the tip of needle to make an equatorial dent in the zona and trophoblast layers. Stem cell injection pipette arrowed. (**d**) Injection pipette cleanly inserted into the blastocoel cavity. (**e**) Stem cells (arrowed) deposited in the blastocoel cavity, here adjacent to the inner cell mass. (**f**) Injection pipette withdrawn leaving cells in the blastocoel cavity. (**g**) Collapsed blastocysts. Degree of collapse depends on the time which has elapsed since injection. Stem cells (arrowed) still visible in some cases. (**h**) Re-expanded injected blastocysts. Stem cells have mostly integrated into inner cell mass, though a few are still visible. ICM = Inner Cell Mass; Z = Zona pellucida; T = Trophoblast; SC = Stem Cells; B = Blastocoel cavity.

(ii) Orientate an expanded blastocyst by expelling jets of medium from the holding pipette. Grasp the embryo by suction so that it is held in either one of the two alternative orientations (see *Figure 11a* and *b*). Note that applying too much suction to the blastocyst will cause it to collapse. When immobilized, lower the blastocyst just sufficiently so that it is brought into contact with the floor of the injection chamber. This will restrict the movement of the blastocyst still further.

(iii) The standard inverted microscope gives a two-dimensional image and therefore does not provide sufficient information in the third dimension. This information is only attained by fine focus. To determine the precise plane of the injection pipette raise the tip and move it horizontally so that it just contacts the blastocysts to cause a small dent (but without damaging the zona or trophoblast) (see *Figure 11c*). Attain an equatorial position on a 'trial and error' basis. It is often convenient to inject the embryo at a 'junction' between trophoblast cells; cellular damage is minimized, penetration is easier and the amount of debris which adheres to the needle is reduced (see *Figure 11b*).

(iv) When the final positioning is complete, push the pipette into the blastocoel cavity (*Figure 11d*). The speed of the movement is important and has consequences on the success of the injection. If entry is too slow and hesitant, the pressure in the blastocoel will be released and the embryo will collapse before the pipette has penetrated. If the entry movement is very fast and uncontrolled, the pipette will enter the blastocoel but it is easy to accidentally damage the end on the opposing holding pipette, or damage the inner cell mass (ICM) component of the embryo.

(v) When the pipette is clearly within the blastocoel expel the cells from the injection pipette by increasing the pressure (*Figure 11e*). This should be performed using a controlled action as a sudden surge of pressure can force the expelled cells out of the blastocoel through the point of entry of the injection pipette.

(iv) When the cells have been expelled, slowly withdraw the pipette (see *Figure 11f*). Retain the blastocyst by suction on the holding pipette, change to a lower power objective, and move the embryo to a remote location in the injection chamber. Release the blastocyst by reducing the suction. Injected blastocysts collapse in a rather alarming manner (*Figure 11g*), but they soon regain their original shape and form when incubated at 37°C in culture medium (see *Figure 11h*).

Repeat the procedure on successive blastocysts, placing successful embryos in one group, and segregating failed embryos in another group. Using this procedure, and with practice, injections can be accomplished at a rate of 30–40 per h and as many as 100 embryos can be injected without changing the micro-instruments. Although the procedure can be very straightforward, in practice problems do arise. Their nature, and possible remedies are listed in *Table 4*.

Place the embryos in the injection chamber in groups of about 15–20. Transfer the successfully injected embryos back to culture drops and incubate at 37°C. Re-expansion of the blastocysts normally takes place within 1–3 h. As re-expansion occurs, the introduced cells should be seen adhering to the ICM component of the embryo (see *Figure 11h*). Generally, embryos should be transferred to 2.5 day p.c. pseudo-pregnant foster mothers 1–6 h after injection (see Section 8). If they are left too long after injection the re-expanding embryos begin to hatch from the zona pellucida. This makes handling

Table 4. Problems encountered during blastocyst injection.

Problem	*Reason*	*Solution*
1. Meniscus does not move when micrometer is turned	(i) Blockage in pipette	Replace pipette
	(ii) Air in system at any point	Remove pipette, purge using reservoir and replace pipette
	(iii) Leaky joint	Check and replace suspect joints
2. Cells 'rush' in and out of the injection pipette when micrometer is turned	Air in injection system	Remove pipette, purge with reservoir and replace pipette
3. Cells continually sucked into injection pipette	Leaky 3-way tap	Replace tap
4. Debris adhering to pipette	General cell lysis and 'stickiness'	(i) Raise the pipette through oil/medium meniscus
		(ii) Purge pipette with oil
		(iii) If not successful replace pipette
5. Pipette does not penetrate blastocyst before collapse occurs	(i) Jagged or blunt pipette	Replace pipette
	(ii) Point of pipette too long	Replace pipette
	(iii) Technique	Practice!
6. Cells lyse when sucked in and out of pipette	(i) Debris higher up in injection pipette	Replace pipette
	(ii) Pipette too small (ejected cells ellipsoidal)	Replace pipette with larger diameter pipette

Table 5. Typical results obtained following blastocyst injection (from refs 3,4,14–16).

Combination	*Time of analysis*	*% Chimaeras obtained*		*Level of contribution by cultured cells*	
		Normal animals	*Abnormal animals*	*Normal animals*	*Abnormal animals*
Blastocyst-EC	Pre-natal	5	7	0– 33%	0–34%
Blastocyst-EC	Birth	10	2	0– 53%	0–57%
Blastocyst-ES	Birth	35	0	30–100%	–

slightly more difficult and should be avoided where possible. Typical efficiencies of chimaera formation following blastocyst injection are given in *Table 5*. A typical chimaeric adult mouse is illustrated in *Figure 12*.

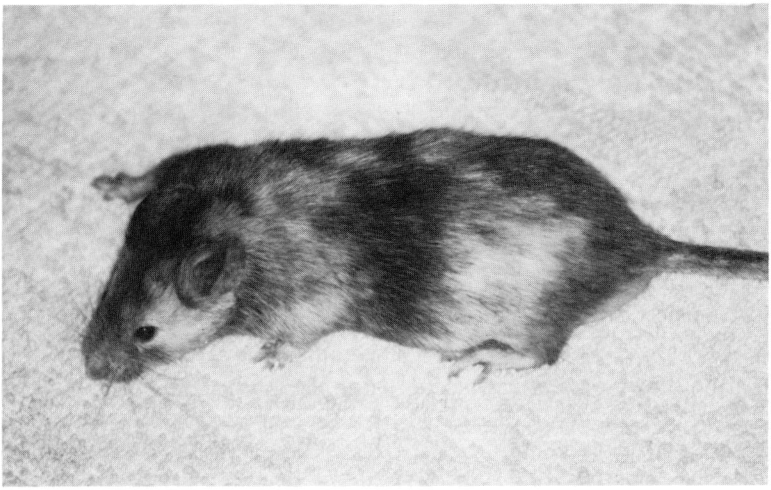

Figure 12. Typical chimaeric mouse.

7. MORULA AGGREGATION

Although the aggregation technique has the advantage of simplicity, the method has two problems.

(i) Because the zona pellucida is removed from the embryo it has to be cultured to the blastocyst stage before being returned to the uterus. The different nutritive requirements of stem cells and cleavage stage embryos places a requirement of an overnight culture step in a 'compromise' medium.

(ii) A single cleavage stage embryo may be unable to properly regulate a group of stem cells so that these combinations may lead to abortive embryogenesis. This problem can be partially overcome by combining two embryos with a single group of stem cells (6).

The equipment used for morulae aggregation is much less sophisticated than that required for blastocyst injection. All operations are performed under a stereo dissecting microscope. Basically, the technique consists of removing the zona pellucida from 8–16 cell embryos. The embryos are then combined either singly or in pairs, with clumps of stem cells in micro-drops. Aggregation will usually occur at 37°C, but the process can be chemically enhanced.

7.1 **Schedule**

The following schedule is found to be convenient.

(i)	Day 1 p.m.	Set up mice for fertile matings.
(ii)	Day 2 a.m.	Plug check female mice for fertile matings (day 0.5 p.c.).
	p.m.	Set up F_1 females with vasectomized males for pseudo-pregnant recipient females.
(iii)	Day 3 a.m.	Plug check the F_1 mice for vasectomized matings.

(iv) Day 4 a.m. Flush 8-cell embryos from fertile matings (day 2.5 p.c.) remove zonae, aggregate with stem cells and culture overnight.

(v) Day 5 a.m./p.m. Transfer blastocysts to F_1 pseudo-pregnant recipients (day 2.5 p.c.).

7.2 Removal of the zona pellucida

The zona pellucida can be removed either with acidified Tyrodes solution or by enzymatic digestion with a solution of pronase. The recipes for these solutions are given in *Table 6*. Use the following procedure. Intact and zona-free morulae are illustrated in *Figure 13a* and *b*.

(i) Using a drawn-out Pasteur pipette, place the morulae in a drop of Tyrodes or pronase (maximum of 10 embryos at a time) under oil.
(ii) Observe the embryos under the stereo dissecting microscope. The zona will disappear quite suddenly (after 20–40 s in Tyrodes, or 2–3 min in pronase).
(iii) Immediately the zonae have dissolved remove the embryos into a drop of Medium 16 supplemented with 4 mg/ml of BSA in groups of 3–4 as prolonged exposure to the agents will damage the embryo. Note that zona-free embryos are very sticky. To minimize any loss caused by sticking, handle the embryos in small numbers (3–5 at a time) and retain them at the very end of the drawn-out Pasteur pipette.
(iv) Wash the embryos by serial transfer through several drops of Medium 16.
(v) Culture the zona-free embryos for 1–2 h at 37°C singly, or in pairs, in microdrops before attempting the aggregation.

7.3 Decompaction

Decompaction is only necessary if embryos have proceeded beyond compaction. In a compacted morula the blastomeres are not distinct, while in a pre- or de-compacted

Table 6. Solutions for removal of the zona pellucida.

Acidified Tyrodes	
For 100 ml	
NaCl	0.800 g
KCl	0.020 g
$CaCl_2.2H_2O$	0.024 g
$MgCl_2.6H_2O$	0.010 g
Glucose	0.100 g
Polyvinyl pyrrolidone (PVP)	0.400 g
Water	100 ml

Adjust water to pH 2.5 with HCl. Filter sterilize and store in aliquots at −20°C.

Pronase (protease from *Streptomyces griseus*; Sigma Chemical Co.)

1. Make up as a 0.5% solution in Medium 16. Add 0.5% PVP.
2. Incubate for 30 min at 37°C for 1 h to remove contaminating nucleases and proteases.
3. Centrifuge to remove insoluble material.
4. Filter sterilize and store in aliquots at −20°C.

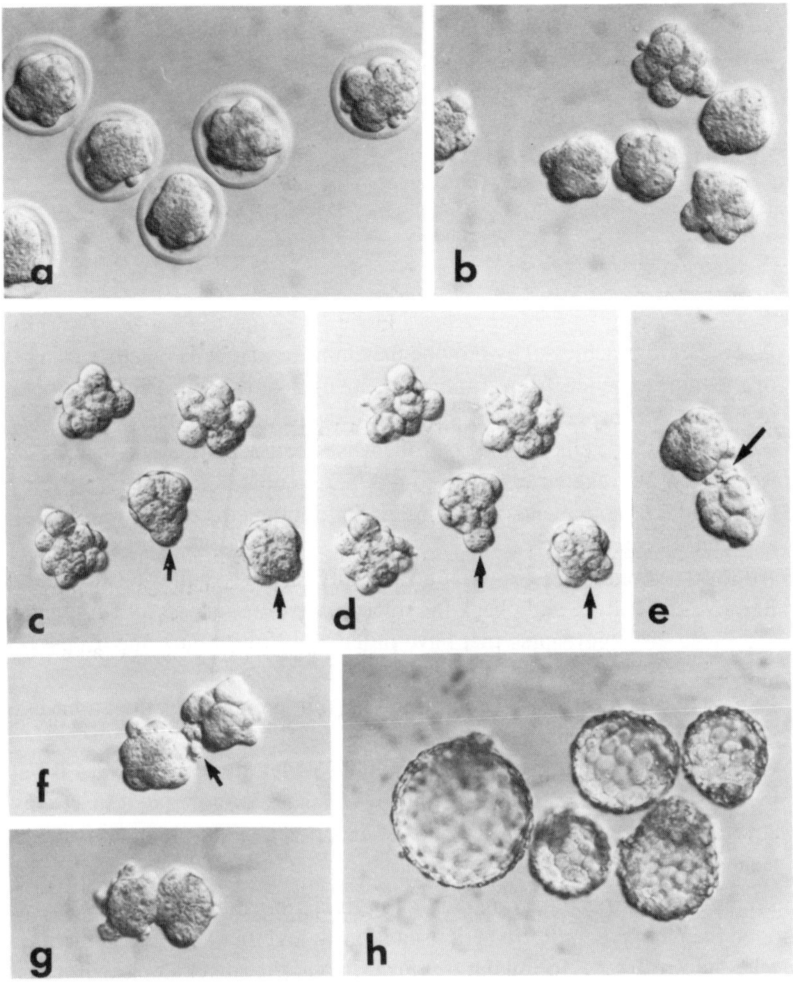

Figure 13. Stages in morula aggregation. (**a**) Group of morulae with zona pellucida intact. (**b**) Group of morulae after zona removed. (**c**) and (**d**) Successive stages in decompaction of compacted morulae—note the differences in the morulae arrowed. (**e**) and (**f**) Position of stem cells (arrowed) when aggregating morulae pairs. (**g**) After a few hours in culture stem cells no longer distinct and morulae are very closely applied and strongly adhering to one another. (**h**) Expanded blastocysts following culture of aggregated morulae pairs.

embryo they are readily discernable. Decompact embryos by incubating them in Ca^{2+},Mg^{2+} free phosphate-buffered saline (PBS) (see *Table 7*), until the blastomeres can be seen to separate. (See *Figure 13c* and *d*).

7.4 **Aggregation**

Proceed as follows.

(i) Prepare clumps of 5–10 stem cells (see Section 5) and place them in micro-drops of Medium 16 + 10% FBS under oil.

(ii) Place the zona-free 8–16 cell embryos singly, or in pairs, in the micro-drops.

Table 7. Calcium and magnesium free PBS, pH 7.2.

For 100 ml	
NaCl	0.800 g
KCl	0.020 g
Na_2HPO_4	0.115 g
KH_2PO_4	0.020 g
EDTA (0.2 mM)	0.04 ml of a 500 mM stock solution
Water	100 ml
Sterilize by autoclaving.	

(iii) Aggregation is achieved by moving the embryos and cells together using streams of medium expelled from a mouth controlled drawn-out Pasteur pipette. This technique is most effective if the culture drops are kept at 37°C. *Do not* touch the embryos or cells at this stage because they readily stick to glass.

(iv) Return the aggregated embryos to a 37°C incubator, taking particular care to avoid sudden movements which might disrupt the contact between the embryos and cell clumps.

(v) After 30–40 min the embryos are usually strongly adhering to one another. Remove the culture and check the micro-drops for successful aggregation. Re-aggregate any aggregates that have fallen apart and return the drops to the incubator.

(vi) Inspect the culture for a second time 30 min later. All of the embryos should now be strongly adhering to one another.

(vii) Culture the embryo aggregates overnight in Medium 16 + 10% FBS at 37°C in 5% CO_2 in air. Transfer the embryos which have aggregated successfully and developed to the blastocyst stage to 2.5 day p.c. pseudo-pregnant foster mothers (see Section 8).

The stages in the procedure for pairs of morulae are illustrated in *Figure 13e–g*.

Although the embryos and cells will usually stick together at 37°C, sometimes cooling of the micro-drops, following removal from the incubator, will reduce the adhesiveness of the embryos preventing many of them from aggregating. If this proves to be the case either restrict the number of micro-drops in each culture dish to 4–5 drops, or only attempt to aggregate a small proportion of the embryos in the dish at a time. The speed of cooling can also be reduced by aggregating the embryos in a warmed room or on a warmed microscope stage. The efficiency of aggregation may be improved either mechanically or chemically. Mechanically this can be simply achieved by reducing the volume in the micro-drop cultures; bringing the oil/medium meniscus down on the embryos forces them together. Alternatively some experimenters have successfully used phytohaemagglutinin (0.2 μg/ml) to 'glue' the embryos and stem cells together. Once the embryos have stuck they should be washed free of phytohaemagglutinin by serial transfer through several drops of Medium 16 and cultured to the blastocyst stage overnight. (See *Figure 13h*).

The position of attachment of the stem cells should be between blastomeres, when a single embryo is used, or between embryos when two embryos are used. The importance of exact positioning is not, however, clear because provided compaction has not

Table 8. Sample results of stem cells–morulae aggregation experiments (from refs. 7,8,17).

Combination	*Time of analysis*	*% Implanted or born*	*% Chimaeric*
Morula–EC	Pre-natal	32	56
Morulae[a]–EC	Pre-natal	51	40
Morulae[a]–EC	Birth	28	25
Morulae[a]–ES	Birth	36	20

[a]Pairs of morulae aggregated with a single group of stem cells.

Table 9. Surgical instruments.

Large dressing scissors
Iris scissors
Semkins forceps (blunt)
Watchmakers forceps (No. 5)
Needle holders
Clip fixing forceps
7.5 mm Michel clips
Suture silk (with attached needle)

proceeded too far, the stem cells will attain an 'internal' position with respect to the blastomeres as a result of the differential adhesiveness between the blastomeres and stem cells.

Typical results from aggregation experiments are given in *Table 8*.

8. SURGICAL PROCEDURES

Manipulated embryos are surgically transferred to the reproductive tract of pseudo-pregnant females so that they can continue development *in vivo*. Pseudo-pregnant mice are generated by mating with vasectomized males. The transfer of embryos to the female reproductive tract and the vasectomy of male mice are simple surgical procedures which can be accomplished in a relatively short time. It is important to obtain a good set of surgical instruments from veterinary or hospital suppliers (cheaper instruments sold by general laboratory suppliers are not usually very satisfactory). The basic surgical requirements are listed in *Table 9*. In general it is not necessary to take elaborate precautions with sterile techniques. The operations are performed on the bench under general anaesthesia. When performing these techniques on mice it is quite possible, with a little practice, to minimise the bleeding by avoiding cutting obvious blood vessels and keeping the incisions very small. It is much easier to perform the operation when bleeding does not occur because the internal structures are not obscured. Generally, the incisions made in the body wall should be sutured, especially if the incision is fairly long. This is not strictly necessary if the mother is going to be sacrificed at mid-term. Skin incisions are clipped with Michel wound clips. Clips are necessary because the mice will remove silk sutures before the wound has healed.

8.1 **Anaesthesia**

The recommended anaesthetic is Avertin. The dose administered is calculated to anaesthetize the mouse for the shortest possible time, usually 15–20 min. The Avertin stock is a solution of 1.0 g 2,2,2-tribromoethanol dissolved in 0.5 g tertiary amyl alcohol. Store the stock solution in the dark in a glass container. Dilute the Avertin to the final working concentration by dissolving 1.2 ml of the stock in 100 ml of boiling PBS. Shake the suspension until the Avertin dissolves totally. Aliquot and store frozen at −20°C. Once thawed the diluted Avertin should be used within 7–10 days. The dose administered is 0.02 ml/g body weight injected intraperitoneally. In some cases, particularly when relatively old mice are used, selective absorption of the anaesthetic into fat reduces its potency. If this happens a dose equivalent to one and a half times the normal dose can be safely administered.

8.2 **Transfer of manipulated embryos to recipient foster mothers**

Transfer of manipulated embryos to recipient foster mothers usually takes place on the same afternoon following blastocyst injection or the following afternoon following morulae aggregation.

At the blastocyst stage the embryos are found in the uterus, so it is to here that they are returned. Although it is acceptable to use pseudo-pregnant mice at the same developmental stage as the blastocysts, that is at 3.5 days p.c., in practice recipients are best used at 2.5 days p.c. This gives the transferred embryos an opportunity to re-synchronize with the response of the uterus following micro-surgery and the period of *in vitro* culture.

The most significant loss of operated embryos occurs because of recipient failure (the foster mother does not maintain the pregnancy). There are a number of factors which affect the frequency of failure. Firstly, the choice of recipient strain. As discussed in Section 2, F_1 (C57BL × CBA), or other F_1 hybrids, are ideal. Secondly the 'Brewster effect' (the smell of strange males), a particular problem in badly ventilated animal houses, may cause the females to re-absorb the embryos. Thirdly, surgical trauma can be a significant problem, particularly if the operator is inexperienced. Generally, minimizing the period of anaesthesia and developing a smooth surgical routine will effectively eliminate this problem.

The number of embryos transferred to a female can effect the outcome of the transfer if pregnancy is allowed to proceed to term. It is necessary to transfer a minimum of five blastocysts per uterine horn. If fewer than this are transferred, embryonic growth is often too rapid, resulting in a large foetus which can exceed the nutritive capacity of the placenta and result in pre-natal death. Additionally, large embryos are difficult for the mother to deliver and will often die during or shortly after birth. If the recipient mother only delivers a few pups, lactation can be reduced, again causing death of offspring. These factors make it important to 'crowd' embryos to their normal density. If low numbers are being transferred it is quite acceptable to transfer them to a single uterine horn. The maximum number transferred to each horn is eight (allowing for up to three embryos to fail). Transfers which exceed this number should be made bilaterally unless the recipient is going to be sacrificd at mid-term. It is not advised to exceed 12 embryos per recipient. Where the number of pseudo-pregnant foster mothers allows, the best experimental outcome is achieved with unilateral transfers of 6–7 embryos.

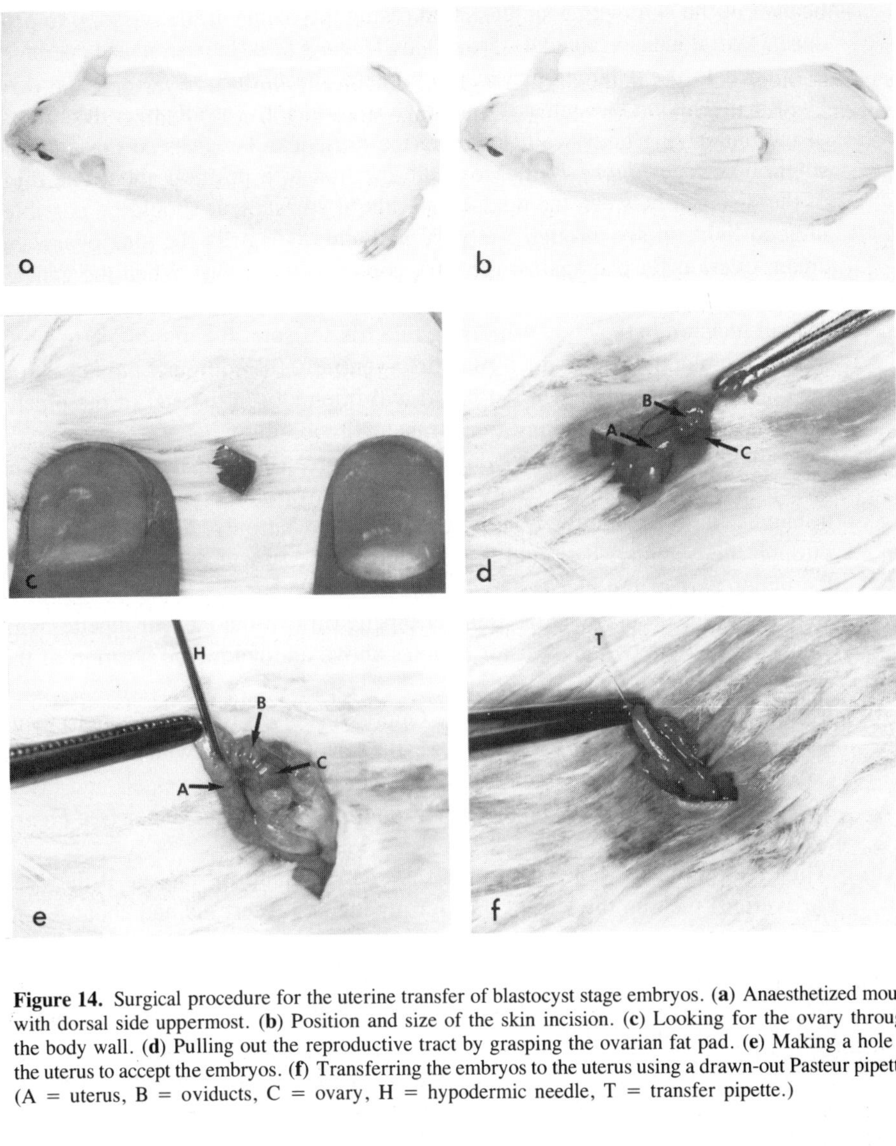

Figure 14. Surgical procedure for the uterine transfer of blastocyst stage embryos. (**a**) Anaesthetized mouse with dorsal side uppermost. (**b**) Position and size of the skin incision. (**c**) Looking for the ovary through the body wall. (**d**) Pulling out the reproductive tract by grasping the ovarian fat pad. (**e**) Making a hole in the uterus to accept the embryos. (**f**) Transferring the embryos to the uterus using a drawn-out Pasteur pipette. (A = uterus, B = oviducts, C = ovary, H = hypodermic needle, T = transfer pipette.)

Once the uterine transfer procedure has become routine, application of this technique should result in 70–80% of operated F_1 hybrid strain mothers giving birth. This may be reduced to 40–50% of females if inbred or outbred strains are used. Uterine transfers are performed by the following technique and the procedure is illustrated in *Figure 14*.

(i) Weigh the mouse and anaesthetize it by intra-peritoneal injection of Avertin anaesthetic. Place the mouse dorsal side uppermost on an absorbent paper towel and swab the back with a liberal quantity of 70% ethanol (*Figure 14a*).

(ii) Grasp the skin on the midline with blunt forceps at the midpoint of the back and raise the skin away from the body wall. Holding the dressing scissors vertically cut the skin with a single action (using the points of the scissors) to produce a lateral incision about 4–5 mm long (*Figure 14b*). Inexperienced operators may need to use a larger incision to proceed successfully.

(iii) Use the incision in the skin as a 'window' by sliding it ventrally over the surface of the underlying body wall. Both uterine horns can be accessed through this single hole (see *Figure 14c*). Move the incision to a position about one third of the distance between the dorsal and ventral surfaces. It should be possible, in good light, to see through the body wall and distinguish the pink ovary and cream ovarian fat pad against the dark contents of the gut. When the general position has been located use fine forceps to grasp the slippery body wall. Make a small incision in the body wall using fine iris scissors. It is important to avoid cutting the blood vessels which run dorso-ventrally (bleeding not only obscures the contents of the abdomen but blood will often clot in the end of the pipette carrying the embryos making their transfer impossible).

(iv) Hold the cut edge of the body wall with fine forceps, insert the blunt Semkins forceps into the abdominal cavity and grasp the ovarian fat pad. It is important to handle the reproductive tract by the associated fat only. Gently pull the fat through the opening. It should be followed by the ovary, oviducts and finally the upper part of the uterus (*Figure 14d*).

(v) Under the microscope load the transfer pipette (drawn-out Pasteur pipette) with embryos. Include two marker air bubbles above and below the position of the embryos in the capillary.

(vi) Under the stereo dissecting microscope (approximately ×10 magnification) focus on the exposed uterine horn. Gently grasp the top of the uterus with blunt forceps. Carefully select an area which is relatively devoid of blood vessels and insert a 25-gauge hypodermic needle through the uterine wall into the lumen (see *Figure 14e*).

(vii) While looking at the hole (its position is often difficult to relocate if the eyes are averted) pick up the loaded transfer pipette and insert the end about 3 mm into the uterine lumen (*Figure 14f*). Gently blow the embryos into the uterus using the air bubbles to monitor the transfer. After transfer check that the embryos have been transferred by expelling the remainder of the fluid in the pipette back into the culture drop.

(viii) Grasp the edge of the incision with fine forceps and gently push the uterus, fat pad and ovary back into the peritoneal cavity using the blunt forceps.

(ix) If required repeat the process on the other side. Close the incision in the body wall with a single suture and clip the skin with one or two Michel clips.

(x) When the foster mother(s) regain consciousness transfer them to clean cages in groups of two to four. They will litter 17–18 days later.

About 3 days prior to birth separate the pregnant mothers into individual cages. This is necessary to reduce the loss of pups by any infanticidal mothers. If the appropriate pigmentation markers have been used chimaeras may be scored at birth by eye colour or 3–4 days later by the presence of pigment in the coat.

8.3 Vasectomy of male mice

Vasectomy is performed on male mice at about 6 weeks of age. It is important to keep

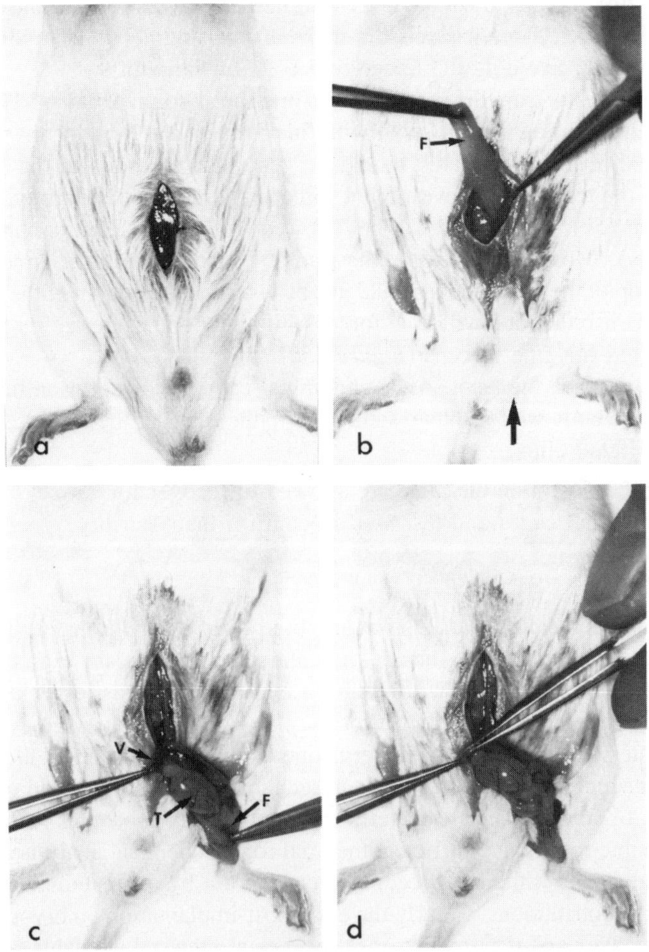

Figure 15. Vasectomy of male mice. (**a**) Position and length of the incision. (**b**) Pushing up testis (large arrow), grasping fat pad and exposing the testis by pulling at the associated fat pad. (**c**) Position of the vas deferens. (**d**) Cauterizing vas deferens using fine forceps. (F = fat pad, T = testis, V = loop of vas deferens.)

the mouse fairly warm during and after the procedure, because while the abdomen is open extensive heat loss will occur. An adjustable lamp with a 100 W bulb is convenient for this purpose.

Vasectomy is performed by the following procedure. The technique is illustrated in *Figure 15*.

(i) Weigh and anaesthetize the mouse using Avertin.
(ii) Lie the mouse on its back and swab the abdomen with 70% ethanol.
(iii) Part the hairs along the mid-ventral line.
(iv) Using blunt forceps lift the skin clear of the body wall and using dressing scissors cut along the mid-ventral line, in the position indicated in *Figure 15a*. Keep the points of the scissors up.

(v) Using fine forceps grasp the body wall in the mid-ventral line and raise it away from the gut. Using iris scissors make an incision about 1 cm along the mid-ventral line, avoiding the underlying coils of the gut.
(vi) Push one testis from the scrotal sac up into the abdominal cavity. Hold one edge of the incision in the body wall with fine forceps and look for the white fat pad attached to the testis. This should lie ventral to the coils of the gut.
(vii) Grasp the testis fat pad with blunt forceps and gently pull through the incision (*Figure 15b*). The fat pad will be followed by the testis (*do not* handle the testis directly). Move and orientate the testis by manipulating the attached fat pad.
(viii) The loop of the vas deferns should just be visible at the bottom of the testis (*Figure 15c*). Grasp the loop with fine forceps, lift clear and cauterize it off completely using red hot fine forceps (*Figure 15d*).
(xi) Gently replace the testis in the abdominal cavity and repeat on the other side.
(x) Close the body wall with two sutures. Clip the edges of the skin incision together with Michel clips.

Following the operation the mice are allowed to recover for 2 weeks before being set up in individual cages and test bred to confirm their sterility.

9. ANALYSIS OF CHIMAERAS

Chimaeras may be taken for analysis at any time after uterine transfer. The timing of the analysis clearly depends on the experimental design. Thus, for example, genes expressed during embryogenesis will require a mid-term analysis, especialy if analysis of the extra-embryonic components is necessary. Live-born chimaeras may be sacrificed at birth or later for examination. If germ line chimaerism is one of the aims of the experiment the chimaeras can only be analysed biochemically at the end of their useful breeding life. In practice the ease and efficiency with which chimaeras can be generated, makes it possible to produce sufficient material for biochemical analysis at all the pre- and post-natal stages while test breeding is carried out by other individuals to test for any germ line contributions. Clearly the early post-implantation embryos are the most difficult to recover and the amount of experimental material available for analysis is limited. A study of the very early differentiation events of a cell line is best carried out entirely *in vitro* (see Chapter 2, Section 3 and Chapter 4, Section 6) if biochemical studies are required. Early *in vivo* events are, however, amenable to lineage and immunohistochemical analysis. Dissection of the early post-implantation stages is detailed in Chapter 1. Later stages of embryogenesis are more amenable to study, in terms of the dissection involved, the quantity of material available and the ease of identification of chimaeras. After day 10.5 the embryos are relatively large, the decidual swelling is readily dissected from the uterus and further dissection using fine forceps yields the embryo, membranes and placenta.

Identification of chimaeras and assessment of the overall level of contribution from the cultured component in the mosaic individual is always an important consideration. Typically a population of manipulated embryos might include some 30–80% chimaeras (100% has been observed) with contributions from the cultured cells ranging from 5% to 75% between individuals (higher contributions have been observed in our hands).

9.1 Pigmentation markers

In general, pigmentation markers are the usual method of identification as chimaeras can be scored by simple observation. Identification can take place 4–5 days after birth when the first signs of skin pigmentation are evident. Pigmentation markers can also be used pre-natally because the embryonic eye begins forming pigment around day 10 of gestation. This can be detected with the naked eye if the contribution is heavy or with a dissecting microscope if the contribution is at a low level. Eye pigmentation can also be detected at birth through the closed eyelid of the animal. We routinely use albino embryos and introduce stem cells originally derived from a pigmented strain of mouse. It is also possible to utilize the *agouti* locus as a visual marker of chimaerism, although this marker cannot be applied to pre-natal or early post-natal individuals. Agouti hairs are normally detectable about 10 days after birth. The typical combination when this locus is used is to use a Black non-Agouti host embryo (*BB aa*) and introduce cells originally from a Black Agouti (*BB AA*) mouse strain. Chimaerism is seen as areas of brown hairs on a black background. In this case the chimaerism is assayed in the epidermal cells of the hair follicle rather than the melanocytes or pigmented retina. Although, in principal, there are a number of other pigmentation loci which could be utilized (reviewed ref. 18), in practice many of these are quite complex and the outcome in a chimaeric association may not be very easy to interpret. The author would recommend utilizing one of the two systems described above unless the coat colour genetics are well understood or form a necessary part of the experimental design.

9.2 GPI markers

Where coat or eye colour are not available as markers to distinguish chimaeras early in gestation (before pigmentation forms in the eye), or where pigmentation markers are required on both the host embryo and introduced cells, isoenzyme analysis is used to distinguish chimaeric from non-chimaeric tissues. Although there are a number of candidate enzymes for this type of analysis, in practice the choice is restricted to ubiquitously expressed enzymes that can be electrophoretically separated and stained. For most studies GPI is the enzyme of choice. It occurs in three forms in laboratory mice: GPI-1^a and GPI-1^b are common while GPI-1^c is relatively rare. The GPI isozymes are very stable during extraction and electrophoresis and may be stained in a relatively short time. The staining reaction for the GPI enzyme is based on the system shown in *Figure 16*. The method is a very sensitive and rapid micro-method which uses cellulose acetate plates, although the isozymes may also be separated on starch gels. Using the assay described here the fastest GPI isozymes run about 15 mm in a cathodal direction with a separation of approximately 4 mm between the GPI 1^a and 1^b bands and slightly more between the 1^b and 1^c bands. GPI gels are run according to the following protocol. A typical GPI gel from a chimaeric mouse is shown in *Figure 17*.

(i) Soak cellulose acetate plates (Titan III Zip Zone cellulose acetate plates; Helena Laboratories), in soaking buffer (25 mM Tris, 200 mM glycine [ammonia free], pH 8.5) for 20 min, taking care not to trap air bubbles when immersing the plate.

(ii) Prepare samples for analysis by freeze/thawing tissue samples and diluting them with distilled water. The sensitivity of the technique is such that considerable

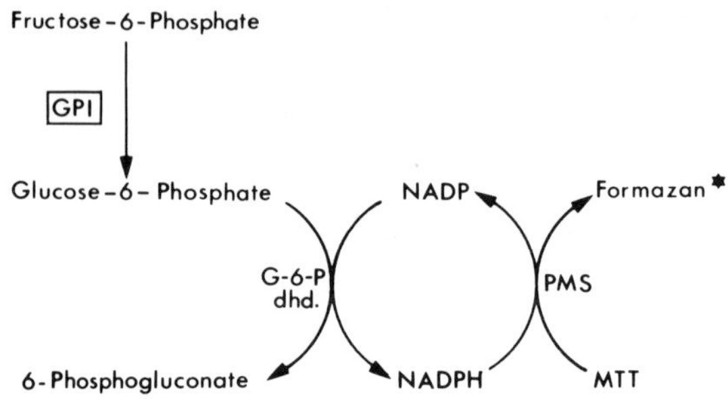

Figure 16. Staining system for GPI isozymes.

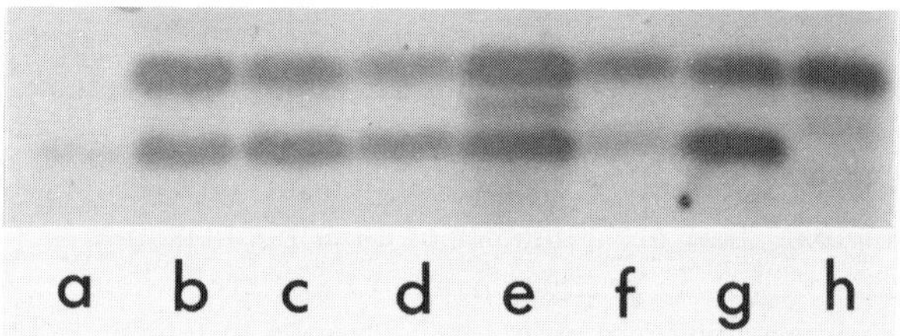

Fig. 17. GPI separation gel. A typical GPI separation gel analysing contributions to tissues in a chimaeric mouse. Injected ES cells are GPI − I^aI^a and the host embryo is GPI type I^bI^b. Hybrid band in lane e is the result of heterodimerization between the two different isozymes in this syncitial tissue. a = aa standard; b = kidney; c = thymus; d = eye; e = skeletal muscle; f = brain; g = liver; h = bb standard.

Table 10. Visualization stain for GPI isozyme separation gel.

	Volume required	*Stock solution*
Tris-HCl	2.0 ml	(24.2 g/l, pH 8.0)
Mg-acetate	0.1 ml	(54.1 g/l)
Fructose-6-phosphate	0.1 ml	(100 mg/ml)
NDP (Na_2 salt)	0.1 ml	(10 mg/ml)
Glucose-6-phosphate dehydrogenase	5.0 μl	(140 U/ml)
Methylthiazolium tetrazolium (MTT)	0.1 ml	(10 mg/ml)
Phenazine methosulphate (PMS)	0.05 ml	(2.5 mg/l)

dilution of each sample is usually required to avoid the production of misleading trailing bands.

(iii) Apply up to 16 samples per plate using a gel loader which applies eight 0.5 μl samples in 5 mm wide tracks. Two parallel sets of samples may be applied to the same plate. Run control standards in the peripheral tracks.

(iv) Run the plate at 180 V for 30−40 min at 4°C with wicks trailing into reservoirs of running buffer (100 mM Tris, 750 mM glycine, pH 8.5).

(v) Remove the plate and place it onto absorbent tissue.

(vi) Make up the stain as indicated in *Table 10* omitting the glucose-6-phosphate dehydrogenase (G6PD). Just before use add the G6PD and 5 ml of warmed agar (2% solution of agar in water). Mix well and pour the stain over the cellulose acetate plate.

(vii) The stain is light sensitive so, to avoid a high background, staining is performed in the dark. Invert a light-proof box over the plate. The isozymes are usually visualized in a few minutes.

9.3 **Analysing contributions to the germ line**

An important consideration is whether it is worth assaying for germ line chimaerism. In practice there are a number of criteria which, if satisfied, will minimize wasteful breeding experiments.

(i) The cell line from which chimaeras have been constructed should be euploid. The maintenance of euploidy is a problem that a number of workers have encountered. It is generally worth spending time analysing the karyotype of clones before constructing chimaeras if germ line contributions are crucial to the outcome of the experiment (see Chapter 4, Section 6 for karyotype procedure).

(ii) The efficiencies of chimaera formation should be high, the contributions to tissues should exceed 50% by GPI analysis and the mosaicism should be wide spread and fine grained (contributions should include every organ). If cloned lines are used these should be demonstrated to form chimaeras with a similar efficiency to the parental line.

(iii) Preferably the cell line used should be XY for a number of reasons. Firstly, the XY composition is stable in culture compared with XX lines which tend to suffer deletion or loss of one X chromosome (19). Secondly, a sex distortion phenomenon produces more male than female chimaeras. Where sex conversion has occurred and the animal is fertile it will transmit its culture-derived genome only (4). Thirdly, the analysis is much faster. This is important when there is only a low percentage contribution to the germ line. Females will take their entire breeding lives (about 1 year) to produce 100 offspring, while males can sire this number of offspring in 4–5 weeks.

If all of these criteria are satisfied then test breeding should demonstrate any germ line chimaerism relatively rapidly. If an XX cell line is used it may be possible to demonstrate a contribution if the first two criteria are satisfied, though large numbers of chimaeras should be generated and breeding programmes will be protracted. It is worth while noting that only three female germ line chimaeras have ever been reported and that these were generated in the same laboratory (20–22).

9.3.1 *Breeding programmes*

The assay for germ line chimaerism is an exercise in mouse breeding and simple genetics. As with many other aspects of the system, the most straightforward assay for germ line chimaerism involves the use of pigmentation markers.

In the system used in our laboratory, pigmented cells are normally injected into albino host blastocysts. The resultant chimaera can breed either from its embryo-derived albino

Table 11. Breeding results from chimaeras constructed between an XY pigmented cell line (*CC*) and an albino host embryo (*cc*) following mating to an albino animal (*cc*).

Sex of embryo	*Sex of chimaera*	*Genotype of germ cells*		*Genotype of progeny*	*Germ line contribution?*
		Host-derived	*Stem cell-derived*		
Female	Female	c	[a]–	cc	No
Female	Male	[b]	C	Cc	Yes
Male	Male	c	C	Cc,cc	Yes
Male	Male	c	–	cc	No

[a]The introduced cells are XY and unable to form functional oocytes. This is confirmed by the genotype of the progeny.
[b]The host embryo is XX. The resulting chimaera is male due to the incorporation of the culture-derived XY cells. The XX cells are unable to form functional sperm.

Table 12. Breeding results from chimaeras constructed between an XY albino cell line (*cc*) and a pigmented F_1 host embryo (*Cc*) following mating to an albino animal (*cc*).

Sex of host embryo	*Sex of chimaera*	*Genotype of germ cells*		*Genotype of progeny*	*Germ line contribution?*
		Host-derived	*Stem cell-derived*		
Female	Female	C,c	[a]–	Cc,cc	No
Female	Male	–	c	cc	[b]Yes
Male	male	C,c	c	Cc,cc	[c]Possible

[a]The chimaera is female. As the introduced cells are XY they are unable to form functional oocytes, all the albino progeny (*cc*) result from the transmission of the *c* marker from the host embryo.
[b]The absence of transmission of the dominant host-derived pigmentation marker (*C*) indicates that these animals are germ line chimaeras and transmit only sperm derived from the cultured cells.
[c]In practice, since the contributions are low in mosaic germ line animals, an excess of albino (*cc*) progeny is not a convincing demonstration of germ line transmission. In order to score for germ line contribution it is necessary to use an additional marker, e.g. GPI to distinguish the origins of the germ cells.

component (recessive) or its culture-derived pigmented component (dominant). The presence of one dominant and one recessive allele means that germ line chimaeras can be simply scored by the presence of pigmented progeny in the litter following mating to an albino animal. The expected pattern of the segregation of the pigmentation marker, for various categories of chimaeras, is shown in *Table 11*. In the absence of the pigmentation marker system use the GPI isozyme assay. We routinely use stem cells which are homozygous for the GPI-1^c isozyme and blastocysts that carry the GPI-1^a or 1^b markers. Any GPI-1^c component detected in the F_1 generation demonstrates a germ line contribution.

In practice several combinations of markers can be used for both constructing chimaeras and for breeding analysis, but care should be taken to ensure that they can be distinguished. The fact that chimaeras can be detected by pigmentation markers does not necessarily mean that the breeding analysis will be straightforward. In some combinations it may be impossible to answer the question of whether germ line transmission has occurred when single markers have been used. The example results given in *Table 12* demonstrate this point.

10. ACKNOWLEDGEMENTS

I am indebted to Professor M.H.Kaufman for teaching me many aspects of the embryo culture and surgical techniques described here. I would like to thank Kate Littlechild for typing the manuscript and Brian Curtis for photographic work. The author is a Beit Memorial Fellow.

11. REFERENCES

1. Gardner,R.L. (1968) *Nature,* **220**, 596.
2. Martin,G.R. and Evans,M.J. (1975) *Proc. Natl. Acad. Sci. USA,* **72**, 1441.
3. Papaioannou,V.E. and Rossant,J. (1983) *Cancer Surv.,* **2**, 165.
4. Bradley,A., Evans,M.J., Kaufman,M.H. and Robertson,E. (1984) *Nature,* **309**, 255.
5. Bradley,A. and Robertson,E.J. (1986) In *Current Topics in Developmental Biology 20,* Moscona,A.A. and Monroy,A. (eds.), Academic Press, New York, p.357.
6. Stewart,C.L. (1980) *J. Embryol. Exp. Morphol.,* **58**, 289.
7. Stewart,C.L. (1982) *J. Embryol. Exp. Morphol.,* **67**, 167.
8. Fujii,J.T. and Martin,G.R. (1983) *J. Embryol. Exp. Morphol.,* **74**, 79.
9. Mullen,R.J. and Carter,S.C. (1973) *Biol. Reprod.,* **9**, 111.
10. Champlin,A.K., Dorr,D.L. and Gates,A.H. (1973) *Biol. Reprod.,* **8**, 491.
11. Whittingham,D.G. (1971) *J. Reprod. Fertil. Suppl.,* **14**, 7.
12. Gardner,R.L. (1971) *Adv. Biosci.,* **6**, 279.
13. Gardner,R.L. (1978) In *Methods in Mammalian Reproduction.* Daniel,J.C.,Jr. (ed.), Academic Press, New York, p.137.
14. Papaioannou,V.E., McBurney,M.W., Gardner,R.L. and Evans,M.J. (1975) *Nature,* **258**, 70.
15. Illmensee,K. and Mintz,B. (1976) *Proc. Natl. Acad. Sci. USA,* **73**, 549.
16. Rossant,J. and McBurney,M.W. (1982) *J. Embryol. Exp. Morphol.,* **70**, 99.
17. Stewart,C.L., Vane,R.M. and Wagner,E.F. (1985) *EMBO J.,* **4**, 3701.
18. Silvers,W.K. (1978) *The Coat Colors of Mice.* Springer Verlag, New York.
19. Robertson,E.J., Evans,M.J. and Kaufman,M.H. (1983) *J. Embryol. Exp. Morphol.,* **74**, 297.
20. Stewart,T.A. and Mintz,B. (1981) *Proc. Natl. Acad. Sci. USA,* **78**, 6314.
21. Stewart,T.A. and Mintz,B. (1982) *J. Exp. Zool.,* **224**, 465.
22. Illmensee,K. (1978) In *Genetic Mosaics and Chimaeras in Mammals.* Russell,L.B. (ed.), Plenum Press, New York, p.3.

CHAPTER 6

Introduction of DNA into embryonic stem cells

ROBIN H.LOVELL-BADGE

1. INTRODUCTION

The ability to introduce DNA into embryonic stem cells is of critical importance for a wide range of experimental approaches that are aimed at providing an understanding of some aspects of embryogenesis and developmental gene regulation. Although this chapter is primarily a review of methodology it is pertinent to discuss briefly the types of experiment that have been done, or are envisaged, as this is relevant to the choice of technique and cell line to be used. This chapter will not attempt a comprehensive description of all of the techniques that are now available for DNA-mediated gene transfer in cultured cells but will concentrate on a few key methods that have met with some success when used with stem cell lines, and refer briefly to others that may be worth trying for particular experimental applications. Also, this is not really the place to describe molecular biological techniques; there are already a number of excellent manuals covering these aspects. However, some simple methods for DNA and RNA preparation that have proven very reliable, and a protocol for RNase protection assays are described. The latter seems, at present, to be the most suitable and sensitive technique available for detecting gene transcription, which will in many cases be an essential objective in the transformation experiment.

2. AIMS

These can be divided roughly into two categories. Firstly, experiments where one wishes to study the expression of the introduced gene, or the effect of that expression on differentiation or development and secondly, experiments where the main aim is to eliminate the expression of an endogenous gene. The former will require regulated expression of the introduced gene, the latter could involve constructs making antisense RNA or simply the interruption of genes through integration.

The differentiation of EC cells *in vitro*, especially the relatively controlled differentiation of F9 cells to parietal and visceral like endoderm with retinoic acid (see Chapter 2), has been very useful to investigate the regulation of genes expressed in these cell types. Thus, alpha fetoprotein (AFP) minigene constructs (1) and an H-2 gene (2), have been introduced into F9 cells, and both shown to be regulated correctly on differentiation of the cells to endoderm. For AFP in particular this approach has been very helpful in defining enhancer-like elements 5′ to the gene. F9 cells have also been used in studies on promoter and enhancer usage. This research stems from the observation that a number of viruses fail to replicate in early mouse embryos. EC cells generally show the same phenomena, although there are subtle differences between lines, and have therefore been used in experiments designed to analyse this block. This appears to be at the level of transcription of specific viral genes, e.g. the polyoma and SV40 early genes (3). The situation with SV40 is still not well understood as the T-antigen promoter and enhancer will function in early embryos and stem cells when linked to other genes (4–6 and author's unpublished observations) and T-antigen itself can be detected in F9 cells after transfection (3). In the case of polyoma it has been possible to select mutant viruses that will grow in EC cells (PyEC), and a change in the early gene enhancer was shown to be responsible. The mutant enhancer will increase expression of thymidine kinase (*tk*) and chloramphenicol acetyltransferase (CAT) genes in F9 cells (7). It seems that there is a transacting factor present in EC cells that is responsible for repression of viral enhancers (8). This factor appears functionally similar to the adenovirus E1A gene product, and it is suspected that it is responsible for regulating the expression of sets of genes in the early embryo. Some EC cells, F9 in particular, may not provide a good model of normal embryogenesis but it is much easier to study promoter usage where cellular differentiation can be relatively tightly controlled.

EC cells that can contribute to chimaeras have also been used as recipients in gene transfer experiments (9,10). In principle these cells should allow the study of any gene as they should be able to contribute to all tissues in the chimaeras. However, ES cells are considerably better for this purpose, and have been used successfully to study expression of a *tk*/*neo* construct introduced via a retroviral vector (11), and the human type II collagen gene introduced by transformation (5,12). Because ES cells may contribute to the germ line they may also allow the transmission of the introduced sequences to the offspring of the chimaera. ES cells can be used in this way to determine the *cis*-DNA requirements for correct developmental regulation of genes. Normally, however, the method of choice for this type of experiment would be to produce transgenic mice by direct injection of DNA into the pronuclei of fertilized eggs (13–16). However, it is envisaged that the stem cell/chimaera approach will be very useful in particular cases. These are:

(i) to study expression of genes whose expression might compromise development, e.g. oncogenes (17) or antisense constructs (18);
(ii) where it is necessary to characterize some aspect of the integration event before differentiation, such as site, copy number, or expression, in particular when this event is likely to be a rare, such as homologous, integration (19);
(iii) for techniques such as chromosome-mediated gene transfer (20) which would be technically very difficult to do with direct injection, but which are quite standard for cells *in vitro*.

3. TRANSFORMATION TECHNIQUES

The choice of technique will ultimately depend on the nature of the experiment. The various physical methods currently available can be ranked according to their simplicity and efficiency, but unfortunately there is an inverse correlation between the two (see *Table 1*). The biological methods, essentially those employing viral vectors, are both highly efficient and relatively simple to carry out, but they are not sufficiently well

Table 1. Physical methods to introduce DNA into cultured cells.

Calcium phosphate/DNA precipitation	Simple; no special apparatus needed. Some cell types refractory	Very inefficient 10^{-5} to 10^{-6}
Electroporation	Simple; apparatus required although this need not be sophisticated. Good for cell types resistant to other methods	Inefficient 10^{-4} to 10^{-5}
Protoplast fusion	Complex procedure. Introduces bacterial as well as plasmid DNA. The latter as supercoils	Moderately efficient 10^{-3} to 10^{-4}
Microinjection	Technically the most difficult. Sophisticated apparatus required	Very efficient 10^{-2} to 1

understood. This leads to design problems. While it is likely that at least some of these will be solved in the near future, there are also a number of constraints imposed by the infection route which will be more intractable (see Section 5.1).

3.1 Physical methods

3.1.1 *Calcium phosphate/DNA precipitation*

This is the most commonly used technique, partly because it was the first to be used successfully with mammalian cells but also because of its relative simplicity. Most cell types show reasonable transformation frequencies. However, sometimes these can be so low as to be unworkable, and the method is not very suitable for cells that grow in suspension. It has often been used with EC and ES cells, and although frequencies are low compared with fibroblasts, it is possible to generate sufficient transformants for most purposes. The technique is described in detail in Section 4.

3.1.2 *Electroporation*

This is a relatively new technique made attractive by its simplicity, improved transformation frequencies compared with calcium phosphate/DNA precipitation and success with cell types resistant to other transformation methods. The technique is described in ref. 21. Briefly, cells are suspended in buffer containing the transforming DNA and placed between two electrodes in a simple chamber. A brief high voltage shock is then applied across the chamber. This should ideally be supplied through a series of capacitors but a high voltage powerpack with automatic cutout can deliver the correct type of pulse. The DNA enters via pores which open up in the cell membranes; however, there is normally appreciable cell death. The critical parameters are the strength and duration of the potential difference and the concentration of DNA. Titration curves should be determined for a range of values of these against both cell viability and transformation rate. It is likely that the optimum conditions will vary according to cell type. Electroporation has been shown to work for ES cells (M.Evans, personal communication), but ideal conditions have not yet been established.

3.1.3 *Protoplast fusion*

Although this is a moderately efficient technique (22), it has a number of drawbacks. In addition to those mentioned in *Table 1*, the method works poorly unless the bacteria contain a high plasmid copy number, and it is impossible to transfer genes free of vector.

3.1.4. *Microinjection*

This is the most efficient 'physical' method. As long as the cells can be immobilized sufficiently well the technique can be applied for most cell types. Stable transformation rates of >20% can be obtained with mouse fibroblasts (23–25, and author's unpublished observations). At these frequencies it is not necessary to use selection; following injection of linearized pSV2*neo* into HD14 ES cells, G418 resistant colonies can be recovered at a frequency of 2% (author's unpublished observations). The difference between the transformation rates for fibroblasts and ES cells is similar to that found with calcium phosphate/DNA precipitation with the same plasmid, and probably reflects either a basic difference in integration rate between the two types of cell or comparatively poor expression of the *neo* gene in the ES cells. However, it should be noted that ES cells are more difficult to inject as they are smaller, require finer pipettes, are more susceptible to damage and can accept less volume of the DNA solution. Microinjection techniques are described in Section 6.

3.2 **Viral vectors**

A number of viruses are currently being exploited and developed as vectors for introducing foreign genes. These include SV40, polyoma, adenoviruses, bovine papillomavirus (BPV) and a variety of murine ecotropic and amphotropic retroviruses. They all offer the advantage that it should be possible to infect every cell in a population. BPV vectors may in addition be useful as they can remain extrachromosomal. To date, only retroviral vectors have been used successfully in ES cells. Some of these vectors and the problems associated with them are described in Section 5.

3.3 **Types of experiment**

3.3.1 *Transient expression*

Transient expression of genes often occurs after their introduction with the majority of the techniques mentioned above. This is due mainly to transcription from unintegrated copies and typically reaches a maximum about 48 h after transformation. Although the expression can sometimes be inappropriate, the presence of the correct cell type-specific promoter elements (enhancers) can significantly increase transcription. This has been exploited in a number of assays to map the precise position of these elements and to look for transacting factors associated with them. The most commonly used assay employs the CAT gene linked to the test sequence (see refs. 3 and 26 for methods). The advantage of the CAT assay is its sensitivity and ease of quantification; however, other enzymes or antigen markers have been used. It should be noted that supercoiled DNA is more efficient than linear for the purpose of these assays (27).

Most of the physical methods of introducing DNA are suitable for use with transient expression assays, although with microinjection one is limited to *in situ* detection methods such as antibody staining. Retroviruses are inappropriate; however other viral vectors such as those based on SV40 or BPV should be ideal.

3.3.2 *Stable transformation*

(i) DNA-mediated gene transfer. Stable transformation only occurs after the incoming DNA has integrated into a host chromosome. Occasionally there is a considerable delay before this happens, leading to poor growth in selective medium, or loss of a colony on passaging. Although it tends to be rare, it is also possible for integrated sequences to be lost. It is, therefore, essential to check for the exogenous sequences if transformed ES cells are taken out of selection as will happen, for example, if the cells are used to generate tumours or chimaeras. Most of the methods can be used for stable transformation or co-transformation (see Section 4.2 for other considerations).

(ii) Chromosome-mediated gene transfer. Whole chromosomes or fragments can be taken up quite efficiently by cells if applied as calcium phosphate precipitates together with a glycerol shock (20). These may integrate intact into an endogenous chromosome, but sometimes rearrangements or internal deletions are found. Selectable marker plasmids can be introduced at the same time as the chromosomes and are normally found to be integrated at the same site.

(iii) Homologous integration. This was thought to be very rare, but recent experiments have shown homologous recombination events to occur in 0.1−0.5% of transformants (19,28). The parameters affecting the frequency of homologous integration are not well understood at present. It seems likely that electroporation or microinjection will be the most suitable methods.

3.4 Selective markers

With the exception of infection by retroviral vectors and perhaps microinjection, all other transformation techniques require the use of some method of selection to identify the rare transformants in a culture population. A number of these selection methods are given in *Table 2*.

Recessive selection systems have the disadvantage that they require the recipient cells to have been mutated prior to the transformation. This is a particular problem with ES cells as they may lose developmental potential or undergo chromosomal rearrangements that would prevent them from forming viable gametes. A further complication is that the reversion rate must also be lower than the transformation rate.

Dominant selection systems are therefore much more attractive. Most of those used with EC and ES cells rely on bacterial genes that confer resistance to antibiotics. The bacterial genes have to be driven by eukaryotic promoter sequences and it is important that these are 'strong' enough to give sufficient gene product in the stem cells. Most current work uses the SV40 T-antigen promoter as in the pSV2 series of vectors, the Herpes simplex *tk* gene promoter, or a combination of the two. It should be emphasized that the SV40 promoter will work efficiently in embryonic stem cells. G418 selection for *neo* expression is the most popular system for embryonic stem cells as very

Table 2. Commonly used selective markers.

Gene		*Selective agent*	*Comments*
Recessive			
Thymidine kinase	*(tk)*	HAT medium[b]	Back selection with 5-bromodeoxyuridine
Hypoxanthine phosphoribosyl transferase	*(hprt)*	HAT medium	Back selection with 6-thioguanine
Dominant			
Dihydrofolate reductase	*(dhfr)*	Methotrexate	Selection increased in steps to give amplification, or used as recessive marker
Xanthine guanine phosphoribosyl transferase[a]	*(gpt)*	Mycophenolic Acid (MPA) (plus HAT)	
Tn*5* aminoglycosidase phosphotransferase[a]	*(neo)*	G418	
Hygromycin B phosphotransferase[a]	*(hph)*	Hygromycin B	
Chloramphenicol acetyltransferase	*(cat)*	Chloramphenicol	Only for few cell types; main use as reporter gene
Cell surface antigens		Fluorescence activated cell sorter (FACS)	
Oncogenes		Growth characteristics	

[a]These are prokaryotic genes, therefore they require eukaryotic promoters and polyadenylation signals, e.g. the SV40 early gene promoter and enhancer as in pSV2*neo*, or the MoSV LTR as in pY3 (containing *hph*). Other commonly used promoter sequences come from the Herpes simplex thymidine kinase gene, the Rous sarcoma virus LTR and the mouse metallothionein-1 gene.
[b]HAT medium is usually 100 μM hypoxanthine, 10 μM methotrexate (or 46 nM aminopterin), 16 μM thymidine, in normal growth medium.

low levels of *neo* gene product confer resistance. While mycophenolic acid (MPA) selection for *gpt* expression has been demonstrated with some EC lines (4) it was impossible to find appropriate levels for ES cells. Hygromycin selection has not been tried with ES cell lines.

4. THE CALCIUM PHOSPHATE/DNA PRECIPITATE TECHNIQUE

4.1 **Cells**

Feeder cells take up the precipitate quite efficiently (unpublished observations), probably more so than ES cells. The transformation rate is therefore improved significantly by using cells that have been passaged onto gelatin-coated dishes in the absence of a feeder layer. Most feeder 'dependent' lines (ES and EC) will maintain a largely undifferentiated phenotype for one passage on gelatin only, but they should not be maintained under these conditions for longer periods of time. Feeder cells can be added at any time, so it is advisable to add feeder cells to the cultures after the precipitate has been removed; this step is included in the protocol given in *Table 4*. High serum concentrations also seem to inhibit uptake of precipitates. This could be explained by a reduction in the rate of pinocytosis, but it may simply be that the very fine precipitate fails to sink onto the cells. For this reason the protocol includes changes in serum concentration. For feeder independent lines and those not grown in high serum, both these parts of the protocol can be ignored.

The cells should be maintained at their optimum pH and temperature. Do not add more than 0.5 ml of precipitate per 5 ml of culture medium. The calcium phosphate precipitate will retard cell growth slightly, and so should not be left on the cells for an extended period. It has been found that cell viability is not affected after 24 h; however, if problems are encountered, reasonable transformation rates can be obtained after a 4 h incubation in combination with a glycerol shock step. The stem cells are killed slowly by selective agents such as G418. Therefore it is important to passage the cells between adding the precipitates and initiating selection, otherwise the cells may overgrow and 'pile up' and be lost by differentiation. The necessity for subculture means, however, that each original transformed cell may give rise to more than one colony.

For feeder-dependent cell lines it is essential to have feeder cells resistant to the selective agent. For dominant selection procedures it is simple to derive transformants containing the appropriate marker. The protocols given in *Tables 3* and *4* work well for fibroblasts, although it is necessary to adjust the cell numbers so that they are in an actively dividing phase when the precipitate is added. STO cells are more convenient for this purpose than primary fibroblasts.

Table 3. The preparation of calcium phosphate/DNA precipitates.

Prepare the following solutions.

Solution A

For each dish dissolve DNA in TE (1 mM Tris, 0.1 mM EDTA, pH 7.9) to give 5 μg DNA per 187.5 μl TE. Add 62.5 μl 1 M $CaCl_2$ to give a total volume of 250 μl containing 250 mM $CaCl_2$ and 20 μg/ml DNA.

Solution B

For each dish use 250 μl 2 × Hepes-buffered saline (2× HBS):280 mM NaCl, 50 mM Hepes, 1.5 mM Na_2HPO_4, pH 7.12.

Method

1. Add solution A dropwise to solution B in a clear plastic centrifuge tube (e.g. Falcon 2001 F), *while* bubbling air. This is easily done using a 1 ml plugged plastic pipette (e.g. Falcon 7521 F) connected via an inline Millipore filter to a pipette aid or mouth tube. Bubble air through the solution fairly vigorously. A fine precipitate should form immediately. (If the precipitate is stringy then the DNA may not have been properly dissolved; if no precipitate forms then the pH is probably wrong.)
2. Leave the precipitate to stand for 45 min at room temperature.
3. Resuspend the precipitate by pipetting and add 500 μl of the suspension to each 6 cm dish. Add the precipitate dropwise over the area of the dish and gently 'swirl' to ensure a uniform distribution.
4. Return the cells to the incubator.

Table 4. Cell culture protocol for the calcium phosphate/DNA precipitation technique.

Day 0.	Plate out ES cells onto 6 cm gelatinized dishes (without added feeder cells) at a density of 2.5×10^6 cells per dish, in culture medium (DMEM) containing 15–20% fetal calf serum and 5×10^{-5} M β-mercaptoethanol.
Day 1.	(i) After 18–24 h inspect the cell cultures and change the medium for one containing only 10% serum and incubate for 1–2 h. (ii) Add the calcium phosphate/DNA precipitate (see *Table 3*) and re-incubate. [Note: If a glycerol shock is to be used, 4 h after adding the precipitate treat the cells with 15% glycerol in DMEM for 4 min and rinse with Hepes buffer (25 mM Hepes pH 7.1, 134 mM NaCl, 5 mM KCl, 0.7 mM Na_2HPO_4), or medium, three times to remove all traces of glycerol and then reincubate in normal medium.]
Day 2.	After 24 h check the cells. The precipitate should be visible on the cells. Aspirate off the medium and replace with normal medium containing high serum and 5×10^5 G418 resistant feeder cells per dish.
Day 3.	Passage the cells expanding each 6 cm dish into 3×10 cm gelatinized dishes supplemented with 3×10^6 fresh G418 resistant feeder cells per dish. Use normal culture medium supplemented with the appropriate concentration of G418, e.g. 0.3 mg per ml (concentrations of G418 are discussed in Section 4.3).

Change the medium every 2 days, maintaining selection. It is not normally necessary to add fresh feeders but if they become sparse the feeder layer may be supplemented at any time.
Stem cell colonies growing in the selective medium should be visible after a further 7 days.

4.2 DNA

Linearized DNA is considerably more efficient than supercoils at giving stable transformation (see ref. 13 for discussion). Cutting within the vector, or at the vector/insert junction, is also more likely to lead to integration events where the gene of interest is not interrupted. Ideally the selectable marker gene should be covalently linked to the sequence of interest; however, a consistent finding is that the frequency of successful co-transformation is in the region of 70% (see *Table 5*). Consequently, unless there is a specific reason for knowing the precise linkage, it is not generally worth redesigning the input DNA.

The quality of the DNA is very important. It is not advised to use chloramphenicol amplified plasmid as this may contain ribonucleotides, and is therefore more prone to nicking. The DNA must be clean and free from traces of phenol. Gel-purified fragments occasionally give problems. If repeated precipitation and washing of the DNA does not help, either try changing the batch of agarose or use an alternative method of purification such as a column or sucrose gradient.

Apart from the various points mentioned above, the main factors affecting transformation frequency are the amount of DNA, presence of carrier DNA and the quality of the precipitate. A titration curve with pSV2*neo* on HD14 ES cells shows increasing numbers of G418 resistant colonies per dish with increasing amounts of input DNA. However, the frequency per μg of pSV2*neo* reaches a plateau with about 1 μg. This is illustrated in *Table 6*, which also compares transformation frequencies between ES cells and fibroblasts. The amount of DNA will affect the nature of the precipitate, and this may well have some subtle effect on the titration curve. The use of carrier DNA will improve the efficiency. Heterologous genomic DNA (e.g. salmon sperm DNA) was frequently used in early transformation experiments. However, because co-trans-

Table 5. The effect of co-transforming DNA on transformation rate.

Cell line	*Cell type*	*G418 resistant colonies per 10^6 cells per μg pSV2neo* (a) 1 μg *pSV2neo*	(b) 0.5 μg pSV2neo 5 μg cosHcol.1	*Percentage in (b) containing cosHcol.1*
HD14	ES	0.63	2.55	70
CP3	ES	0.6	2.01	67
C11-D	fibroblast	2.01	4.8	75

Transformation using the calcium phosphate/DNA precipitation technique was carried out as in the text. pSV2*neo* was linearized with *Eco*RI, cosHcol.1 with *Sal*I. CosHcol.1 is a cosmid containing the human type II collagen gene and is approximately five times the length of pSV2*neo*. The two transforming DNAs were therefore present in similar molar amounts. The presence of cosHcol.1 was determined by Southern blot analysis.

Table 6. The effect of increasing amounts of pSV2*neo* on transformation frequency.

Cell line	*Cell type*	*pSV2neo (μg)*	*Colonies per dish*
HD14	ES	2.5	16
HD14		1.0	14
HD14		0.5	8
STO	fibroblast	2.5	286
STO		1.0	169
C11-D	fibroblast	2.5	69
C11-D		1.0	67
C11-D		0.5	40

Transformation was carried out using the calcium phosphate/DNA precipitation technique as described in the text. A constant volume of precipitate was added to each dish (0.5 ml/6 cm dish).

fected DNA tends to integrate together, this means that the DNA of interest will be flanked by unknown heterologous sequences. The integration site and its effect will therefore be meaningless. For most purposes sufficient numbers of transformants are obtained not to warrant the use of such carrier. However, in co-transformation experiments, the co-transforming DNA will itself act as carrier, and will significantly increase the transformation frequency. Thus, optimum frequencies were obtained with 0.5 μg pSV2*neo* when it was precipitated together with 5 μg of cosmid DNA; and these frequencies were 2–4 times those found with 1 μg pSV2*neo* alone (*Table 5*). A fine calcium phosphate precipitate works best. Although it may be affected by the concentration of DNA the factors that are most critical are the pH and freshness of the solutions, and the rate of 'bubbling' (Section 4.3).

The ratio of co-transforming to selectable marker DNA should be such that there are at least equivalent, and preferably greater, molar quantities of the former.

4.3 The preparation of calcium phosphate/DNA precipitates and cell transformation protocol

All solutions should be made up fresh, using tissue culture grade water and sterilized by filtration. The DNA solution should be sterile because of ethanol precipitation, but

Table 7. Use of cloning rings.

1.	Aspirate medium from the dish, and gently wash the cells with PBS. Remove all the PBS by aspiration, but take care not to let the cells dry out, especially if it is intended to continue culturing the dishes.
2.	Pick up a cloning ring with a pair of bent forceps, smear its bottom edge with sterile (baked) silicone grease and place it over an individual colony, pushing down to make a good seal with the surface of the dish.
3.	Add two drops of trypsin with a Pasteur pipette to the cells within the cloning ring, and leave for 5 min.
4.	With a small volume of medium in a Pasteur pipette take up the trypsinized cells. It is possible to detach them by pipetting up and down but take care not to dislodge the ring.
5.	Transfer the cells to a 1 cm diameter well of a multiwell dish (gelatinized and with an appropriate number of feeder cells).

if in doubt this too can be filtered; however, allowances may need to be made for some losses due to binding of the DNA to the filter. The protocol detailed in *Table 3* employs 5 μg DNA per 6 cm dish.

Transformation rates are determined by comparing the number of colonies relative to the number of cells on day 3. If the cells are not required Giemsa staining clearly reveals the stem cell colonies against the background of feeders. Wash the cells with PBS and fix them with 100% methanol. Stain for 2 min with 4% (v/v) Giemsa in pH 6.8 buffer (Gurr, buffer tablets, 33199), rinse off the excess stain with buffer (or tap water), invert the plates and leave them to dry.

The colonies may grow rapidly and should be picked within 3 or 4 days before they pile up excessively. They may be picked and trypsinized with a drawn out glass pipette as described in Chapter 3, or by using cloning rings (*Table 7*). Cloning rings may be made out of plastic, porcelain or stainless steel, and should be about 5 mm in diameter, 4 mm in height, with a wall thickness of 1 mm and a flat bottom edge. They can be conveniently made by sawing up an appropriately sized stainless steel tube.

It is normal to find that a proportion of colonies do not survive this first passage. These colonies probably represent unstable transformants where the exogenous DNA is episomal. However, by the time the cells have been expanded to give a confluent 10 cm dish, transformation is normally stable and it is no longer necessary to maintain selection.

Different cell types have varying sensitivities to G418, and batches of G418 may differ considerably in their activity. It is therefore worth testing each batch with each cell type to determine the optimum concentration to use for selection. This should be the minimum required to kill all cells, both in a cloning efficiency test and during normal passage. As a rough guide, with one particular batch, the concentration of G418 that completely inhibited growth of ES cells, STO fibroblasts and clone-1D L cells were found to be 0.3, 0.5 and 1.0 mg/ml, respectively.

5. RETROVIRAL VECTORS

There has been much recent work to develop retroviral vectors suitable for introducing genes into stem cells in general. These include those of adult animals such as the haemopoetic lineages as well as early embryonic cells. Some of this work is urged on by the need to develop highly efficient techniques for use in gene therapy, but the following properties of retroviral vectors give them some advantages over other transformation techniques for experimental purposes.

(i) They are highly efficient – all cells in a population can be infected.
(ii) Each provirus integrates at a single site, unlike other methods which tend to give tandem repeats.
(iii) It is possible to adjust titres so that each infected cell contains only one integrated provirus. Alternatively, with high titres or repeated infection there can be multiple copies, each at a unique site. This latter means that it is possible to use the vectors as mutagens for specific genes where selection can be applied. This has been done for the hypoxanthine phosphoribosyl transferase (HPRT) gene in F9 cells (29).
(iv) The viral LTRs may protect the sequences within from any effect of integration site making them good for constitutive expression.
(v) The properties of the LTRs may make the vectors powerful insertional mutagens.

However, there are also a number of real or potential disadvantages.

(i) Size constraints imposed by the viral packaging system mean that the vectors would be unsuitable for many genes and for others it would be necessary to use 'intronless' constructs. 6 kb is probably the upper limit that can be inserted in present vectors. Long range effects (chromatin structure etc.) which appear to be required for the correct regulation of expression in particular tissues for genes such as AFP simply could not be accommodated in the vectors that are available at present.
(ii) It remains to be demonstrated whether tissue specific promoters can work correctly when they are confined within a retroviral vector. The properties of the LTRs may make expression constitutive.
(iii) Because retroviruses and the vectors derived from them actively integrate by virtue of their LTRs (30), it is unlikely that they will be useful for attempts to obtain homologous integration.

5.1 Vector design

A description of the biology of retroviruses is beyond the scope of this chapter but it is worth mentioning a few points on the design of vectors and their relative merits.

Apart from their role in integration, the LTRs of retroviruses are strong promoters, at least in many differentiated cell types. Many vectors therefore use an LTR to drive expression of sequences placed within them (31). However, LTRs such as that of the Moloney murine leukaemia virus (M-Mul LV) are very inefficient in EC cells (32,33). Moreover, when retroviruses or vectors are introduced into early embryos or EC cells the LTR generally becomes inactive and there is no expression in the stem cells or differentiated derivatives (34). Mice carrying retroviruses introduced during preimplantation stages will become viraemic, but this is probably due to expression in a restricted range of tissue types late in development. This pattern of expression would seem to vary according to the site of integration of each provirus. This means that retroviral vectors which use an LTR to drive, for example, *neo* expression, are very inefficient at giving G418 resistant colonies of EC or ES cells, even though most cells will contain integrated copies of the vector (35). It is possible that the rare transformants selected in this way have different properties, although they may be able to contribute to chimaeras (11). Alternatively this may be another phenomenon related to integration site (36). If the plan of the experiment is to use the vector simply as an insertional mutagen then lack of expression is probably unimportant. However, if expression of a non-selectable gene is desired these simple vectors are inappropriate. Three approaches have been tried to solve this problem.

(i) A strong enhancer that can function in EC cells can be inserted into the 5′ LTR, replacing the M-MuLV enhancer. This has been done with the PyEC mutant enhancer from polyoma (37).

(ii) It is possible to select retroviral mutants that can actively grow in EC cells. These viruses show mutations in the 5′ LTR. These viruses can then be used to derive vectors (32).

(iii) An internal promoter can be included to drive expression from linked genes with the LTRs being relied upon only for integration. The Herpes simplex *tk* promoter either alone (11) or in combination with the SV40 enhancer (6) have both been used successfully.

For many purpoes it would be desirable to have two active promoters – for example, using one to drive a selectable marker gene. The LTR can be used efficiently in differentiated cells, but in EC or ES cells this means using a rare transformant. It is not known whether two internal promoters could be used, but it may be worth trying a combination of one with a mutant LTR.

Retroviral vectors carrying the *SupF* bacterial suppressor tRNA gene are particularly useful as they enable rapid isolation of cellular sequences at the site of proviral insertion (38).

An important consideration for any vector is the number of infectious particles it can give. 'Reconstruction' can have a profound affect on the efficiency of viral RNA synthesis and packaging making some vectors practically useless for embryonic cells which require very high titres.

5.2 **Helper cells**

The development of ψ-2 cells (39), which contain an integrated copy of pMOV-ψ^-, a retrovirus with a 350 bp deletion of the sequence required for packaging, has been extremely important as it allows production of helper-free vector. The cells permit efficient packaging of genomic RNA from any retroviral vector that carries ψ but which lacks any of the other packaging functions supplied by pMOV-ψ^-.

5.3 **Methods**

5.3.1 *Helper cells*

(i) Transfect the vector DNA into actively growing ψ-2 cells using calcium phosphate/DNA precipitation. The protocol described in *Table 4* can be easily adapted for these cells.

(ii) Clone and expand a number of transformants.

(iii) Check the titre of each clonal derivative by infecting 3T3 fibroblast cells with serial dilutions of the culture supernatant (see infection protocol, Section 5.3.3).

Note: Very rarely recombination may occur between the incoming vector and the pMOV-ψ^- in the ψ-2 cells to give infectious virus. Such a recombination is evidenced by an extremely high titre.

(iv) Choose the ψ-2 producer with the highest titre for infecting ES cells. Ideally the titres should be $>10^4$ and ideally up to 10^6 colony forming units (c.f.u.) per ml of culture supernatant.

5.3.2 *Preparation of supernatant*

(i) Collect culture medium from confluent ψ-2 producer cell cultures following a 12–24 h incubation.
(ii) Remove any contaminating ψ-2 cells by low-speed centrifugation for 10 min at 1000 r.p.m. and/or filtration through a 0.45 μm nitrocellulose filter.
(iii) Supplement the medium as appropriate for use with the recipient cells, and add polybrene (Sigma Chemical Co.) to 2 μg/ml. The latter helps virus dispersal. Store the polybrene as a filtered 100× stock solution in PBS at 4°C.

The supernatant is normally used fresh but it is possible to concentrate viral particles with an ammonium sulphate precipitation step (39) by the following protocol.

(i) Add an equal volume of saturated ammonium sulphate (previously neutralized with NH_4OH) dropwise to filtered, virus containing medium. All solutions and procedures should be carried out at 4°C.
(ii) Collect the precipitate by centrifugation and redissolve it in medium to give the required concentration, which can be up to 50 times that in the original supernatant.

5.3.3 *Infection protocols*

(i) *Titration on 3T3 cells*

(1) Construct a series of 10-fold dilutions of supernatant in culture medium plus polybrene to give a range from neat to 10^{-9}.
(2) Add 1 ml of each dilution to a 10 cm plate containing 5×10^5 3T3 cells.
(3) Incubate for 2 h at 37°C. Aspirate the viral supernatant and add 10 ml of fresh medium and incubate for 48 h before adding selective medium.
(4) Count the surviving colonies after 10 days (see Section 4.3), and determine the number of c.f.u. in 1 ml of the original supernatant.

(ii) *Embryonic stem cells*

(1) If the titre is very high ($>10^5$ c.f.u./ml), EC or ES cells can be infected simply by replacing the culture medium with supernatant for 12–24 h.
(2) If the titre is low ($<10^5$ c.f.u./ml), or if multiple proviral insertions per cell are required either concentrate the virus by ammonium sulphate precipitation (see Section 5.3.2), or repeatedly infect the cell culture by replacing the medium with fresh supernatant every 6–8 h over a number of days. Alternatively, the stem cells can be grown on feeder layers of producer cells. Mitomycin-C treat the ψ-2 cells and prepare feeder layers (see Chapter 4, Section 2). Maintain the stem cells on ψ-2 feeders for one passage. It is not necessary to add polybrene.

(3) If selection is to be used, passage the cells into the selective medium 24–48 h after infection.

6. MICROINJECTION

Microinjection is the most efficient of the physical methods for introducing DNA into cultured cells. Up to 100% of injected cells can show transient expression and, depending on cell type, about 20% can be stably transformed. However, it requires some considerable investment in equipment. A degree of skill is also necessary, but good quality micromanipulators and microscope considerably reduce operator problems.

6.1 **Equipment**

6.1.1 *Microscope*

There are essentially two different ways of organizing a microinjection set up depending on whether an upright (23) (see *Figure 1*) or an inverted microscope (24,25) is chosen. The upright version requires the use of a covered chamber which makes the construction of the microinstruments more complex; however, the optics can be superb. An inverted microscope is perhaps easier to use, as the cells can be injected in a Petri dish with simple micropipettes, but at some expense with regard to optical quality.

Recent developments in micromanipulator design are leading to a wider choice of suitable microscope. Most micromanipulators are free-standing and this imposes a requirement on the microscope to have a fixed stage, as it is not possible to focus without also moving the microinstruments. Some of the major microscope manufacturers see a fixed stage as an outmoded concept and a number of models have recently been withdrawn from the market. However, there is now an option to have micromanipulators

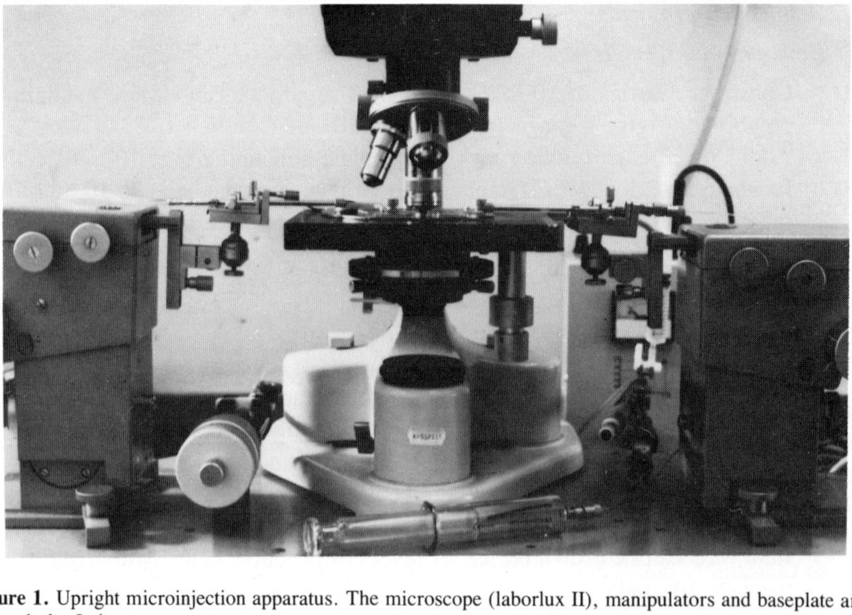

Figure 1. Upright microinjection apparatus. The microscope (laborlux II), manipulators and baseplate are all made by Leitz.

that clamp onto the stage. Consequently, these can be used with any microscope. However, at present, the movement they give is too smooth for many applications and free-standing manipulators are still generally preferred.

An example of a suitable upright microscope is the Leitz Laborlux II. This is an old design, but it can be adapted to take most modern optics. It is also very convenient to use with the Leitz mechanical manipulators. Zeiss Jena make a similar microscope.

Zeiss, Olympus and Nikon all make good fixed stage inverted microscopes suitable for injection. They can be used with Leitz micromanipulators by raising the latter on solid metal supports.

The microscope should be fitted with phase contrast and, if possible, with interference contrast optics. The latter works best when a glass chamber is used. The condenser and objectives should have long working ranges. Suitable objectives are 10×, 32× and 100× (oil immersion). A diamond scriber is also useful. 10× and 16× eyepieces are sufficient for most purposes.

6.1.2 *Other equipment*

(i) *Micromanipulators.* For good control it is necessary to have a joystick for movement in the horizontal plane. Leitz, DeFonbrune and Narishige all make suitable micromanipulators, although the author's personal preference is for the Leitz mechanical micromanipulators. For cell injections a single micromanipulator is sufficient, although a second is always useful.

(ii) *Pipette puller.* Good quality pipette pullers are supplied by Campden Instruments, Brown Flaming, Narishige and Kopf.

(iii) *Microforge.* DeFonbrune (available in the UK from Oxford Microinstruments) and Narishige make suitable instruments. It is useful to have an additional, high power, objective, for example Leitz UTK 50/0.63:L32/0.40.

(iv) *Injection system.* A number of alternatives are available to provide pressure for injecting DNA solutions. These vary considerably in their sophistication and cost. The simplest is a large air filled glass syringe. This has the advantage that flow is stopped and cannot be reversed simply by releasing pressure; however, it can be difficult to control the amount injected. A hydraulic system (*Figure 2*) can be used to have both continuous or intermittent flow, and it offers good control over the amount injected, but practice is required to judge when flow has stopped. Gas-pressure injection systems are now supplied by a number of manufacturers, e.g. Zeiss. These make it possible to inject the same volume (assuming constant pipette tip dimensions and flow) into each cell. Backflow can be prevented by a 'holding pressure'.

(v) *Glass capillaries.* It is essential to have thinwall capillaries with an internal fibre. The latter aids filling of the micropipette. Suitable capillaries are GC100TF-10 (10 cm length, 1 mm o.d., 0.8 mm i.d., borosilicate glass) from Clark Electromedical Instruments. To clean them, soak the capillaries overnight in 40% nitric acid, rinse thoroughly with distilled water, and finally with acetone. Leave to dry and siliconize with dimethylchlorosilane solution. Rinse once with acetone to remove excess silicone, wash thoroughly with distilled water followed by acetone, and bake dry. Siliconizing prevents loss of DNA through absorption onto the glass, and helps keep the outside free of cell debris.

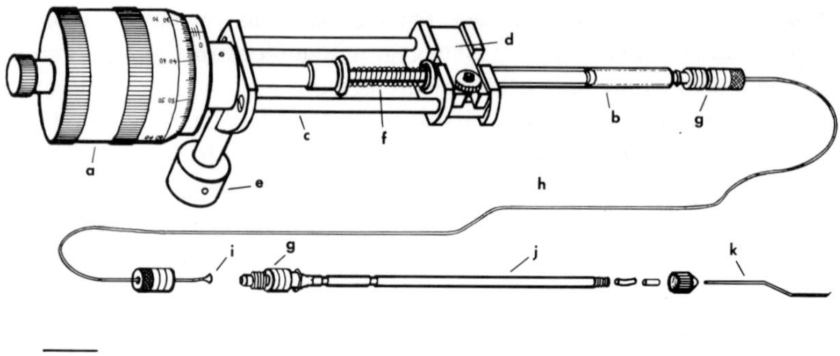

Figure 2. Hydraulic injection system. (a) Mitutoyo 0–25 mm (0.005 mm) micrometer, (b) 1 ml Hamilton gas tight syringe, (c) holding device for micrometer and syringe, (d) telflon clamp to hold syringe (hinged top piece secures with hinged bolt and nut, the syringe resting in a V-shaped bottom piece), (e) bolt-on attachment to baseplate (the holding device fits into this and is prevented from sideways movement with a screw), (f) spring, (g) Clay Adams tubing adaptor, (h) catheter tubing, (i) flange produced over flame, (j) Leitz instrument tube, (k) micropipette. The injection system is filled with water. (The bar at bottom left = 2 cm.)

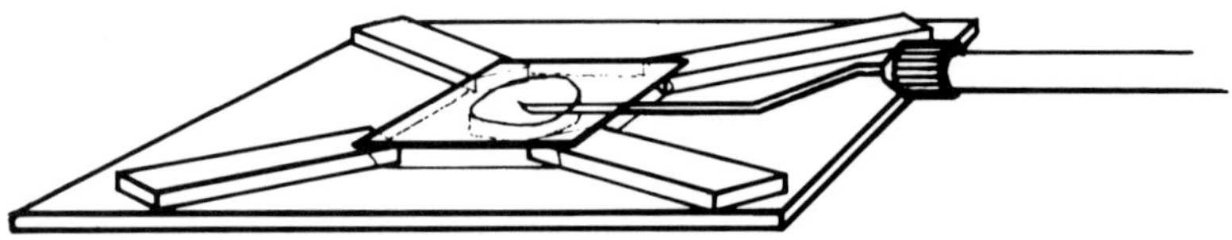

Figure 3. Injection chamber. (See text for details.)

(vi) *Connection systems.* The most suitable system is one that is rigid and can withstand fairly high pressures. A suitable system is shown in *Figure 2*. It is composed of a Leitz instrument tube into which the glass capillaries fit. The Luer end of a 'plastic tubing adaptor' [Clay Adams, B(7541); Horwell] is attached with Araldite to the other end of the instrument tube. Polythene catheter tubing [Clay Adams, Intramedic PE60 (7415)] is threaded through the other half of the tubing adaptor, its end is then flanged over a small flame, and the adaptor reassembled. This should give an air tight seal. The injection device, which normally has a Luer end, is connected to the tubing with another adaptor. If a hydraulic system is used it should be filled with autoclaved water (paraffin oil is messy), taking care to avoid introducing air bubbles.

(vii) *Injection chamber.* A simple design is shown in *Figure 3*. It consists of a 2 × 3 inch glass slide onto which four glass supports 2 mm in thickness have been stuck with silicone aquarium glue. The coverslip is held in place with (baked) vacuum grease. Use coverslips that are 22 × 22 mm, 0 or 1 thickness. For cell attachment they must be clean. Drop them individually into 40% nitric acid and leave to soak overnight. Rinse thoroughly with distilled water and finally rinse twice with Aristar grade acetone. Oven sterilize in stainless steel or porcelain coverslip racks.

To clean the chambers after use, remove as much grease and oil as possible with soft tissues; scrub with a soft sponge in strong detergent (e.g. Decon or 7X); rinse thoroughly in tissue culture grade water to remove all traces of detergent; leave to dry and then oven sterilize.

(viii) *Vibration-free conditions*. It is important to have vibration-free conditions for injection. The amount of vibration will depend on the local environment (nearby plant or equipment, type of building fabric, main roads, trains, tectonic activity) and there are a number of more or less sophisticated solutions. There is a problem if it is not possible to see the pipette tip at high magnification. Normally at least a heavy, sturdy table (e.g. concrete, marble or steel) is recommended. This should be situated just away from the wall, and if necessary positioned on isolation pads made out of a cork/rubber composite such as 'Tico S' (James Walker & Co, Woking, Surrey, UK) to cut out high frequency vibration. In addition, a 1 cm thick aluminium plate just larger than the base plate for micromanipulation set up is placed on top of three groups of three squash balls (yellow dot) positioned on the table. The micromanipulation set up then rests on this. If necessary a second aluminium plate and set of squash balls can be added to the pile. With this arrangement it is possible to stamp on the floor next to the table without any movement of the pipette tip. Alternatively, a commercial isolation table can be used (e.g. 'Micro-G', from Moore and Wright, Sheffield, UK).

6.2 **Injection procedure**

6.2.1 *Cells*

To obtain high transformation frequencies it is necessary to inject the DNA into the cell nucleus. Large flat fibroblastic cells are the simplest to inject as it is easy to observe the pipette within the nucleus and the flow of DNA solution. Such cells can be injected when grown at low density if the fate of individual injected cells is to be followed, or at higher densities when the number injected is important. With confluent monolayers it is possible to inject at a rate of about one cell every 2 or 3 sec as long as the pipette holds out, and almost all cells will be injectable.

ES cells, on the other hand, are far from ideal. They are small cells and fairly high magnifications, such as 512× or greater, are required to adequately monitor the injection (even though their nucleus/cytoplasm ratio is high). When grown as single cells in the absence of feeders they tend to round up. This makes observation difficult, especially under phase contrast, and injection is complicated because the cells are easily dislodged by the pipette. An additional problem with single cells grown in this way is that a high proportion of them differentiate to endoderm. The presence of a confluent feeder cell layer will help solve this but again affects visibility, making it difficult to both locate and inject the ES cells. In practice compromise works best, with feeders present at sub-confluent levels (10–20%).

If plated at higher densities and allowed to grow, the ES cells will form clumps which are often more than one cell thick. This can again cause problems of visibility and it is difficult to inject anything but the outermost cells. This leaves the majority uninjected and drastically slows down the rate of injection. Treatments that increase the adhesivity of the substrate (lectins etc.) can promote flattening but often affect cell viability. However, reducing the amount of serum in the medium is a simple and moderately effective solution, with both low and high density cultures, and this is described below.

With an inverted microscope the ES cells can simply be plated out onto gelatinized plastic tissue culture dishes as normal (although glass is essential with interference contrast optics). With an upright microscope it is better to use a chamber, such as that in *Figure 3*, with the cells growing on the coverslip. To confine the cells to the centre of the coverslip use cloning rings (with a very little silicone grease) until the cells have attached.

6.2.2 *DNA*

The DNA should be linearized for stable transformation or injected as supercoils for transient expression.

(i) After caesium chloride gradient purification, restriction enzyme digestion, gel electrophoresis, or other isolation techniques, extract the DNA with equal volumes of phenol, phenol/chloroform, and chloroform twice; add NaCl to 250 mM and 2.5 volumes ethanol (Aristar) and leave at −20°C to precipitate.

(ii) Centrifuge at 10 000 r.p.m. for 10 min, wash the pellet three times with 70% ethanol and dry briefly under vacuum.

(iii) Dissolve the DNA in 0.1× TE to give 250 μg/ml (see Section 7, *Table 8*, for details of solutions). The concentration of DNA should be checked by absorbance at 260 nm or by gel electrophoresis.

(iv) The purity of the DNA is very important. Traces of phenol and other organic compounds can be very toxic, and other impurities generally affect survival and transformation rates.

The procedure outlined above is minimal, and if problems are encountered repeat the precipitation or both the extraction and precipitation steps. For more elaborate alternative methods such as columns, caesium chloride gradients or selective absorption see ref. 41.

High transformation rates can be obtained when as little as one copy (on average) of the DNA sequence is injected per cell nucleus. Increasing the number of copies injected will improve the transformation rate a little, but the main effect is to increase the number of integrated copies. It seems that both homologous recombination and ligation occur before integration to give head-to-tail concatamers, normally at one or a few sites (see ref. 28).

Just before injection, dilute the DNA in water to the appropriate concentration. 5 μg/ml of a 5000 bp DNA is approximately equivalent to 1 molecule/femtolitre (1 fl = 10^{-15} l or 1 μm^3). Depending on the cell type the volumes injected will be 1−20 fl. Being small, ES cells will lyse if injected with more than 5 fl, so 1 molecule/fl is an ideal concentration.

To pellet particulate matter that could block the microinjection pipette, centrifuge the DNA at 10 000 r.p.m. for 10 min at room temperature.

6.2.3 *Injection procedure with an upright microscope and chamber*

(i) Place coverslips into a Petri dish. Lightly grease a cloning ring and place onto centre of coverslip. Prepare appropriate dilutions of ES cells in DMEM without serum but with 10^4 feeder cells per ml, and add 100 μl to each ring. With 5 mm diameter cloning rings, 10^5 cells will give an almost confluent layer, but 10^4 cells will be well separated from each other. Incubate for 2−4 h at 37°C to allow the cells to attach, remove the cloning rings and either set up on a chamber ready for injection or flood the dish with medium containing 5% serum and re-incubate.

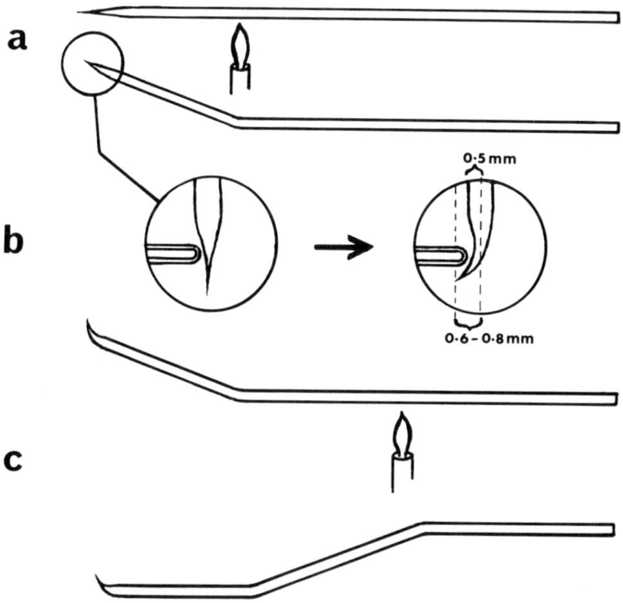

Figure 4. The manufacture of microinjection pipettes for use with a chamber. See text for details. Steps (a) and (c) are carried out over a small flame from a microburner; step (b) uses the microforge.

(ii) Take a chamber and smear a little sterile silicone grease onto each support at the centre. Use fine curved forceps to pick up a coverslip with attached cells, invert it and place it onto the chamber supports, pushing down gently onto the grease at each corner. Add Hepes-buffered DMEM containing 5% serum to cover the area of the cells and to span from coverslip to chamber floor. Surround the medium with either light paraffin oil (BDH, selected non-toxic batches, sterilized by filtration) or with silicone oil (e.g. Rhodorsil, 47v50; oven sterilized). If what is now the topside of the coverslip is wet with medium then dry it with sterile cotton buds, wipe with distilled water and with 70% ethanol and dry once more. Put the assembled chamber on the microscope and focus on the cells at low power (×10 objective). It is not necessary to have the microscope set up in a sterile environment as the chamber and oil provide adequate protection.

(iii) The injection pipette needs to have a fairly complex shape, as shown in *Figure 4*. The 'bayonet' shape is to overcome the diameter of the instrument tube, which would otherwise hit the microscope stage, and the upward bend at the tip allows the cells to be reached. Pull the capillary on the pipette puller, to give a very fine tip. The diameter must be less than 0.3 μm, however, this is below the limits of resolution of the light microscope. As a rough guide, if it is possible to measure the tip diameter it will be too large. The pipette should have a fairly wide angled final taper, such that it is about 5 μm in diameter at 20 μm from the tip. Put a bend of about 30° into the pipette, about 25 mm from the tip, over a small flame. Fit the pipette into an instrument tube up to the bend and attach this onto the microforge (see Chapter 5 for description on use) with the tip pointing straight down. Then bend the tip, either by applying heat close to one side so that it bends towards the filament, or by stroking with the filament at low heat to bend it away. The tip should be 0.6–0.7 mm from the axis and at an angle of about 60° compared to the adjacent capillary shaft. Remove the pipette from the holder and add the second bayonet bend over the flame (with the same angle as the first), making sure that all the bends are in the same plane and the bayonet is parallel. It is best to use the injection pipettes immediately as, on storage, they tend to become blunt and accumulate dirt.

(iv) Draw out a siliconized glass capillary over a flame to give a shaft about 6 cm long and 0.3 mm in diameter. Attach this to a mouth piece or rubber bulb and fill with about 2 µl DNA solution. Insert this capillary into the injection pipette and expel the DNA at the tip. Using a long 30 guage metal needle (Hamilton, non-standard needle; N730: 90 mm length: point 3: 90030) attached to a glass syringe, backfill the pipette with silicone oil or Fluorinert FC77 (3M Company) from the DNA at the tip, taking care not to leave any air bubbles. This provides an interface between the DNA and the water or air in the injection system.

(v) Assemble the pipette into the instrument tube attached to the injection device and place on the micromanipulator such that the tip is pointing upwards and the main shaft is parallel to, and almost touching, the chamber floor. Advance the pipette into the chamber taking care not to break the tip on the coverslip. Position the pipette tip so that it is in the light path from the microscope condenser. Focus down onto the pipette tip. Raise alternatively the focus and the pipette until the cells are in focus and the tip is just out of focus below them.

(vi) It is important to be able to judge the flow rate of DNA from the tip. With a good pipette, rigid tubing and no air bubbles, the pressure required to expel the DNA solution is not very great, although with long DNA molecules (such as cosmid clones) viscosity is a problem. Injection can be carried out with either intermittent or constant flow with any injection system. If intermittent flow is used care should be taken to avoid sucking back medium which will dilute the DNA. The hydraulic system is the most difficult in this respect. There is no simple way to calibrate the flow rate. The microscope stage can be moved so that the pipette tip is in the oil surrounding the medium, pressure applied and the DNA should form a drop around the tip (but only if the glass had been well siliconized). A drop with a diameter of 2 µm is about 4 fl. Usually it is necessary to inject a few cells, adjusting the flow rate until one can still see DNA enter the nucleus but without lysis. With constant flow, the amount injected can be moderated by the length of time the tip is within the nucleus as well as by flow rate. With intermittent flow it is shape of the pressure pulse that is critical. The commercial gas pressure injection systems are designed to give a constant pulse and so are ideal for injecting uniform amounts. With a foot pedal controlling such a system it is possible to inject cells at rate comparable to constant flow injection.

(vii) It is useful to practise on a few feeder cells before injecting ES cells, but remember that the former will tolerate much greater volumes. With the pipette tip still out of focus move the stage and select an appropriate cell, which is large, well attached and with a distinct nucleus. Change to a higher power objective (×32). Raise

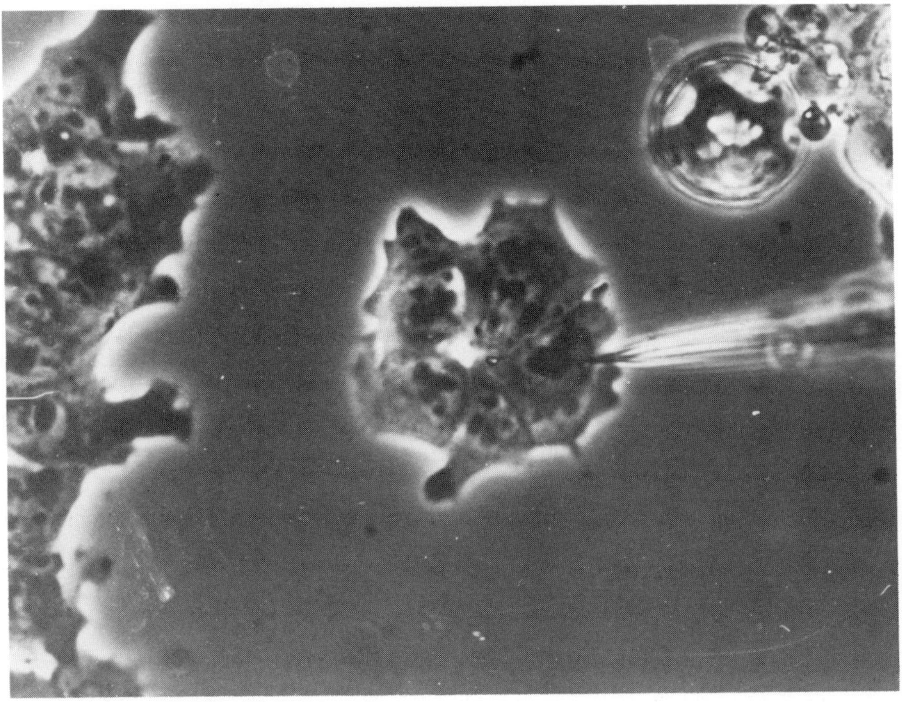

Figure 5. Microinjection of ES cells.

the pipette slowly until it appears to be within the nucleus, taking care not to push through onto the coverslip. If the pipette is flowing you should see a change in the nucleus due to the different refractive index of the DNA solution, and a brief expansion in size. The pipette tip should be moved out of the cell as soon as possible. If too much solution is injected you may see the nucleus burst and the cell will rapidly bleb and lyse. It is useful to see the difference when cytoplasm is injected instead of the nucleus, again by choosing a suitably flat feeder cell.

(viii) If it is necessary to move the stage to select another cell, first lower the pipette until it is out of focus and well clear of the cell layer.

(ix) When sure of the flow rate begin injecting ES cells (shown in *Figure 5*). They are much more sensitive than the fibroblasts, and will often lyse. They are also not so well anchored to the substrate, and if the pipette is withdrawn at the wrong angle they will detach. It may be necessary to increase magnification. With this system of injection it is possible to use a 100× oil immersion lens.

(x) To dismantle the chamber after injection, withdraw the pipette and transfer the chamber to a sterile flow cabinet. Wipe the coverslip with cotton buds dipped in 70% ethanol and allow to dry. With sterile forceps, carefully take off the coverslip, and place it in a 30 mm Petri dish with the cells uppermost. Add 2 ml normal medium and incubate. The cells can be left to grow on the coverslips, with more feeders if required, or trypsinized and replated after a short recovery period of 1 or 2 h. If selection is to be used add selective medium 24–48 h after injection.

6.2.4 *Injection procedure with inverted microscope and Petri dish*

The procedure to be followed is identical in most respects to that described above, with the following modifications.

(i) Seed the cells at the appropriate density either in a suitable Petri dish, or, as above, onto coverslips. It is not necessary to confine the cells to the centre with a cloning ring.
(ii) Change the medium to Hepes-buffered DMEM with 5% serum. Cover with light paraffin oil or silicone oil. This is not essential but it helps prevent drying out and contamination.
(iii) Pull the injection pipette; it does not require any bends. Load the pipette with DNA, backfill with oil and assemble into an instrument tube, as described. Fit the tube on the micromanipulator such that it is raked at an angle of 40–60°.
(iv) Move the pipette tip into the light path and lower it until it is just in the medium. Focus on it at low power. Lower alternatively the focus and pipette tip until the cells are in focus and the pipette just out of focus.
(v) Inject as described above.

6.2.5 *Hints and troubleshooting*

(i) It is useful first to practise injecting with water or with a fluorescent marker as these are free from the viscosity problems associated with DNA. It may also be useful to try silicone oil, because, although it is viscous and requires higher pressures, it is simple to see whether the drop of oil is within the nucleus, cytoplasm or between the cell and substrate.
(ii) If a cell or cell debris becomes impaled on the pipette it can usually be removed by taking the pipette through the oil/medium interface. Alternatively, a glass microneedle attached to a second micromanipulator can be used.
(iii) If there is a clear area around the pipette tip then it has not penetrated the plasma membrane. This may be because the tip is blunt, or because the pressure of liquid flowing from the tip is too great and forms a cushion around it.
(iv) If the tip is at too shallow an angle and it is being moved only in a vertical direction it may tear the cell membrane, either on entry or withdrawal. Change pipette or use the joystick and vertical controls at the same time to enter and leave the cell at the correct angle.
(v) If the tip is too vertical it is likely to be broken on the cell substrate.
(vi) If the pipette blocks, change it. With larger cells it is often possible to chip off the end and remove the blockage, but ES cells are unlikely to tolerate the resulting larger tip.
(vii) If a pipette breaks in the instrument tube on dismantling, it can be pushed out with a length of piano wire.
(viii) A diamond scribe 'objective' can be used to circle individual injected cells so that they can be followed in subsequent culture.

(ix) When injecting all or most cells on a coverlsip or dish a squared eyepiece particulate is useful. Start at one corner, e.g. top left, inject all cells in that square, then in the adjacent square to the right etc, until the other side is reached. Then move the stage to the square below and scan back from right to left. Do not be tempted by a cell just outside the square! An alternative method when using cells grown on coverslips is to etch a grid onto them. Dip the coverslips into soft wax, allow a thin layer to harden on them. Score the desired grid with a razor blade. Dip the coverslips into 10% hydrofluoric acid (HF) for five minutes and then rinse in distilled water. Remove the wax with zylene. Alternatively, to make the grid, stretch plastic mesh of the appropriate size (e.g. 2 mm) over a group of coverslips on a flat surface. Use a fine lacquer spray to cover the glass with an HF resistant layer. Remove the mesh; there should be a clear grid. Cover the other side of the coverslip completely with lacquer. Treat with HF as above and then remove the lacquer with acetone.

7. BIOCHEMICAL TECHNIQUES

7.1 DNA preparation

The solutions and reagents for the DNA preparation techniques are given in *Table 8*.

7.1.1 *DNA preparation from tissues*

(i) Place the tissue into a sterile plastic 25 ml universal tube containing 4 ml of SB at 4°C. Homogenize with a polytron, or equivalent, for 5–10 sec on ice. (Note: wash the homogenizer thoroughly between samples.)

(ii) Add 224 μl of 4 M NaCl and 60 μl of proteinase K. Mix gently, and then add 1.2 ml of 10% SDS. Mix gently by rocking and incubate at 37°C for >4 h or overnight.

(iii) Transfer the lysate to a phenol/chloroform resistant container. [14 ml polypropylene capped tubes from Falcon (2006) are very convenient for this purpose.] Extract three times with phenol and twice with chloroform (Analar).

Table 8. Solutions for DNA preparation from tissues and cells.

1.	Suspension buffer (SB); 100 mM EDTA, 50 mM Tris, pH 8.0.
2.	4 M NaCl.
3.	Proteinase K: stock solution 20 mg/ml in H_2O. Store at −20°C.
4.	10% SDS.
5.	TE for dialysis: 10 mM Tris, 2 mM EDTA, pH 8.0 (can be made up as a 20× stock solution).
6.	Phenol, equilibrated with TE and containing 0.1% 8-hydroxyquinoline. (It is important that the pH is between 7 and 8; either dissolve in TE and wash several times with TE, or dissolve in an equal volume of 20× TE and wash three times with water.) Store at 4°C.
7.	RNase A: stock solution 20 mg/ml in H_2O. Store at −20°C.
8.	2 M sodium acetate, pH 5.5.
9.	Homogenization buffer: 250 mM sucrose, 10 mM Tris-HCl, 2 mM $CaCl_2$, 30 mM NaCl, pH 7.4.
10.	10% (w/v) Triton X-100.

Note: It is important to keep all solutions, glassware etc. sterile and separate from those used for plasmids or phage.

(iv) The DNA can be ethanol precipitated at this point (see steps viii–x) and is normally pure enough for restriction enzyme analysis with most enzymes. Otherwise, it should be dialysed overnight against four changes of TE.
(v) Add NaCl to 200 mM and RNase to 100 μg/ml and incubate at 37°C for 1–2 h.
(vi) Add proteinase K to 100 μg/ml and incubate at 37°C for 1 h.
(vii) Extract twice with phenol and twice with chloroform.
(viii) Add one tenth volume sodium acetate, transfer to a suitable volume container (e.g. 50 ml Falcon tube, 2070F) and cool to 4°C.
(ix) Add 2.5 volumes Analar ethanol, pouring it from a height. High molecular weight DNA should precipitate as a clump and rise to the surface because of trapped air bubbles.
(x) Hook out the DNA with a heat sealed and bent Pasteur pipette. (Note: do not spool the DNA). Wash the DNA through 5 ml 70% ethanol three times (e.g. in Sterilin plastic bijous, 129A). Drain off most of the ethanol against the side of the container, leave the DNA to dry in the air briefly (about 1 min) and dissolve it in 0.5 ml 0.1 × TE in a sterile Eppendorf tube.
(xi) Take an aliquot for determining the concentration by optical density (OD) (1 OD_{260} = 50 μg/ml; OD_{280} should be ~1.8.) Store the remainder of the DNA solution frozen at −20°C.

7.1.2 *DNA preparation from tissue culture cells*

This alternative method, which involves making a crude preparation of nuclei, is useful for cells with a high glycogen content (such as liver cells) or those with a large 'glycocalyx'. EC and ES cells have many complex glycolipids and glycoproteins on their surfaces and these can come through the extraction procedure with the DNA.

(i) Suspend the tissue culture cells by scraping them from the surface of the dish with a 'policeman'.
(ii) Pellet the cells (10 min, 1000 r.p.m.) and resuspend in cold (4°C) PBS. Transfer the suspension to a 14 ml polypropylene tube and re-pellet.
(iii) Suspend the cells in 5 ml homogenization buffer at 4°C.
(iv) Add 0.5 ml 10% Triton X-100 and mix by gentle rocking for 30 sec.
(v) Centrifuge for 5 min at 700 *g* at 4°C.
(vi) Aspirate the supernatant, resuspend the nuclei in SB and continue with the protocol in Section 7.1.1, omitting the homogenization step.

If this method is to be used for organs the tissue should be gently homogenized with a loose fitting glass Dounce, in homogenization buffer plus Triton at 4°C. Then continue with step (v) above.

7.2 **RNA extraction**

With any method of RNA extraction the main concerns are the inactivation and removal of RNase and the elimination of DNA. Solutions for extraction of RNA are listed in *Table 9*. The protocol for the extraction of RNA is given in *Table 10*. This method can be used for organs, either fresh or after snap freezing in liquid nitrogen, or for

Table 9. Solutions for isolation of RNA.

Use diethyl pyrocarbonate (DEP) treated and autoclaved water in the preparation of all solutions. Add DEP to 0.01% (v/v), mix well and autoclave. Note: DEP is rapidly inactivated by Tris and so cannot be used to eliminate RNase from solutions containing it. (**Caution:** DEP is highly toxic.)

Prepare the following solutions.

1. 6 M urea/3 M LiCl: 180.2 g Aristar urea; 63.6 g Analar LiCl. Make up to 500 ml with sterile H_2O. (Do not autoclave.) Store at 4°C.
2. Phenol (Analar) containing 0.1% 8-hydroxyquinoline and saturated with 10 mM Tris, pH 8.0 (see *Table 6*).
3. Proteinase K: stock solution 20 mg/ml in H_2O, store at −20°C.
4. 1 M Tris-HCl, pH 7.5.
5. 10% SDS.
6. 2 M sodium acetate pH 5.5.

Table 10. Extraction of RNA.

1. Place tissue (fresh or frozen) into 50 ml plastic centrifuge tubes (e.g. Falcon 2070F), with 10 ml of 6 M urea/3 M LiCl. Tissue culture cells should be harvested either by scraping with a 'policeman', followed by a cold PBS wash, pelleting and addition of urea/LiCl, or, after a PBS wash, by direct lysis on the plate with about 2 ml urea/LiCl per 10 cm dish.
2. Homogenize with a polytron, or equivalent, at its maximum speed setting for 1 min, on ice then leave the suspension to precipitate in the fridge at 4°C overnight.
3. Transfer the suspension into a 15 ml Corex tube. Spin down the precipitate for 20 min at 10 000 r.p.m., at 0°C in a Sorvall HB4 swing-out rotor.
4. Pour off the supernatant. Vortex briefly to resuspend the precipitate. Add 8 ml urea/LiCl and centrifuge as before.
5. Repeat step (4). Drain the pellet thoroughly.
6. Dissolve the precipitate in 6 ml 10 mM Tris pH 7.5, 0.5% SDS, and add 15 μl proteinase K (to give a final concentration of 50 μg/ml). Transfer to a 14 ml polypropylene tube (Falcon 2006) and incubate at 37°C for 30−120 min.
7. Phenol extract three times. Chloroform extract three times.
8. Add 0.6 ml (i.e. approximately one tenth the volume) 2 M sodium acetate pH 5.5 and transfer to a 30 ml Corex tube. Add 17 ml (i.e. approximately 2.5× the volume) Analar ethanol. Leave to precipitate at −20°C overnight or at −70°C for in excess of 1 h.
9. Centrifuge for 30 min at 10 000 r.p.m. at 0°C in Sorvall HB4 swing-out rotor.
10. Pour off the supernatant carefully. Vortex briefly to resuspend the precipitate. Add 10 ml of 80% ethanol. Spin again for 20 min.
11. Repeat step 10. Drain well and dry the precipitate briefly under vacuum.
12. Dissolve the RNA pellet in a total of 0.4 ml H_2O and transfer to a sterile Eppendorf tube. The concentration of RNA can be calculated by optical density. Take 4 μl of the RNA solution, add to 496 μl H_2O and take readings at 235, 260 and 270 nm (1 OD_{260} = 40 μg; 260/280 = 2, 260/235 = 2 if the RNA is pure). Store the RNA solution frozen, or as an ethanol precipitate, at −70°C.

tissue culture cells. It works by selective precipitation of most RNA species (there is poor recovery of small RNAs such as tRNA). Protein and DNA are left in solution; however it is essential to shear the DNA, hence the long period of homogenization. The volumes quoted in the method in *Table 10* work well with a range of tissue masses, from about 20 to 500 mg. Adjust the volumes throughout for very small or large amounts.

Table 11. Solutions for *in vitro* transcription.

1.	10 × transcription buffer (10 × TB): 400 mM Tris-HCl pH 7.5, 60 mM $MgCl_2$, 100 mM spermidine. Store at 4°C.
2.	20 × rNTP's: 5 mM each of ATP, CTP, GTP and 250 μM UTP, made from stocks of 20 mM ATP, CTP, GTP and 1 mM UTP mixed in equal proportions. Store stock solutions and mix at −20°C.
3.	100 mM Dithiothretiol (DTT). Store at −20°C.
4.	1 mg/ml RNase free DNase. Store at −20°C.

7.3 **RNase protection assays**

A specific RNA polymerase is used to transcribe a 'template' sequence from a strong bacteriophage promoter. The template should be a genomic or cDNA subclone that is diagnostic for the mRNA to be assayed. The sequence is placed adjacent to the promoter in a direction opposite to its normal orientation in the genome, so that the transcripts are complementary to the mRNA. ^{32}P- or ^{35}S-labelled ribonucleotides are included in the transcription reaction to generate very high specific activity RNA transcripts. These 'antisense' labelled transcripts are then hybridized in solution to the RNA sample to be assayed. RNase will digest only single-stranded RNA, leaving double-stranded RNA intact. RNase digestion is then used to remove all single-stranded RNA which will include any non-complementary RNA species, unhybridized segments of either the specific mRNA or probe, and any excess probe, to leave only double-stranded complementary RNA. This is then run on a denaturing polyacrylamide gel and autoradiographed to reveal protected fragments of the labelled RNA strand.

The choice of template is important. Ideally the probe it generates should be longer than the RNA to be detected so that there will be a reduction in size after RNase digestion. Short probes, however, generally work better than long ones, presumably because of the increased likelihood of premature termination with the latter. Probes of 100−1200 bases have worked well for us, but a 3 kb probe gave a smear rather than a discrete band. The choice of restriction site at which to linearize the SP6 vector is also important. In particular enzymes that give a 3′ overhang, such as *Pst1*, should be avoided as the polymerase can transcribe the wrong strand as well leading to a high background due to 'self protected' bands. Blunt ends can also show this effect. In any analysis it is useful to run tracks of the original transcript and of a control with no added RNA to check the size and quality of the probe and the extent of self-protection.

7.3.1 *In vitro transcription of SP6 plasmids*

The following protocol, which has been derived from references 42 and 43, has been written for *in vitro* transcription from SP6 promoters, but it will work with T4 and T7 promoters by simply substituting the appropriate RNA polymerase.

The solutions used are listed in *Table 11*.

(i) Mix in the following order at room temperature:

	μl
H_2O	28.3
10 × TB	5
100 mM DTT	5
RNasin	1.7 (40−50 units)

10 × rNTP's	5
1 μg DNA template	1
^{32}P UTP	3
SP6 polymerase	1 (10–20 units)

(ii) Incubate for 1 h at 37°C.
(iii) Add 1 μl of DNase and incubate for 10 min at 37°C.
(iv) Add 5 μl of 10% SDS and 5 μl of 100 mM Tris, 100 mM EDTA, pH 8.
(v) Load onto a Sephadex 650 column equilibrated in 10 mM Tris, 10 mM EDTA and 1% SDS, pH 8 (TES2), and collect the excluded fraction in an Eppendorf tube. (A spinning column is more convenient.) Take 1 μl for determining c.p.m. by Cerenkov counting.
(vi) Add an equal volume of phenol/chloroform (1:1), vortex briefly, centrifuge for one min in an Eppendorf centrifuge and take off the aqueous phase to a fresh tube.
(vii) Add 50 μl of TES2 to the organic phase, vortex and spin.
(viii) Pool the aqueous phases. Add one tenth volume of 2 M sodium acetate pH 5.5 and 2.5 volumes of ethanol. Leave to precipitate at −70°C for 1 h or on dry ice for 15 min.
(ix) Spin down the precipitate at 10 000 *g* for 15 min at 0°C. Wash the precipitate twice with 80% ethanol, drain well and dry briefly in a vacuum dessicator.
(x) Dissolve the labelled probe in the appropriate volume of hybridization buffer (see below).

7.3.2 *RNase protection assay*

The solutions used are listed in *Table 12*.

(i) Make up hybridization buffer as follows: 800 μl of Formamide, 100 μl of Pipes/EDTA and 100 μl of 4 M NaCl (to give 80% formamide, 40 mM Pipes, 400 mM NaCl, 1 mM EDTA, pH 6.7).
(ii) Dissolve the hot probe in hybridization buffer at 2–5 × 10^5 c.p.m./μl.
(iii) The RNA sample to be assayed, normally 1–20 μg, should either be in 1 μl of H_2O or ethanol precipitated and dried briefly. Dissolve the RNA in 30 μl hybridization buffer.
(iv) Add 1 μl probe to each RNA solution. Heat shock for 1 min at 65°C.
(v) Incubate at 45°C for over 8 h or overnight.
(vi) Add 300 μl of RNase solution. Incubate for 1 h at 30°C.
(vii) Add 20 μl of 10% SDS, 2.5 μl of proteinase K (20 mg/ml) and 1 μl of yeast tRNA (10 mg/ml). Incubate for 10–60 min at 37°C.

Table 12. Solutions for RNase protection assay.

1.	Formamide (crystallized and deionized). Store at −70°C.
2.	40 mM Pipes, 10 mM EDTA pH 6.7.
3.	4 M NaCl.
4.	RNase digestion buffer: 10 mM Tris, 5 mM EDTA, 300 mM NaCl, pH 7.5.
5.	RNase solution: 40 μg/ml RNase A (Sigma type III), 2 μg/ml RNase T (Sigma) in solution (4). Heat for 5 min in boiling water and allow to cool at room temperature. Make up fresh just before use.

Table 13. Solutions for gel electrophoresis.

1.	40% Acrylamide: 190 g of acrylamide, 10 g of bisacrylamide, make up to 500 ml with H_2O. Deionize by stirring with 20 g of MB-1 (Biorad) ion-exchange resin and filter. Store at 4°C.
2.	5 × TBE: 54 g of Tris base, 27.5 g of boric acid, 4 ml of 250 mM EDTA pH 8.0, make up to 1 litre with H_2O. Store at room temperature.
3.	Loading buffer: 4.2 g of urea, 2 ml of 5 × TBE, 1 ml of 1% bromophenol blue, 1 ml of 1% xylene cyanol, make up to 10 ml with H_2O. Aliquot and store frozen at −70°C.
4.	TBE gel mix, for 100 ml: 46 g of (Aristar) urea, 31 ml of H_2O, 20 ml of 5 × TBE, 15 ml of 40% acrylamide. Use fresh.
5.	Gel fixative: 10% methanol, 10% acetic acid.

(viii) Extract once with phenol/chloroform and precipitate the aqueous phase with 1 ml ethanol.

(ix) Centrifuge at 10 000 r.p.m. for 15 min at 0°C. Wash the precipitate twice with 80% ethanol and dry briefly in vacuum.

7.3.3 *Gel electrophoresis*

The solutions used are listed in *Table 13*.

Use a sequencing gel apparatus. The two glass plates should be very clean. Siliconize one of them by wiping with dimethylchlorosilane solution (in a fume cupboard), allow to dry, rinse with tap water, distilled water and then dry with soft tissues. The siliconizing will last several runs.

Run an aliquot of the original transcript and end-labelled restriction digest fragments of known size (e.g. *Hae*III cut Lambda DNA), denatured in loading buffer, as size markers and as standards for quantifying bands (either by densitometry or scintillation counting). To prevent uneven running or 'smiling', clamp a 3 mm aluminium plate onto one glass plate after loading the gel.

(i) Assemble the glass plates and spacers.

(ii) Briefly de-gas 100 ml of TBE gel mix, add 200 μl of 25% ammonium persulphate (made up fresh in a plastic container) and 200 μl of TEMED and mix gently by swirling. Pour between the glass plates, taking care to avoid bubbles. Insert the comb and clamp the glass together over it with bulldog clips. Leave to set for at least 1 h.

(iii) Dissolve the RNA pellets in 6 μl loading buffer. Denature for 5–10 min at 90°C. Centrifuge briefly to bring all the sample to the bottom of the Eppendorf tubes.

(iv) Assemble the gel apparatus. Fill the reservoirs with running buffer. Carefully remove the comb.

(v) Draw out 50 μl siliconized microcaps (Drummond) over a flame to give loading pipettes with a tip approximately 30 × 0.5 mm. With a hand operated suction device load the samples directly into the wells.

(vi) Run the gel at 40–44 Watts. In a 6% denaturing acrylamide gel, bromophenol blue runs at a position roughly equivalent to 26 bases and xylene cyanol to 106 bases.

(vii) After running, dismantle the apparatus. Lift off the siliconized glass plate leaving the gel attached to the other. Carefully pour gel fixative over the gel in a tray and leave for 5–10 min. Carefully lift out the glass plate with the gel still attached, stand it upright on absorbant paper and allow to drain briefly.

(viii) Lie the plate flat and place a piece of Whatman 3MM paper on top of the gel; press gently and peel back the paper. The gel should stick to the paper. Cover the gel with Saran wrap and, with an additional sheet of 3MM underneath, dry it on a vacuum gel dryer.

(ix) When dry, remove the cling film and expose the gel to pre-flashed X-ray film.

8. ACKNOWLEDGEMENTS

I would like to thank all those in the laboratories of Martin Evans, Jacques Jami and Ann McLaren for their help and support in the establishment of the techniques described. I am also indebted to Angelika Schnieke for the RNA preparation method, Kathryn Cheah for the RNase protection assay, and the late Elaine Diacumakos for much of my knowledge on microinjection. Anne Bygrave deserves a special thank you for her practical help over the last three years.

9. REFERENCES

1. Scott,R.W., Vogt,T.F., Croke,M.E. and Tilghman,S.M. (1984) *Nature,* **310**, 562.
2. Rosenthal,A., Wright,S., Cedar,H., Flavell,R. and Grosveld,F. (1984) *Nature,* **310**, 415.
3. Gorman,C.M., Rigby,P.W.J. and Lane,D.P. (1985) *Cell,* **42**, 519.
4. Bucchini,D., Lasserre,C., Kunst,F., Lovell-Badge,R.H., Pictet,R. and Jami,J. (1983) *EMBO J.*, **2**, 229.
5. Lovell-Badge,R.H., Bygrave,A.E., Bradley,A., Robertson,E., Evans,M.J. and Cheah,K.S.E. (1985) *Cold Spring Harbor Symp. Quant. Biol.,* **50**, 707.
6. Rubenstein,J.L.R., Nicolas,J-F. and Jacob,F. (1984) *Proc. Natl. Acad. Sci. USA,* **81**, 7137.
7. Linney,E. and Donerly,S. (1983) *Cell,* **35**, 693.
8. Jones,N.C. (1986) *Nature,* **321**, 202.
9. Wagner,E.F. and Mintz,B. (1982) *Mol. Cell Biol.,* **2**, 190.
10. Kondoh,H., Takahashi,U. and Okada,T.S. (1984) *EMBO J.,* **3**, 2009.
11. Stewart,C.L., Vanek,M. and Wagner,E.F. (1985) *EMBO J.,* **4**, 3701.
12. Lovell-Badge,R.H., Bygrave,A.E., Bradley,A., Robertson,E., Tilly,R. and Cheah,K.S.E. (1986) Manuscript submitted.
13. Brinster,R.L., Chen,H.Y., Trumbauer,M.E., Yagle,M.K. and Palmiter,R.D. (1985) *Proc. Natl. Acad. Sci. USA,* **82**, 4438.
14. Ornitz,D.M., Palmiter,R.D., Hammer,R.E., Brinster,R.L., Swift,G.H. and MacDonald,R.J. (1985) *Nature,* **313**, 600.
15. Chada,K., Magram,J., Raphael,K., Radice,G., Lacy,E. and Costantini,F. (1985) *Nature,* **314**, 377.
16. Krumlauf,R., Hammer,R.E., Tilghman,S.M. and Brinster,R.L. (1985) *Mol. Cell. Biol.,* **5**, 1639.
17. Hanahan,D. (1985) *Nature,* **315**, 115.
18. Izant,J.G. and Weintraub,H. (1984) *Cell,* **36**, 1007.
19. Smithies,O., Gregg,R.G., Boggs,S.S., Koralewski,M.A. and Kucherlapati,R.S. (1985) *Nature,* **317**, 230.
20. Pritchard,C. and Goodfellow,P.N. (1986) *EMBO J.,* **5**, 979.
21. Potter,H., Weir,L. and Leder,P. (1984) *Proc. Natl. Acad. Sci. USA,* **81**, 7161.
22. Schaffner,W. (1980) *Proc. Natl. Acad. Sci. USA,* **77**, 2163.
23. Anderson,W.F., Killos,L., Sanders-Haigh,L., Kretschmer,P.J. and Diacumakos,E.G. (1980) *Proc. Natl. Acad. Sci. USA,* **77**, 5399.
24. Capecchi,M.R. (1980) *Cell,* **22**, 479.
25. Graessman,M. and Graessman,A. (1983) *Methods in Enzymology,* Academic Press, NY, Vol. 101, p. 482.
26. Gorman,C.M., Moffat,L.F. and Howard,B.H. (1982) *Mol. Cell. Biol.,* **2**, 1044.
27. Weintraub,H., Cheng,P.F. and Conrad,K. (1986) *Cell,* **46**, 115.
28. Thomas,K.R., Folger,K.R. and Capecchi,M.R. (1986) *Cell,* **44**, 419.
29. King,W., Patel,M.D., Lobel,L.I., Goff,S.P. and Nguyen-Huu,C. (1985) *Science,* **228**, 554.
30. Panganiban,A.T. (1985) *Cell,* **42**, 5.
31. Cepko,C.L., Roberts,B.E. and Mulligan,R.C. (1984) *Cell,* **37**, 1053.
32. Seliger,B., Kollek,R., Stocking,C., Franz,T. and Ostertag,W. (1986) *Mol. Cell. Biol.,* **6**, 286.
33. Linney,E., Davis,B., Overhauser,J., Chao,E. and Fan,H. (1984) *Nature,* **308**, 470.

34. Jahner,D., Stuhlmann,H., Stewart,C.L., Harbers,K., Lohler,J., Simon,I. and Jaenisch,R. (1982) *Nature,* **298**, 623.
35. Stewart,C.L., Stuhlmann,H., Jahner,D. and Jaenisch,R. (1982) *Proc. Natl. Acad. Sci. USA,* **79**, 4098.
36. Sorge,J., Cutting,A.E., Erdman,V.D. and Gautsch,J.W. (1984) *Proc. Natl. Acad. Sci. USA,* **81**, 6627.
37. Van der Putten,H., Botteri,F.M., Miller,A.D., Rosenfeld,M.G., Fan,H., Evans,R.M. and Verma,I.M. (1985) *Proc. Natl. Acad. Sci. USA,* **82**, 6148.
38. Lobel,L.I., Patel,M., King,W., Nguyen-Huu,C. and Goff,S.P. (1985) *Science,* **228**, 329.
39. Mann,R., Mulligan,R.C. and Baltimore,D. (1983) *Cell,* **33**, 153.
40. Robinson,W.S., Pitkanen,A. and Rubin,H. (1965) *Proc. Natl. Acad. Sci. USA,* **54**, 137.
41. Maniatis,T., Fritsch,E.F. and Sambrook,J. (1982) *Molecular Cloning,* Cold Spring Harbor Laboratory Press, NY.
42. Zinn,K., DiMaio,D. and Maniatis,T. (1983) *Cell,* **34**, 865.
43. Melton,D.A., Krieg,P.A., Rebagliati,M.R., Maniatis,T., Zinn,K. and Green,M.R. (1984) *Nucleic Acids Res.,* **12**, 7035.

CHAPTER 7

Experimental analysis of teratocarcinoma cell multiplication and purification of embryonal carcinoma-derived growth factor

JOHN K.HEATH

1. INTRODUCTION

This chapter is concerned with methods for experimentally analysing the growth of teratocarcinoma cells and their differentiated progeny, and with techniques for identifying and characterizing the expression of growth regulatory molecules by embryonal carcinoma (EC) cells. Interest in this area has arisen from a desire to understand growth control mechanisms in early development, using teratocarcinoma cells as a model, and from the realization that EC cells, and EC cell-derived cell lines provide a useful source of potential embryonic growth regulatory factors. This analysis of EC cell growth control has almost exclusively focused on feeder-independent cell lines for the obvious reason that the presence of feeder cells precludes controlled analysis of growth requirements, as the feeder cells themselves act as a source of undefined growth regulatory agents. Recent experimental findings in this area are reviewed elsewhere (1).

In this chapter, the methods that have been developed for the analysis of PC13 EC cell growth and differentiation are described. The emphasis is on the underlying principles of the specific experimental techniques so the reader can tailor the methods to their own requirements.

2. CULTURE AND DIFFERENTIATION OF EC CELLS IN SERUM-CONTAINING MEDIA

2.1 Introduction

The starting point for any experimental analysis of cell multiplication is the growth of cells in serum-containing media, which in most cases provides a baseline from which to assess the effects of experimental manipulations on EC cell proliferation. This section should be read in conjunction with Chapter 2 which describes the routine culture of EC cell lines other than PC13.

The objective of successful growth in serum-containing media is to obtain maximum yield of viable cells in the absence of selection for cells with aberrant growth phenotype. The three major variables for culture in serum-containing media are the tissue culture surface, the basal medium and serum and the sub-cultivation regime.

2.2 The tissue culture surface

Although some EC cells can be propagated successfully on untreated tissue culture plastic, most feeder-independent cell lines, such as PC13, are cultured on gelatin-treated tissue culture surfaces. Plating PC13 EC cells on untreated dishes rapidly yields cells with abnormal morphology and poor growth rates. The methods for preparing the gelatin solution and for treating tissue culture plastic with gelatin are described in *Table 1* and *Table 2*. Gelatin-treated tissue culture surfaces can be made in bulk and stored before use.

Table 1. Reagents for retinoic acid-induced differentiation of PC13 EC cells.

Culture media (DMEM:F12:FBS)

50:50 (v/v) Ham's Medium F12 (F12) and Dulbecco's modified Eagle's medium (DMEM) supplemented with 10% by volume of batch tested FBS. The medium should be pre-equilibrated at 37°C, 5% CO_2 in air before exposure to cells.

Gelatin stock

Dissolve 1 g of gelatin (Sigma Chemical Co., swine skin type 1) in 1 litre of Analar water, split into 100 ml aliquots and sterilize by autoclaving. Store at 4°C.

Trypsin/EDTA stock

Dissolve 1.25 g of trypsin (Difco, or Worthington 3 × crystallized), 0.2 g of EDTA (tetrasodium salt) in 500 ml PBS. Filter sterilize and store at 4°C.

Retinoic acid stock

10^{-2} M all-*trans* retinoic acid (Eastman) dissolved in dimethyl sulphoxide (Analar, BDH), split into 100 μl aliquots and stored in the dark at −70°C.

Polybrene stock

1 mg/ml of polybrene (Sigma Chemical Co.) dissolved in PBS. Filter sterilize and store at 4°C. Add to DMEM:F12:FBS to make 2 μg/ml final concentration.

Table 2. Retinoic acid-induced differentiation of PC13 EC cells.

1. Coat the tissue culture surface with gelatin. Incubate the tissue culture flasks or dishes with sufficient gelatin stock (*Table 1*) to completely cover the culture surface and incubate for at least 30 min at 4°C. Longer incubations are acceptable, provided the gelatin solution does not dry out. Aspirate the gelatin solution before use. Tissue culture plastic may be gelatin treated in bulk and stored indefinitely before use.
2. Dissolve retinoic acid stock and polybrene stock (*Table 1*) in DMEM:F12:FBS to a final concentration of 10^{-6} M RA, 4 μg/ml polybrene in half the intended final volume of culture media.
3. Rinse the EC cell monolayers with ice-cold PBS, add 2 ml/10 cm^2 of ice-cold trypsin/EDTA solution (*Table 1*), leave for abut 10 s, aspirate and add a second aliquot of the trypsin/EDTA solution and aspirate, leaving a thin film of trypsin solution covering the cells.
4. Monitor cell detachment under the microscope. When all the cells have detached (~45−60 s) add pre-equilibrated DMEM:F12:FBS, pellet the cells, resuspend in DMEM:F12:FBS and count. Resuspend the cells at twice the intended final concentration in half the intended final volume of DMEM:F12:FBS, mix with the pre-equilibrated DMEM:F12:FBS:RA:polybrene medium (giving a final concentration of 5×10^{-7} M RA, 2 μg/ml polybrene).
5. Plate out the cell suspension at a final density of 4×10^3 cells/cm^2 culture surface and 10^4 cells/ml.
6. Culture at 37°C in 5% CO_2 in air for 5 days.

2.3 Basal medium and serum

PC13 cells in our laboratory are routinely propagated in a basal medium comprising Ham's F12 (F12) and Dulbecco's modified Eagle's medium (DMEM) basal media mixed in equal proportions (F12:DMEM; *Table 1*). Although the optimal nutrient media requirements for EC cell propagation have not been systematically investigated the F12:DMEM mixture offers good pH stability and contains potentially important trace metal supplements. The use of F12:DMEM is particularly advisable where serum-free culture experiments are planned since it allows EC cells to be plated into serum-free media without metabolic adaptation.

EC cells are routinely grown in basal medium supplemented with 5 – 10% foetal bovine serum (FBS). The major factors supplied by serum seem to be nutrient factors (see Section 3.3) which are required in relatively large amounts since lower serum concentrations almost invariably lead to rapid impairment of cell viability. The source of serum, unfortunately, remains an uncontrolled feature of EC cell culture. It has been empirically found in many laboratories that different batches of serum differ wildly in their ability to support EC cell growth, although, in general, FBS has proved superior to newborn or adult bovine sera. Before embarking on long-term propagation of EC cells, as many serum batches as possible should be tested for the ability to support EC cell growth at 'clonal' density (10 – 50 cells/cm^2), paying particular attention to plating efficiency (number of colonies/number of cells plated) and growth rates, and selecting the best. It is advisable to allow cells to grow for 20 – 30 generations in 'promising' sera to check for changes in morphology and growth rates, and to check the performance of candidate sera in retinoic acid (RA)-induced differentiation (Section 2.5) before purchase. Do not adulterate expensive FBS with cheaper, but untested, sera for anything other than short-term experiments which will be terminated.

2.4 Sub-cultivation regime

The continued passage of EC cells can lead to selection for cells with atypical (usually rapidly growing) growth characteristics which can cause serious problems when comparing experimental results separated in time. EC cells, like many other malignant cell types, are also prone to rapid depletion of basal medium nutrients and release of toxic products at high cell number/medium volume ratios, which leads to rapid compromise of cell viability and concomitant opportunity for cell selection. These problems can be partly controlled by paying strict attention to sub-cultivation regimes. Different EC cell lines (and different sub-clones from the same parental cell line) differ in their rate of proliferation. By measuring the population doubling rate, and cell density at confluence, of a new stock of cells it is possible to devise a sub-culture regime which will allow 4 – 5 population doublings per passage, and to sub-culture at one population doubling before confluence. By adhering to the same sub-cultivation regime, the investigator will be alerted to changes in the growth rate of the cell population indicative of poor growth conditions, differentiation or cell selection. Discard cell stocks which have overgrown and acidified the culture medium. It is, in any case, advisable to discard EC cell stocks after about 20 passages in order to avoid undetected changes in the cell population.

2.5 **Retinoic acid (RA) induced differentiation**

A wide variety of conditions exist for the promotion of EC cell differentiation *in vitro*, many of which are detailed elsewhere in this volume (see Chapter 2 and Chapter 4). However the use of RA (2) as a differentiation-inducing agent offers considerable advantages to the investigator of the developmental regulation of PC13 EC cell proliferation:

(i) It induces the differentiation of EC cell lines which do not differentiate 'spontaneously', allowing homogeneous EC cell populations to be compared with homogeneous differentiated cell populations.

(ii) It allows a degree of control over the timing of differentiation which can be critical in certain types of experiments.

(iii) Under the correct conditions (see below) it is extremely 'efficient', in that very few EC cells are left undifferentiated after exposure to RA. This aspect is especially critical in experiments where differentiated cell proliferation is the subject of investigation.

(iv) Careful analysis (3) has shown that RA *induces* cell differentiation rather than selects a pre-differentiated sub-population.

(v) The use of RA allows differentiation to occur in cell monolayers which offers more experimental opportunities for the analysis of cell multiplication than differentiation in aggregates.

The reagents and techniques for RA-induced differentiation of PC13 EC cells are detailed in *Table 1* and *Table 2*, respectively. Analysis of the optimal conditions for RA-induced PC13 EC cell differentiation has revealed a number of critical variables for 'high efficiency' differentiation.

(i) Different PC13 sub-clones differ widely in their susceptibilty to RA-induced differentiation, as measured by the number of colony-forming EC cells left after 5 days exposure to RA.

(ii) RA-induced differentiation is more efficient at low cell number/medium volume ratios since active RA seems to be rapidly depleted from the culture medium on exposure to cells. Optimal differentiation seems to require prolonged (12–24 h, ref. 3) exposure of the cell population to RA which can be achieved by exposing the cells to a relatively large volume of medium.

(iii) RA can decompose, yielding toxic and inactive side products. This can be circumvented by storing RA stocks at −70°C (or lower) in the dark and discarding immediately after use. Do not store RA solutions at 4°C or room temperature for more than a few days.

(iv) It has been found empirically that the charged polymer polybrene significantly enhances the efficiency of RA-induced differentiation, although it does not act synergistically with RA and has no differentiation-inducing activity on its own. This effect is particularly striking in the case of PC13 EC cell sub-clones which respond poorly to RA-induced differentiation. The basis of the polybrene effect is obscure. The optimal concentration of polybrene is important since it is toxic at concentrations above 10 μg/ml and ineffective below 1 μg/ml.

(v) RA rapidly associates with serum proteins in the conditions of cell culture and so has to be employed at relatively high (i.e. non-physiological) concentrations.

The optimal concentrations of RA required for efficient differentiation need to be determined empirically for each cell line but usually lie between 10^{-7} and 10^{-6} M. It is important to use the lowest concentration required for efficient differentiation, since higher concentrations may be toxic.

(vi) Although RA effects on cell morphology, gene expression and cell multiplication can be detected within 24 h (4) it is usual to allow differentiation to proceed for 4–5 days to ensure complete differentiation of the cell population before experimental analysis. Since differentiated PC13 EC cells (PC13END) are not capable of indefinite multiplication (4) the results of studies on PC13END cell multiplication will depend upon the 'age' of the cells at the time of analysis.

The efficacy of RA-induced EC cell differentiation can be assessed by a variety of criteria. The most appropriate (and sensitive) measure in the case of studies on differentiated cell multiplication is the proportion of undifferentiated EC cells in a differentiated cell population. This is most easily measured by re-plating a sample of the differentiated cell population back into serum-containing media lacking RA at 'clonal' (10–50 cells/cm^2) density, culturing for 10–12 days followed by fixation (methanol/acetic acid, 3:1 v/v, 15 min) and staining with Giemsa. Since PC13 EC cells, unlike their differentiated progeny, grow progressively, contaminating EC cells can be quantified by colony counting, yielding a measure of the proportion of viable EC cells in the original sample.

The major problems likely to be encountered by the investigator are excessive cell death and/or toxicity where the RA stock, polybrene concentration and serum batch are the most likely culprits, or poor differentiation efficiency where the cell density or RA concentration may be sub-optimal.

3. SERUM-FREE CULTURE OF EC CELLS

3.1 **Introduction**

The objective of serum-free culture is to remove the requirement for serum substitution in cell culture and to replace serum with an optimal mixture of defined macromolecular additives. There may be several reasons for adopting this approach:

(i) Serum-free culture is a very powerful technique for identifying exogenous requirements for cell multiplication, differentiation and gene expression. Serum contains a complex mixture of bioactive molecules which can completely obscure or inhibit cellular responses to exogenous factors.

(ii) Serum-free medium is very useful for the analysis of bioactive molecules secreted by cell lines *in vitro*. It is particularly useful when trying to identify activities which are also present in serum, such as growth and differentiation factors. Serum-free media generally contain a limited number of defined components and have a low protein concentration compared with serum-containing media and thus provide a favourable source for attempting the purification of bioactive molecules (see Section 5).

(iii) In particular cases serum-free media provides an economical alternative to serum-containing media, which is particularly relevant when generating large quantities of conditioned media for analytical purposes (Section 5).

3.2 Development of serum-free media

There are two approaches to the development of serum-free media for a particular cell type. The first, pioneered by Ham and his colleagues (reviewed in ref. 5) involves progressively lowering serum concentrations until cell viability or growth is impaired and then restoring the activity of serum by either optimizing the basal nutrient medium or adding back defined macromolecular additives. The cycle of serum depletion and re-optimization is repeated until the serum is withdrawn completely. This approach has the major advantage of being comprehensive but is extremely time consuming in practice. The second, and more common method, is to start with basal media supplemented with a limited number of macromolecular additives (usually derived by 'guesswork' or from consulting the literature) and to restore growth and function by optimization of the macromolecular additives and additional supplementation. All the serum-free media for EC cells have been developed by this second approach.

The development of serum-free culture conditions for EC and other cell types has led to a number of general findings:

(i) There is not necessarily one unique formulation for any given cell type; different macromolecular additives may have similar overall biological effects, or high concentrations of one additive will substitute for lower concentrations of a second. This can be used to advantage.

(ii) The requirements for low density, or clonal cell multiplication, are invariably more complex than those for growth at high density. This results partly from 'conditioning' effects which are dependent upon cell density, and partly from factors such as medium toxicity (see Section 3.3.4) which can become acute at low cell density.

(iii) The basal nutrient media plays an important role in successful serum-free culture. Many commercial basal media will not support growth under serum-free conditions. This is partly due to the lack of essential trace elements such as selenium, zinc, etc., which would be provided by serum and partly because of inappropriate nutrient balance. In general, media specifically developed for serum-free culture such as Ham's F12 or the MCDB series are recommended.

(iv) The purity of media additives is important, in that trace levels of contamination can be responsible for striking biological effects leading to the erroneous identification of the importance of particular macromolecular additives.

It is important to realise, therefore, that successful serum-free culture is operationally defined by the investigator's research objective. Thus, analysis of exogenous requirements for cell multiplication will require more stringent attention to reagent purity and the biological relevance of particular macromolecular supplements than simple short-term culture of cells at high density in the absence of serum.

Current serum-free culture techniques are limited in that there is no 'sure-fire' combination that will work for any cell type. Some cell types may require media additives which may be not be available in pure form, are prohibitively expensive to purchase or prepare, or indeed may not have even been discovered. Published formulations often provide a useful starting point for the development of media for particular experimental purposes. Thus the ECM media formulation (ref 6; *Table 4*) can support the growth of a variety of EC cell and EC cell-derived lines other than PC13, including P19 (Chapter

2) and PYS-2, and has provided the basis for the development of serum-free media for PC13END cells, normal mouse and human fibroblasts (ECM/F; *Table 5*) and human teratocarcinoma-derived cell lines (7).

3.3 Reagents for serum-free cell culture

The reagents for serum-free cell culture are given in *Table 3*.

Table 3. Reagents for serum-free cell culture.

Basal medium (DMEM:F12)

50:50 (v/v) mixture of Ham's medium F12 and DMEM. F12 may be substituted with medium MCDB104. Pre-equilibrate all media at 37°C, 5% CO_2 in air before exposure to cells.

Fibronectin

Make a 10 mg/ml stock solution dissolved in PBS, filter sterilize and store at −20°C. Make a working solution of 1 μg/ml fibronectin periodically by dilution of the stock solution in PBS and store at 4°C.

Transferrin

Make a 10 mg/ml transferrin (Behring) stock solution in DMEM:F12 supplemented with 10^{-4} M ferric chloride. Incubate at room temperature for 30 min to allow iron absorption, filter sterilize and store at 4°C. Make a working solution of 1 mg/ml transferrin periodically by diluting the stock with sterile DMEM:F12.

Insulin

Dissolve 10 mg of insulin in 0.5 ml of 10 mM acetic acid (it may be necessary to warm to 37°C to ensure disolution), make up to 10 ml with 1% BSA (pentex crystalline, Miles) in PBS. Filter sterilize and dispense into 100 μg aliquots in sterile freezing vials and store at −70°C.

Lipoproteins[a]

Prepare these from *fresh* human plasma obtained using the following protocol.

1. Adjust the plasma to a density of 1.21 g/ml by addition of solid KBr (Analar, BDH) (if you overshoot dilute with PBS).
2. Centrifuge for 48 h at 100 000 *g*. The best results are obtained with a swing-out rotor (SW28 or equivalent) but for larger volumes a high capacity fixed angle rotor such as 45Ti is acceptable. The lipoprotein fraction is a cloudy bright orange layer at the top of the tube, which should be separated by a clear zone from the yellow/brown plasma fraction.
3. Take the lipoprotein fraction and adjust the density to 1.25 g/ml with solid KBr.
4. Dispense the density-adjusted lipoproteins into swing-out rotor centrifuge tubes (SW28 or equivalent) such that they occupy one third or less of the tube volume. Carefully overlay the lipoproteins with PBS and fill the tube to the top.
5. Centrifuge for 24 h at 100 000 *g*.
6. The lipoproteins separate into four discrete zones which are readily identified by colour. Unload the tubes from the top with a Pasteur pipette. The uppermost, milky white layer is composed of very low density lipoproteins and chylomicrons. Discard this layer. The second layer is composed of bright orange low density lipoprotein (LDL). Save this fraction. The third layer is composed of bright yellow high density lipoproteins (HDL). Save this fraction. The final layer (located below the KBr/PBS interface) contains plasma contaminants and is discarded.
7. Dialyse the HDL and LDL fractions, separately, overnight against PBS at 4°C, filter sterilize and aliquot in 1 ml or 5 ml lots into sterile freezing vials. Store the vials in liquid nitrogen until required.

   It is important not to expose lipoproteins to light, air or ambient temperatures for long periods.

   The exact yields of HDL and LDL depend on the physiological state of the donor. Normally 1 ml of plasma will yield approximately 5 mg of each of the lipoproteins.

[a]Observe the local regulations regarding the use and disposal of human blood.

Table 4. Culture of EC cells in ECM serum-free medium.

1.	Gelatinize tissue culture vessels (*Table 1*), aspirate gelatin solution and add sufficient fibronectin working solution (*Table 3*) to cover the culture surface, incubate at 4°C for at least 30 min and aspirate. As with gelatinization, fibronectin-coated tissue culture vessels can be prepared in bulk and stored at 4°C. The fibronectin solutions may, if required, be re-used at least once.
2.	Prepare ECM medium by supplementing DMEM:F12 with 5 μg/ml of transferrin (from working solution), 50 μg/ml of LDL, 50 μg/ml of HDL (freshly thawed from frozen stocks, discard excess and do not re-freeze). Equilibrate at 37°C in 5% CO_2 in air.
3.	Trypsinize EC cell monolayers (*Table 2*), leaving the minimum possible volume of trypsin solution on the cells.
4.	When the cells have detached, add pre-equilibrated DMEM:F12 basal media, pellet the cells by centrifugation, and aspirate as much medium as possible.
5.	Resuspend the cells in ECM medium, count the suspension and plate into tissue culture vessels at the appropriate density ($10^3 - 10^4$ cells/cm^2).
6.	Culture at 37°C in 5% CO_2 in air.

3.3.1 *Attachment factors*

Attachment factors are required for the serum-free cultivation of most cell types. The main commercially-available attachment factors are collagen, fibronectin and laminin. PC13 EC cells require collagen (usually provided in the form of gelatin) and either fibronectin or laminin for optimal attachment. This contrasts with human teratocarcinoma-derived cells (7) which do not respond to any of these above but require undefined serum-derived factors for attachment. The methods for pre-treating tissue culture plastic with gelatin and fibronectin are given in *Table 2* and *Table 4*. If large amounts of attachment factors are required, commercial sources may prove prohibitively expensive. Fibronectin is readily prepared from outdated plasma by gelatin–Sepharose affinity chromatography (8) or can be obtained as a side product from hospital plasma fractionation laboratories.

3.3.2 *Transport proteins and macromolecular nutrients*

Macromolecular nutrients seem to be important for the cultivation of most cell types and EC cells in particular exhibit a stringent requirement for macromolecular transport and nutrient factors, principally transferrin (as a source of iron) and lipoproteins (as a source of lipids).

Transferrin is readily available commercially, but is often contaminated with biologically active impurities, which should be evaluated before experiments where the possibility of trace contamination is important. For optimal results transferrin should be pre-loaded with iron before use as described in *Table 3* and is usually employed at concentrations between 5 and 20 μg/ml.

Lipids are most effectively supplied in the form of plasma-derived lipoproteins, which are isolated from fresh human plasma as described in *Table 3*. The storage of lipoprotein preparations is important as they rapidly deteriorate unless stored under liquid nitrogen. Although plasma lipoproteins have to be prepared in the laboratory, they form an essential component for EC cell growth in serum-free medium, and are important for the growth of many other cell types. The lipid composition of plasma lipoproteins, and hence their biological activity, depends on the physiological state of the plasma donors, and this affects the optimal concentration required for EC cell growth. Ideally the optimal concentrations of lipoprotein supplements should be determined for each batch, but should lie between 50 and 100 μg/ml. An inferior, although convenient, substitute for plasma lipoproteins are fatty acid/serum albumin complexes which can be prepared by sonicating linoleic acid and bovine serum albumin (BSA) in phosphate-buffered saline (PBS) at a fatty acid/albumin ratio of 1:1 (w/w). The optimal concentration of the fatty acid/albumin additive has to be experimentally determined but is usually between 1 and 5 mg/ml. At this high protein concentration the possible contamination of commercial albumin preparations may be significant.

3.3.3 *Growth factors*

It is a significant feature of EC cells that they do not require exogenous growth factors and hormones for continued proliferation. A requirement for exogenous growth factors becomes evident upon differentiation. Early media formulations for EC cells (9) used relatively high concentrations of insulin (5 μg/ml) without lipid supplementation. It is now apparent that insulin, presumably by activating nutrient uptake pathways or *de novo* lipid synthesis, can provide limited substitution for exogenous lipids. The effect of high concentrations of insulin is actually due to cross-reaction with EC cell receptors for insulin-like growth factors (IGFs) which can effectively substitute for insulin at lower concentrations (6). PC13END cells are, by contrast, mitogenically responsive to nanomolar concentrations of insulin (10) and IGFs. Insulin is, however, a valuable supplement in the serum-free culture of other cell types. The preparation of insulin stocks is given in *Table 3*.

PC13END cells are responsive to a variety of exogenous growth factors including epidermal growth factor (EGF), acid and basic fibroblast growth factor (aFGF and bFGF), embryonal carcinoma derived growth factor (ECDGF, see Section 5) as well as IGF-I/II and insulin. It is possible that this list will increase as more growth-regulatory agents are discovered. With the exception of ECDGF, whose purification is described in Section 5, these growth factors are commercially available, although commercial FGF is at present an impure mixture of both acid and basic fibroblast growth factors. Commercial growth factors should be carefully checked for biological activity using the growth assay techniques described in *Table 5*, aliquoted in suitable quantities, and stored at −70°C. The optimal concentrations for specific growth factor activity needs to be experimentally determined, but should not exceed 5−10 nM, when contamination of the factor preparations should be suspected. Thawed aliquots of most growth factors are stable at 4°C for several weeks in the presence of suitable stabilizing agents such as 1% ultrapure BSA, or 0.02% Tween 20, or 50% ethylene glycol. The use of plastic tubes for storing hormone stocks is recommended (1.5 ml microfuge tubes, cryovials or equivalents). Fibroblast growth factors and ECDGF are sensitive to acid inactivation and should not be exposed to conditions below pH 4.

Supplementation of media with specific growth factors seems to be an essential requirement for serum-free culture of most non-malignant cell types, and should certainly be investigated when analysing the multiplication of differentiated EC cell derivatives or primary embryonic cell types.

Table 5. Measurement of PC13END and 10T1/2 cell proliferation by incorporation of [^{3}H]thymidine.

General preparation

1. Coat 24 well multi-well plates with gelatin and fibronectin as described in *Tables 2* and *4*.

Preparation of C3H 10T1/2 cells

1. For each multi-well plate prepare 48 ml of ECM/F medium. ECM/F medium is ECM supplemented with 100 ng/ml of insulin (*Table 3*) added from frozen stock. Prepare an additional quantity of medium for washing and diluting the cell suspension.
2. Pre-equilibrate the medium at 37°C, 5% CO_2 in air.
3. Trypsinize the C3H 10T1/2 cells, pellet and resuspend the cells in ECM/F medium (as in *Table 4*).
4. Count the suspension, adjust the cell density to 2.5×10^4 cells/ml in equilibrated ECM/F media and plate 2 ml of cell suspension into each well of the multi-well plate.
5. Incubate the plates at 37°C, 5% CO_2 in air.
6. Following overnight incubation add the required experimental additions (see text).
7. 6 h later add 2 μCi/well [^{3}H]thymidine ([^{3}H]methyl thymidine, 70−90 Ci/mmol, Amersham). The stock solution is 100 μCi/ml made up in PBS.
8. Incubate the cells for a further 18 h before processing for scintillation counting or autoradiography.

Preparation of PC13END cells

1. Prepare PC13END cells by RA treatment (*Table 2*). Take cells for analysis after 5 days exposure to RA.
2. For each multi-well plate prepare 48 ml of DMEM:F12:FBS (*Table 1*). Prepare an additional quantity of medium for washing and diluting cell suspension. Pre-equilibrate the medium at 37°C, 5% CO_2 in air.
3. Trypsinize, pellet, resuspend and count PC13END cell as described for EC cells in *Table 2*, using DMEM:F12:FBS medium.
4. Adjust the cell density to 2.5×10^4 cells/ml and plate 2 ml of cell suspension into each well of the multi-well plates.
5. Incubate the plates overnight at 37°C, 5% CO_2 in air.
6. Aspirate the medium from the wells containing the PC13END cell monolayers and replace with 2 ml/well of pre-equilibrated ECM medium (*Table 4*).
7. Incubate the plates at 37°C, 5% CO_2 in air.
8. Following overnight incubation add the required experimental additions (see text).
9. 6 h later add 2 μCi/well of [^{3}H]thymidine ([^{3}H]methyl thymidine, 70−90 Ci/mmol, Amersham).
10. Incubate the cells for a further 18 h before processing for scintillation counting or autoradiography.

Measurement of [^{3}H]thymidine incorporation by scintillation counting

1. Aspirate the medium from each well and wash with 2 ml/well of ice-cold PBS.
2. Fix the cell monolayers with 1 ml/well of glacial acetic acid:methanol (3:1 v/v) for 10 min at room temperature.
3. Aspirate the fixative and wash each well with 2 ml/well of ice-cold 10% (by weight) trichloroacetic acid in water (10% TCA).
4. Add 2 ml of 10% TCA/well and incubate for 15 min at 4°C.
5. Repeat step 4.
6. Aspirate the TCA solution, and wash each well with 2 ml of water.
7. Aspirate the water and add 200 μl/well of 0.2 M NaOH. Incubate at 37°C for 2 h to solubilize the cell monolayers.
8. Add the 200 μl of NaOH-solubilized cells in each well to 4 ml of aqueous scintillant (Aqualuma plus; May and Baker) in appropriate scintillation vials and count in a suitable scintillation counter.

Measurement of [^{3}H]thymidine incorporation by autoradiography

1. Proceed as described in steps 1 and 2 above.
2. Aspirate the fixative and allow the fixed cell monolayers to air dry.

3. Wash the fixed cell monolayers thoroughly in water and air dry. If required plates can be stored at this point.
4. Heat a water bath with a light proof cover, to 45°C in a dark room. Warm approximately 50 ml of water containing 2% (by volume) glycerol in a beaker, and approximately 5 ml of Ilford K9 emulsion in a 10 ml graduated measuring cylinder (Note, use only the safe light when handling the emulsion). When the emulsion has melted, add an equal volume of the glycerol solution and place back in the water bath for 5–10 min. Stir gently to mix, avoiding the production of of air bubbles in the emulsion mixture.
5. Pipette 250 μl of emulsion into each well and evenly coat the base of each well. After 15–20 s aspirate the excess emulsion with a Pasteur pipette. In practice it is best to work with the wells in groups of four; coat each well with emulsion, aspirate each group of four, change pipettes and proceed with the next row. It may be necessary to practice the coating procedure a few times to produce even coating with the emulsion.
6. Dry the coated plates in the dark at room temperature, for at least 2 h. Put the dried plates in a light proof box (e.g. a 'sandwich' box painted black) containing silica gel, seal the box tightly with tape and store at 4°C for 3 days.
7. Before developing, allow the box containing the coated plates to equilibrate to room temperature (about 2 h).
8. Working in the dark (safe light only), put 0.5 ml of developer (e.g. Kodak D19) in each well and incubate for 5 min. Aspirate the developer and wash with water (plunging the plate into a 5 l beaker of water is a useful method). Add 0.5 ml of fixative solution (e.g. Kodafix, prepared according to the manufacturer's instructions) per well, incubate for a further 5 min, wash with water (as above) and then rise extensively under running tap water for at least 15 min.
9. Invert the plates in a drying cabinet and leave them to dry completely.
10. Wash the wells with 1 ml/well of Giemsa buffer (pH 6.5), aspirate and add 1 ml/well of 10% (v/v) filtered Giemsa stain in Giemsa buffer. Incubate for 15 min, wash extensively in running tap water and dry as above. The staining procedure can be repeated if the initial staining is too faint.
11. Using a microscope count the proportion of labelled nuclei in the population of cells contained in each well.

3.3.4 *De-toxification factors*

Tissue culture basal medium incubated at 37°C in an oxygen-rich atmosphere generates a variety of poorly understood toxic products, especially free radicals of various kinds, which are highly damaging to living cells. The formation and action of these agents is inhibited by serum, and can be negligible at high cell density. These factors can be important when low density growth of cells in serum-free medium is attempted. The solution to this problem is not clear, but substantial improvements in PC13 EC cell proliferation from clonal (10–50 cells/cm^2) density can be made by adding catalase (1 μg/ml), mannitol (10 nM) and glutathione (5 mM) to the media as traps for peroxide and sulphydryl radicals.

3.4 **Technique in serum-free culture**

The principal source of problems in serum-free cell culture is, perhaps surprisingly, tissue culture techniques. Serum contains protease inhibitors and other types of protective macromolecules and has a strong buffering capacity. Culturing cells in the absence of serum makes them acutely sensitive to changes in pH, and damage occurring during trypsinization and re-plating. These problems can be circumvented by pre-equilibrating all tissue culture media to contact the cells at the final intended culture temperature and pH and removing cells from the incubator for the minimal length of time. The use of tissue culture flasks, rather than open dishes, is recommended, particularly when embarking on serum-free cell culture as they provide a closed environment and are less sensitive to fluctuations in gas composition or medium evaporation during culture. The sub-culture of cells for serum-free culture provides a different problem. Considerable cell damage can occur by endocytosis of the proteases used to remove cells from the flask. This can be circumvented by cooling the cells to 4°C before harvesting, using ice-cold solutions and exposing cells to the minimal amounts of protease. This is achieved by washing the cells in trypsin solution and leaving a thin film of trypsin covering the cells during the detachment phase. The use of minimal volumes of trypsin during sub-culture obviates the need for protease inhibitors to stop protease action after detachment. The procedures for sub-culture and harvesting cells for serum-free cell culture are given in *Table 4*.

4. MEASUREMENT OF CELL PROLIFERATION IN VITRO

4.1 **Introduction**

Any analysis of teratocarcinoma cell multiplication has the measurement of cell proliferation as the final objective. A large number of different techniques are available to measure cell multiplication and rate of progress through the cell cycle (see 11 for a complete review and discussion of methods of data analysis). The ideal technique would fulfill the following critiera.

(i) Enable the rapid measurement of large numbers of samples with a fast turn-around time.
(ii) Allow the detection of cell sub-populations with differing cell cycle characteristics.
(iii) Allow for cell death during the period of measurement (visual inspection of cell cultures can give a misleading impression of cell viability).
(iv) Measure the behaviour of a complete population of cells under particular experimental conditions.
(v) Enable determination of the duration and population distribution of different cell cycle phases.

In practice, no single procedure meets all the above criteria, which means that it is important to use a combination of methods to accurately analyse cell proliferation *in vitro*.

Some methods of cell cycle measurement, such as time-lapse video recording or flow cytofluorimetry, require specialized apparatus and are beyond the scope of this Chapter. Nevertheless, time-lapse video recording, in particular, can be a powerful method for analysing population kinetics and cell heterogeneity and has been fruitfully applied to the analysis of EC cell cycle kinetics during differentiation (4). However the simplest, and most convenient, methods available to the investigator are measurement of population doubling rates by cell number determination, and analysis of cell cycle kinetics by incorporation of [^{3}H]thymidine during the S phase of the cell cycle.

4.2 **Measurement of cell numbers**

This is the most straightforward technique to perform in practice, but provides limited information fulfilling only criteria (i) and (iv) above (Section 4.1). The effects of particular experimental manipulations on cell multiplication are determined by measuring the rate of increase of cell number in parallel cultures over a period of time. It is a useful method for simple analysis of EC cell multiplication, because the EC cell cycle transit time is quite constant. The major determinants of EC cell population increase are cell death and differentiation. Cell death can be hard to detect in many cases, although it must be implicated when the rate of population increase appears to plateau or fall. In these situations analysis of cell cycle time by autoradiography must also be considered. A second problem is the existence of sub-populations of cells within the population which possess rapid cell doubling times. These can contribute unduly to the rate of population increase, such as the presence of undifferentiated EC cells in differentiated cell populations. Measuring the growth of mixed cell populations such as spontaneously differentiating teratocarcinoma cultures requires additional means of identifying individual cell types, such as antibody, or enzymic markers.

It is vital, when measuring cell multiplication by the rate of population increase, to ensure a high, and equal, initial plating efficiency amongst the various experimental populations. This may be checked by taking a cell count within the first few hours of plating the cells into experimental conditions and calculating the number of attached cells as a proportion of the number of cells originally plated, or by initially plating the cells in equivalent conditions and then changing to experimental conditions after 24 h culture. It is important to use the low temperature trypsinization techniques described in *Table 2* when analysing PC13END cell growth to ensure high plating efficiencies. PC13END cell plating efficiency is also markedly improved when the cells are initially plated into serum-containing media before exposure to experimental conditions. The initial seeding density should be chosen so that the cells are not allowed to reach confluence during the course of the experiment. This may be determined from knowledge of the population doubling rate determined under optimal growth conditions in serum-containing culture (Section 2.4). Each experiment should be set up in duplicate or triplicate with appropriate control cultures (commonly medium containing 10% FBS) for each intended time point.

Cell numbers are determined after dispersing the cells by incubation in trypsin/EDTA (*Table 1*) at 37°C for 10 min or longer by counting samples using a haemocytometer, or more conveniently, by using an electronic particle counter. It is particularly important to ensure that all the cells have detached from the culture surface and are dispersed into a single cell suspension before counting.

4.3 Measurement of cell proliferation by [^{3}H]thymidine incorporation

The measurement of cell proliferation by [^{3}H]thymidine incorporation is a rapid and flexible technique which depends upon the incorporation of radiolabelled thymidine into newly-synthesized DNA during the DNA synthesis phase of the cell cycle. Incorporated radiolabel can be subsequently detected by scintillation counting of cell populations or by autoradiography and the counting of labelled nuclei. By choosing appropriate experimental conditions the technique can fulfil criteria (i), (ii), (iv) and (v) listed in Section 4.1. [^{3}H]Thymidine incorporation is especially useful when measuring cellular responses to exogenous growth factors in cell types, such as PC13END, where the duration of the cell cycle can be manipulated by the culture environment and the presence of exogenous growth factors. In these situations the response to growth regulatory agents can be detected by measuring the proportion of cells that enter DNA synthesis in a fixed time. In general analysis of cellular incorporation of radiolabel by autoradiography is more informative than measurements obtained by scintillation counting, but is more time-consuming to perform and is not convenient for the analysis of large number of samples. The ideal conditions for measurement of cellular DNA synthesis by scintillation counting are those where the incorporation of radiolabel is linearly related to the number of cells that enter DNA synthesis as determined by autoradiography. This should be checked periodically (see *Table 5* for parallel measurement of [^{3}H]thymidine incorporation by scintillation counting and autoradiography). This condition is achieved where the number of cells in each sample is constant, and the radiolabel is not depleted during culture. Prolonged (more than one cell cycle) labelling should be avoided in scintillation counting experiments to avoid non-linear incorporation due to multiple rounds of DNA synthesis.

The methods for analysis of PC13END and mouse fibroblast DNA synthesis by [^{3}H]-thymidine incorporation are detailed in *Table 5* and could be modified to study any cell type whose cell cycle can be regulated by exogenous growth factors. These methods are based on the use of serum-free medium (ECM/F) lacking growth factors, which results in cessation of proliferative activity in the test cells within a few hours whilst retaining cell viability. DNA synthesis and cell division are re-initiated upon addition of a growth factor (see Section 5) and so the system is extremely sensitive to exogenous growth regulatory agents. The use of 24-well multiwell plates for assaying cell proliferation by [^{3}H]thymidine incorporation is convenient for the parallel analysis of many samples, but the methods can be easily scaled to suite any particular type of tissue culture surface. It is particularly important to choose appropriate controls when carrying out [^{3}H]thymidine incorporation studies by scintillation counting in multiwell plates; each plate should include both positive (usually addition of FBS, or a known mitogen) and negative controls (no additions to ECM/F).

5. IDENTIFICATION AND PURIFICATION OF EMBRYONAL CARCINOMA-DERIVED GROWTH FACTOR

In previous sections methods for the culture and analysis of teratocarcinoma cell proliferation were described. However, teratocarcinoma cells are themselves a source of growth-regulatory molecules which can regulate the growth of heterologous cell types in culture. The best characterized of these growth regulatory agents is ECDGF, a single chain polypeptide of apparent molecular mass 17 500, which is a powerful mitogen for PC13END cells, normal mouse fibroblasts and a variety of other heterologous cell types. In this section methods for the purification and analysis of ECDGF action will be described.

5.1 **Identification and measurement of ECDGF action**

ECDGF can be detected by a variety of methods, the simplest of which is the ability to induce DNA synthesis in appropriate heterologous target cells. ECDGF was originally detected in experiments where quiescent PC13END cells were either exposed to EC cell-conditioned serum-free medium or co-cultured with EC cells (12). It was noted in both these situations that EC cells secreted a soluble activity which could induce PC13END DNA synthesis. Using PC13 EC cell-conditioned ECM medium as a starting source the complete purification of ECDGF was subsquently achieved (13). A number of other EC cell lines, including OC15S1 and P19 have subsequently been found to secrete mitogenic activities which are functionally and structurally similar to PC13EC cell-derived ECDGF. It is not advisable to use feeder-dependent EC cell lines as a source of ECDGF, since many feeder cell lines themselves secrete mitogenic activities.

Before embarking on purification of ECDGF the EC cell line to be employed should be tested for ECDGF expression. Add ECM serum-free medium, which has been exposed overnight to a confluent culture of EC cells and passed through a 4 μm filter, to either PC13END cells, or mouse fibroblast cell lines (such as 10T1/2, *Table 5*, or 3T3) as the experimental addition in the protocol detailed in *Table 5*, and measuring [^{3}H]thymidine incorporation by scintillation counting or autoradiography as described in *Table 5*. It should be found that neat EC cell-conditioned ECM medium has equivalent or superior potency to 10% (by volume) FBS as a mitogenic agent in this assay.

The [^{3}H]thymidine incorporation assay described in *Table 5* provides a fast and accurate method to monitor ECDGF activity during purification. When coupled to scintillation counting as the detection method the large number of experimental samples generated during purification can be readily tested. In general, fibroblast cell lines such as 10T1/2 are routinely used as target cells to detect ECDGF activity, using the protocol described in *Table 5*, although PC13END cells and other target cell types may be preferable in particular experimental situations.

5.2 **EC cell-conditioned medium**

ECDGF is a powerful growth regulatory agent and is maximally active at sub-nanomolar concentrations. Thus, although ECDGF can be readily detected in neat conditioned ECM medium, it is necessary to start with comparatively large volumes of conditioned medium to purify even modest amounts of ECDGF protein. Secondly, yields of ECDGF tend to improve as larger amounts of medium are fractionated, and so it is helpful to start with as large a volume of EC cell-conditioned medium as can be practically or economically accommodated. Thirdly, serum is itself a mitogenic agent and obscures the presence of cell-derived mitogens in conditioned medium. It is consequently necessary to use serum-free media to detect ECDGF action. Thus purification of significant amounts of ECDGF for experimental purposes necessitates the large-scale serum-free culture of EC cells. The yield of ECDGF depends directly on the number of viable cells in the conditioning culture, and ECDGF appears to be degraded or inactivated when cell viability is compromised. Culture conditions should therefore be employed which maximize cell number and viability. Although a variety of commercial solutions to the problem of large-scale mammalian cell culture exist, none has proved as reliable as the simple scale-up of normal cell culture methods. Since the availability of serum-free media additives tends to be the rate-limiting component, cells for the production of conditioned medium are grown up to sub-confluence in medium containing serum, washed and changed to ECM serum-free medium for conditioning. The production of a 5-l batch of PC13 EC cell-conditioned medium as starting material for ECDGF purification is described in *Table 6*.

Table 6. Production of EC cell-conditioned ECM medium.

To prepare 5 litres of conditioned medium.

1. Gelatin and fibronectin treat (*Table 2* and *Table 4*) fifty 175 cm^2 tissue culture flasks.
2. Fill each flask with 70 ml of DMEM:F12:FBS and pre-equilibrate at 37°C in 5% CO_2 in air.
3. Trypsinize EC cell monolayers from five confluent 175 cm^2 flasks (*Table 2*) and resuspend the cells in 100 ml of pre-equilibrated DMEM:F12:FBS.
4. Dispense 2 ml of the EC cell suspension per flask and incubate the flasks at 37°C in 5% CO_2 in air for 2 days or until the cell population is semi-confluent.
5. Aspirate the serum-containing medium, taking care to remove as much of the residual medium as possible.
6. Rinse the cell monolayers with DMEM:F12 (no serum).
7. Add 100 ml of ECM medium (*Table 4*) to each flask.
8. Incubate the flasks for a further 24–48 h at 37°C in 5% CO_2 in air.
9. Decant the medium from the flasks and pool the medium in a plastic container.
10. Rinse each flask with 30–40 ml of Buffer A (*Table 7*) and add the rinse to pooled medium.
11. Centrifuge the pooled medium at 9000 *g* for 30 min to remove detached cells and debris.

5.3 **Purification of ECDGF**

5.3.1 *Principles and problems*

The purification schedule for ECDGF has been developed to be rapid, relatively simple to perform and to maximize protein yield and biological activity. The purification procedure involves absorption of ECDGF from conditioned medium onto carboxymethyl–Sephadex (CM–Sephadex), which removes all the conditioned media additives and the majority of EC cell-derived proteins, followed by elution from the ion-exchange resin with high salt buffers and absorption to the hydrophobic matrix phenyl–Sepharose. ECDGF is eluted from phenyl–Sepharose with the non-polar solvent ethanediol and absorbed to heparin–Sepharose. ECDGF has the valuable property of binding avidly to heparin, which allows a considerable degree of purification in a single chromatographic step, by gradient elution from the heparin–Sepharose with high salt buffers. The final (optional) purification step involves removal of minor contaminating components by high pressure size exclusion chromatography. These chromatographic steps are arranged such that elution from CM–Sephadex is performed under conditions which maximize binding to phenyl–Sepharose, and elution from phenyl–Sepharose is performed under conditions which maximize binding to heparin–Sepharose. The first three steps can therefore be linked together in a continuous operation for routine purification purposes. It is suggested however, that each step is individually tested in initial studies.

The physical characteristics of ECDGF require that certain precautions are observed to minimize losses of protein and biological activity during purification. Although ECDGF binds avidly to laboratory glassware, losses can be avoided by the use of polypropylene tubes for fraction collection and storage. Non-specific losses of ECDGF can occur particularly at the heparin–Sepharose stage, when protein concentrations are low. This can be minimized by including the non-ionic detergent Tween-20 in the final purification steps. ECDGF activity is lost upon prolonged storage of conditioned

Table 7. Buffers[a] for purification of ECDGF.

Buffer A	50 mM Na phosphate, 150 mM NaCl, 5 mM EDTA, pH 6.8
Buffer B	50 mM Na phosphate, 1 M NaCl, 5 mM EDTA, pH 6.8
Buffer C	10% (v/v) solution of ethanediol in *Buffer A*
Buffer D	50% (v/v) solution of ethanediol in *Buffer A*
Buffer E	50 mM Na phosphate, 0.5 M NaCl, 0.02% Tween 20, pH 6.8
Buffer F	50 mM Na phosphate, 1 M NaCl, 0.02% Tween 20, pH 6.8
Buffer G	50 mM Na phosphate, 2 M NaCl, 0.02% Tween 20, pH 6.8
Buffer H	50 mM Na phosphate, 0.2 M NaCl, 0.02% Tween 20, pH 6.8

[a]All buffers should be prepared with Analar grade reagents, or equivalent, and de-ionized distilled water. Buffer H should be prepared with Analar grade or h.p.l.c. grade water and filtered through a 2 μm filter before use.

Table 8. Pre-equilibration and re-generation of low pressure gel matrices.

CM–Sephadex

1. Dissolve CM–Sephadex (Pharmacia C20) in buffer A (*Table 7*) at an approximate concentration of 10 g/100 ml buffer A. Stir gently at 4°C for at least 2 h.
2. Allow the gel to settle and pour off the excess buffer and fines. Add 100 ml of fresh buffer A per 10 ml volume of settled gel and repeat the washing procedure (step 1) until the buffer layer above the settled gel layer is clear.
3. De-gas under vacuum.
4. To re-generate the CM–Sephadex collect the used gel on a sintered glass funnel, resuspend in buffer A (100 ml buffer/1 ml gel) and proceed from step 2 above.

Phenyl–Sepharose

1. Place 20 ml of phenyl–Sepharose (Pharmacia) into a beaker and add 200 ml of buffer A (*Table 7*).
2. Collect the gel on a sintered glass funnel and wash with a solution of 50% propanol water (v/v). Wash through 50 ml of propanol solution per ml of gel. Wash with 50 ml of water/ml of gel. Wash with 50 ml of buffer A per ml of gel.
3. Re-suspend the gel in 5 ml of buffer A per ml of gel and pack it into a 2.5 cm diameter chromatography column.
4. Equilibrate the packed column with two column volumes of buffer B.
5. To regenerate the phenyl–Sepharose unpack the column, resuspend the gel in water and proceed from step 2 above.

Heparin–Sepharose

1. Weigh out 1 g of heparin–Sepharose powder (Pharmacia) and resuspend in 20 ml of buffer A. Allow the gel to swell for at least 2 h at 4°C on a rocking platform.
2. De-gas under vacuum.
3. Pack the gel (~4 ml of volume) into a 1 cm chromatography column.
4. Wash the packed heparin–Sepharose with 40 ml of buffer G followed by 40 ml of buffer H. Repeat this washing cycle at least once.
5. Equilibrate the column with 40 ml of buffer H.
6. To regenerate the heparin–Sepharose proceed from step 4 above.

Table 9. Purification of ECDGF.

CM–Sephadex chromatography

1. Add 20 ml of pre-equilibrated CM–Sephadex slurry (*Table 7*) per litre of clarified conditioned medium (*Table 6*). Stir gently (sufficient to keep the gel in suspension) at 4°C for 2 h.
2. Collect the gel on a sintered glass funnel and wash the gel with about 200 column volumes of buffer A (*Table 7*).
3. Re-suspend the gel in an equivalent volume of buffer A, pack the washed gel into a 5 cm chromatography column, and allow the gel to settle. Open the outlet port and allow the buffer to flow out until the buffer level is just covering the top of the gel bed.
4. Connect the inlet to the column and elute with five column volumes of buffer B at a flow-rate of about 50 ml/h. Either pass the eluate down a phenyl–Sepharose column (see below) *or* collect 10 ml fractions for analysis.

Phenyl–Sepharose chromatography

1. Either elute the CM–Sephadex column directly onto packed and equilibrated phenyl–Sepharose (*Table 8*) *or* pool the active fractions from the CM–Sephadex eluate and load them onto phenyl–Sepharose column. Adjust the flow-rate to about 2.5 column volumes/h.
2. Wash the phenyl–Sepharose with five column volumes of buffer B.
3. Wash the phenyl–Sepharose column with five column volumes of buffer A.
4. Wash the phenyl–Sepharose column with five column volumes of buffer C.
5. Either elute the ECDGF with five column volumes of buffer D onto the heparin–Sepharose column (see below) *or* elute with five column volumes of buffer D and collect 5 ml fractions for analysis.

Heparin–Sepharose chromatography

1. Either elute the phenyl–Sepharose column onto packed and equilibrated heparin–Sepharose (*Table 8*) *or* pool the active fractions from the phenyl–Sepharose eluate and load them onto a heparin–Sepharose column. Adjust the flow-rate to about 20 ml/h.
2. Wash the column with five column volumes of buffer E.
3. Apply a 120 ml (or 30 column volumes) linear gradient using buffer E as the starting buffer and buffer F as the limit buffer. Collect 4 ml fractions for analysis. The ECDGF activity elutes about half way between 0.7 and 0.75 M NaCl.
4. Strip the column with five column volumes of buffer G before regeneration (see *Table 8*).

High pressure size exclusion chromatography

1. Pool the active fractions from heparin–Sepharose chromatography and concentrate by centrifugal ultrafiltration (Centricon 10, Amicon Ltd) to a final volume of 100–200 μl. ECDGF preparations may be stored at 4°C for several weeks at this stage.
2. Connect a guard column and a SW-2000 column in series and equilibrate with at least two column volumes of buffer H, at a flow rate of 0.5 ml/min, until a stable u.v detector baseline is achieved.
3. Filter the concentrated ECDGF solution through a low dead volume 2 μm nitrocellulose filter (e.g. Gelman Acro LC13).
4. Inject the concentrated, filtered, ECDGF solution and elute at a flow-rate of 0.5 ml/min, collecting 0.5 ml fractions. ECDGF elutes with a nominal molecular mass of about 17 000.
5. Assay the fractions for mitogenic activity (*Table 4*). Pool and concentrate the active fractions as in step 1 above.

medium or crude initial fractionation material, and is partially inactivated by exposure to conditions below pH 4. Neutral buffers are consequently used throughout. In general, the best results will be obtained by using fresh conditioned medium, proceeding as quickly as possible through the initial stages of purification and keeping all samples at 4°C.

5.3.2 *Equipment*

The ECDGF purification procedure has been designed so that simple low-pressure liquid chromatography equipment of the type found in many laboratories can be employed through most of the stages. Expensive and specialized high pressure liquid chromatography (h.p.l.c.) equipment is however necessary if a high purity product is required. The minimal requirements are: three low pressure chromatography columns (preferably with adjustable flow adaptors) 5 cm, 2.5 cm and 1 cm diameter, a variable speed peristaltic pump, a fraction collector capable of collecting fractions ranging from 12 ml to 2 ml and a 120 ml gradient maker (either commercial or 'home-made'). H.p.l.c. separations require at least one h.p.l.c. type metering pump (capable of delivering up to 1 ml/min at pressures up to 500 p.s.i.), a u.v. detector with low volume flow cell capable of detection at 210–220 nm, and associated injectors and plumbing. Separations are performed on a Toya-soda TSK SW-2000 (4.5 mm × 60 cm) size exclusion column, with an appropriate guard column, which can be obtained from a variety of specialist h.p.l.c. companies.

5.3.3 *Purification*

The buffers required for purification of ECDGF are listed in *Table 7*. The low-pressure gels have to be washed and equilibrated before use (or re-use) as described in *Table 8*. Pre-cycling is particularly important in the case of freshly swollen heparin–Sepharose

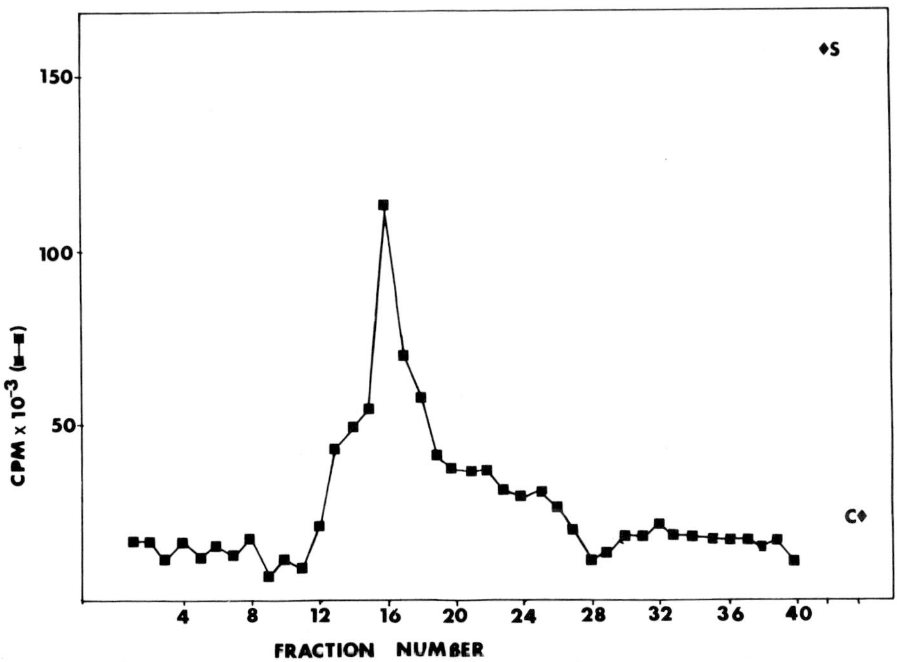

Figure 1. Fractionation of ECDGF activity by chromatography on CM–Sephadex (as described in *Table 9*). 10 ml fractions were collected beginning from the application of Buffer B. 20 μl aliquots from each fraction were tested for [^{3}H]thymidine incorporation into 10T1/2 fibroblasts by scintillation counting. C = [^{3}H]-thymidine incorporation without additions, S = [^{3}H]thymidine incorporation induced by 10% FBS.

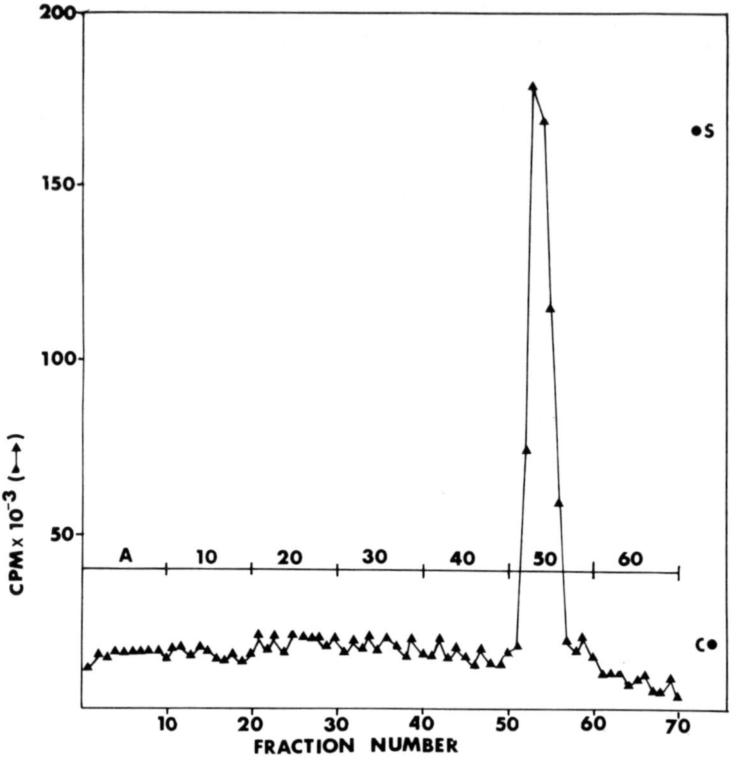

Figure 2. Fractionation of ECDGF activity from CM−Sephadex (*Figure 1*) on phenyl−Sepharose. Pooled fractions from CM−Sephadex eluate were applied to phenyl−Sepharose as described in *Table 9*, but step-eluted with increasing concentrations of ethanediol mixed with buffer A. 5 ml fractions were collected and 10 μl aliquots from each fraction were tested for [^{3}H]thymidine incorporation into 10T1/2 fibroblasts by scintillation counting. C = [^{3}H]thymidine incorporation without additions, S = [^{3}H]thymidine incorporation induced by 10% FBS.

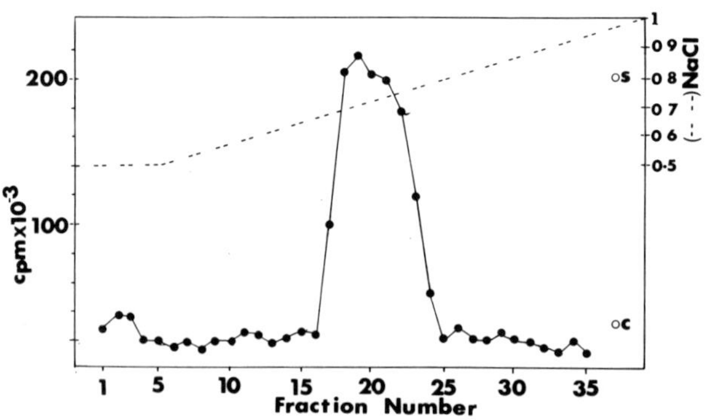

Figure 3. Fractionation of ECDGF activity from phenyl−Sepharose eluate on heparin−Sepharose. Pooled fractions from CM−Sephadex eluate were applied to heparin−Sepharose as described in *Table 9*, and eluted with a gradient from 0.5 to 1 M NaCl. 4 ml fractions were collected and 10 μl aliquots were tested for [^{3}H]-thymidine incorporation into 10T1/2 fibroblasts by scintillation counting. C = [^{3}H]thymidine incorporation without additions, S = [^{3}H]thymidine incorporation induced by 10% FBS.

since the heparin can leak off the column and interfere with biological assays. Although this problem cannot be completely overcome, it can be markedly reduced by vigorous pre-cycling of the gel matrix. Each of the low-pressure column steps are described in *Table 9* and representative results from each of these steps are shown in *Figures 1, 2* and *3*.

At the end of CM–Sephadex, phenyl–Sepharose and heparin–Sepharose chromatography the ECDGF activity is found in about 20 ml of gradient elution buffer at a concentration of about 100–200 ng/ml (depending upon the starting volume of the conditioned medium). Concentration of this dilute solution is readily achieved by centrifugal ultrafiltration to a convenient final volume for further experimentation. Inclusion of Tween-20 is required to prevent protein loss during ultrafiltration, but the Tween-20 is partially concentrated during this step. Although Tween-20 does not interfere with biologial experimentation on whole cells, its presence in ECDGF preparations must be taken into account in biochemical experiments.

The final stage of ECDGF purification is high pressure size-exclusion chromatography (h.p.s.-e.c.) on a TSK-SW2000 column as described in *Table 9*. Representative results are shown in *Figure 4*. This step requires specialized h.p.l.c. equipment, but is required to remove minor contaminants from the ECDGF obtained from the heparin–Sepharose chromatography step, and is essential if further structural study of ECDGF is intended. The presence of contaminating Tween-20 is revealed as a major peak of

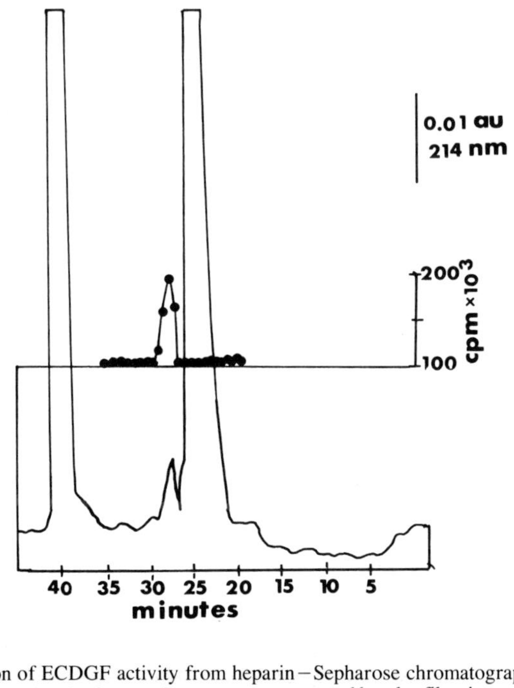

Figure 4. Fractionation of ECDGF activity from heparin–Sepharose chromatography by h.p.s.-e.c. Pooled fractions from the heparin–Sepharose eluate were concentrated by ultrafiltration and applied to an SW-2000 column, and eluted as described in *Table 9*. 10 μl aliquots were tested for [^{3}H]thymidine incorporation into 10T1/2 fibroblasts by scintillation counting.

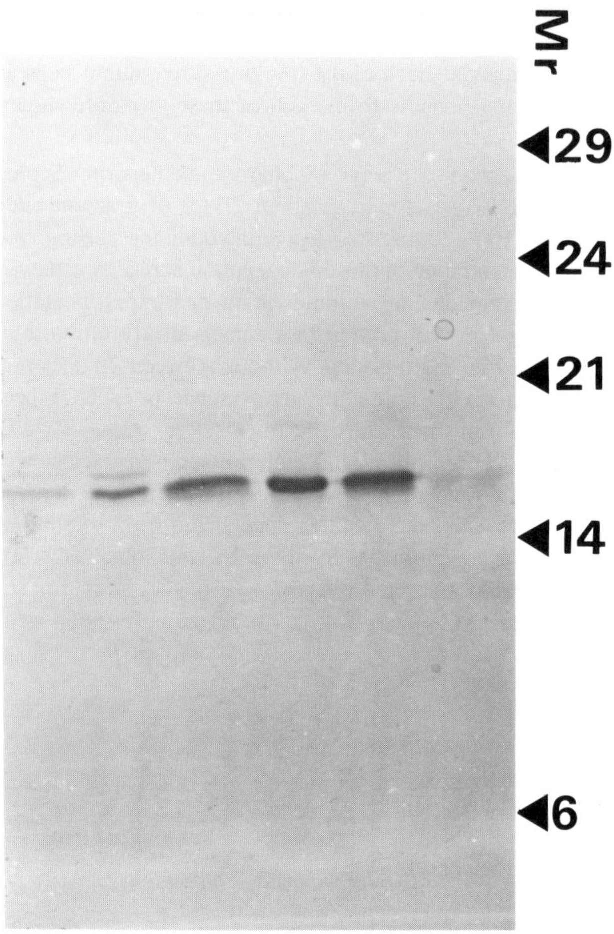

Figure 5. Analysis of ECDGF fractions by electrophoresis. Successive fractions from an h.p.s.-e.c. separations were concentrated by ultrafiltration to 40 μl. 10 μl aliquots were subjected to electrophoresis in polyacrylamide gels (15% acrylamide; 500 μm thickness) followed by silver staining. Mr = relative mobility of marker proteins.

u.v. absorbance eluting just before ECDGF. Following h.p.s.-e.c. the ECDGF is concentrated by ultrafiltration and the purity assessed by electrophoresis in polyacrylamide gel containing sodium dodecyl sulphate (SDS) (*Figure 5*). Concentrated ECDGF stocks are stable for several weeks at 4°C, but should be aliquoted and stored at −70°C for longer periods.

One difficulty encountered during the purification of ECDGF is measuring the low protein concentrations encountered during the final purification steps. Common protein determination methods such as dye-binding assays or amino acid analysis are either insufficiently sensitive, or require the destruction of unacceptably large amounts of sam-

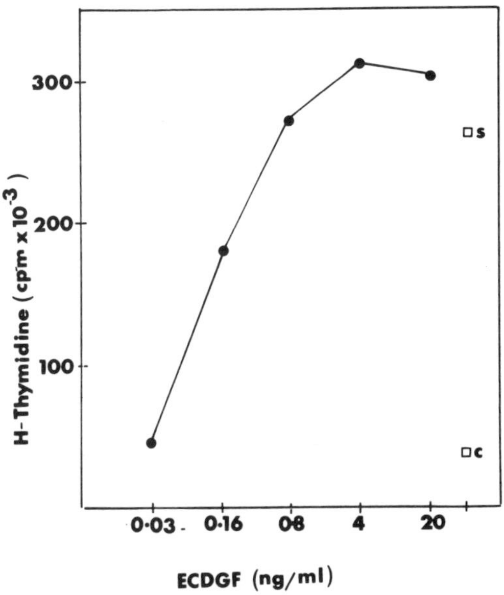

Figure 6. ECDGF concentration-dependent stimulation of [^{3}H]thymidine incorporation. Varying concentrations of ECDGF were tested for [^{3}H]thymidine incorporation into 10T1/2 fibroblasts by scintillation counting. C = [^{3}H]thymidine incorporation without additions. S = [^{3}H]thymidine incorporation induced by 10% FBS.

ple to be of routine use. A major advantage of the final h.p.s.-e.c. step is the visualization of ECDGF by u.v. absorbance (typically 210–220 nm) during the final h.p.s.-e.c. step which provides a convenient non-destructive method of quantifying the amount of ECDGF protein, and compares well with protein determination by destructive methods. Protein absorbance at low u.v. wavelengths must be initially calibrated with known standards of similar apparent molecular mass since the peak heights observed will depend on the characteristics of the u.v. detector and h.p.l.c. configuration employed. A semi-quantitative method of determining ECDGF protein concentration is to run serial dilutions of the ECDGF preparation on an SDS gel with varying dilutions of known amounts of protein standards and comparing band intensities after high sensitivity silver stain methods. In the author's experience this method tends to lead to underestimates of ECDGF protein concentration compared with other methods, and is highly dependent upon the particular staining method employed. For anything other than 'one-off' studies of ECDGF, a reference stock of ECDGF of known protein concentration should be prepared for calibration of biological assays and determination of ECDGF concentration in new preparations and samples. The typical yield of ECDGF is between 500 and 800 ng/l of conditioned medium. A dose–response curve for induction of 10T1/2 mouse fibroblast DNA synthesis (as in *Table 5*) is shown in *Figure 6*. As with other growth factors the optimal concentration for ECDGF-induced mitogenicity must be determined empirically for each target cell type.

6. ACKNOWLEDGEMENTS

I am grateful to many laboratory colleagues for their critical comments on the manuscript, and to Hilary Oakley, Jane Deller and Judy Morton for their help in developing the techniques described herein.

7. REFERENCES

1. Heath,J.K. and Rees,A.R. (1985) *Ciba Found. Symp.*, **116**, 1.
2. Strickland,S. and Mahdavi,V. (1978) *Cell*, **15**, 393.
3. Rayner,M. and Graham,C.F. (1982) *J. Cell Sci.*, **58**, 331.
4. Rayner,M. and Pulsford,J. (1984) *J. Cell Sci.*, **72**, 227.
5. Ham,R. (1981) *Handbook Exp. Pharmacol.*, **57**, 13.
6. Heath,J.K. and Deller,M.J. (1983) *J. Cell Physiol.*, **115**, 225.
7. Engstrom,W., Rees,A.R. and Heath,J.K. (1985) *J. Cell Sci.*, **73**, 361.
8. Engvall,E. and Ruoshalati,E. (1977) *Int.J. Cancer*, **10**, 1.
9. Rizzino,A. and Crowley,C. (1980) *Proc. Natl. Acad. Sci. USA*, **77**, 457.
10. Heath,J.K., Bell,S.M. and Rees,A.R. (1981) *J. Cell Biol.*, **91**, 293.
11. Natchewey,D. and Cameron,I. (1968) In *Methods in Cell Physiology*. Prescott,D. (ed.), Academic Press Inc., New York, Vol. **III**, p.213.
12. Isacke,C.M. and Deller,M.J. (1983) *J. Cell Physiol.*, **117**, 407.
13. Heath,J.K. and Isacke,C.M. (1984) *EMBO J.*, **3**, 2957.

CHAPTER 8

Cell lines from human germ cell tumours

PETER W.ANDREWS, J.WOLTER OOSTERHUIS
and IVAN DAMJANOV

1. INTRODUCTION

The pathology of human germ cell tumours differs in several respects from that of comparable tumours in mice (1). Accordingly, although murine germ cell tumours, particularly teratocarcinomas, offer greater scope for experimental manipulation and direct comparison with embryonic cells, the biology of human teratocarcinomas is more directly relevant to the clinical problem posed by these tumours, as well as to the biology of human embryogenesis which undoubtedly differs in detail, if not overall plan, from embryonic development in the mouse. Nevertheless, the tumours from both species provide complementary model systems, and comparative studies may lead to a better understanding of mammalian embryogenesis in general, and its relationship to oncogenesis, than the study of either system alone. In this chapter, we discuss the techniques and approaches that are used in the experimental study of human germ cell tumours. Because our primary knowledge of these neoplasms comes from their natural history, and because this is essential information for meaningful experimentation, we have begun with a summary of their clinical pathology and the methods used to classify them (Section 2). Readers seeking more detailed information are referred to several monographs and atlases dealing with this subject (2−8). We have then provided a brief review of the cell lines that have already been isolated, and of the general procedures that have proved useful for their derivation (Section 3). However, this lies in the realm of art rather than science, and success in the isolation of new lines is probably more related to the diligence of the investigator than to the particular methods used. Although a wealth of cell lines has already been derived, there are gaps in the available material. For example, no unambiguously identified seminoma cell line has yet been reported, while a pluripotent stem cell with a capacity for differentiation comparable with the best murine stem cells remains elusive. We have tried to indicate how these gaps might be filled. Finally we have described techniques and markers for maintaining and characterizing established human stem cell lines and for inducing and monitoring their differentiation in culture (Section 4).

2. HISTOPATHOLOGY OF HUMAN GERM CELL TUMOURS

Human germ cell tumours arise in the testis, ovary or various extra-gonadal locations. Male and female germ cells are the presumptive cell of origin, and cytogenetic evidence indicates that tumours may develop from pre-meiotic oogonia and spermatogonia and germ cells arrested at various stages of meiosis, and also from post-meiotic and early embryonic post-zygotic cells. However, irrespective of their derivation, the stem cells in germ cell tumours do not have any significant features in common with their presumptive progenitors, the germ cells. The only exception is seminoma, which presumably recapitulates spermatogenesis and has been likened to immature spermatogenic cells. All other germ cell tumours are composed of cells that resemble post-zygotic early somatic cells, and extra-embryonic structures such as yolk sac or trophoblast. Thus although germ cell tumours derived their name from their presumptive cell of origin, they are more closely related histologically to early embryonic structures and their histogenesis can be better understood by studying normal embryogenesis than gametogenesis.

2.1 **Histogenesis and classification of germ cell tumours**

Most of the classifications presently used in the analysis of human germ cell tumours are derived from the basic histogenetic concepts introduced in 1946 by Friedman and Moore and further expanded by Dixon and Moore (3). The basic assumption of these classifications is that germ cell tumours develop from activated germ cells and their derivatives (*Figure 1*). Parthenogenetic activation of ovarian and ephebogenetic activation of testicular germ cells has been described in several vertebrate species, and presumably occurs in humans. Activation of germ cells may be, but is not necessarily, associated with malignant transformation. Activated germ cells that have not undergone malignant transformation give rise to benign tumours composed of differentiated somatic tissues – *teratomas*. If the activation of germ cells occurs hand in hand with malignant transformation, the malignant cells may follow two developmental pathways; either they recapitulate gametogenesis and form *dysgerminomas* (seminomas) or they mimic histogenesis of the early embryo and thus give rise to *teratocarcinomas* and related neoplasms. The malignant stem cells of teratocarcinomas have an embryonic phenotype and are therefore called *embryonal carcinoma cells* (EC cells). Such cells may differentiate into various somatic tissues including skin, nerve tissue, or bone as well as extra-embryonic structures equivalent to yolk sac epithelium or trophoblast. *Teratocarcinomas* are thus typically composed of EC cells and various somatic tissues, yolk sac epithelium

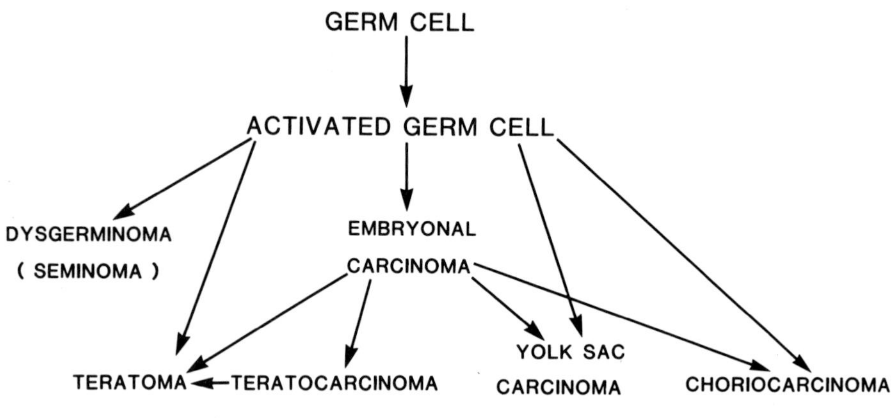

Figure 1. Scheme showing the postulated relationship between germ cells and the various components of germ cell tumours.

Table 1. Comparison of various classifications of germ cell tumours.

WHO 1975 (9)	*Dixon and Moore (3)*	*Mostofi (10)*	*British testicular tumour panel (b)*	*Equivalent animal tumours*
Teratoma Mature Immature	Group III teratoma, pure	Teratoma Mature Immature	Teratoma differentiated (TD)	Teratoma
Teratoma with malignant transformation	Group IV teratoma with carcinoma or sarcoma	Teratoma with malignant area other than seminoma, EC or choriocarcinoma	Malignant teratoma intermediate (MTI)	Not registered
Teratocarcinoma (embryonal carcinoma and teratoma)	Group IV teratoma with EC or choriocarcinoma	Embryonal carcinoma and teratoma (teratocarcinoma)	Malignant teratoma intermediate (MTI)	Teratocarcinoma
Embryonal carcinoma (EC)	Group II embryonal	Embryonal carcinoma adult-type	Malignant teratoma undifferentiated (MTU)	Embryonal carcinoma 'nullipotent'
Choriocarcinoma	Group V choriocarcinoma	Choriocarcinoma pure	Malignant teratoma trophoblastic (MTT)	Trophoblastoma or trophectodermal carcinoma
Yolk sac tumour (endodermal sinus tumour)	Not listed	Infantile embyronal	Yolk sac tumour; orchioblastoma	Yolk sac carcinoma (biphasic, parietal or visceral)
Polyembryoma	Not listed	Polyembryoma	Not listed	Embryoid bodies in ascites tumour

and groups of trophoblastic cells. EC cells unable to differentiate into other cell types form monomorphic tumours which are classified as 'pure monomorphic' embryonal carcinoma, or simply *embryonal carcinoma*.

EC cells capable of differentiation may give rise to somatic tissues and extra-embryonic structures that differ in their malignancy. In some instances, if all malignant stem cells differentiate into non-proliferating somatic tissues, a tumour indistinguishable from an originally benign teratoma will form. If the EC cells differentiate into immature somatic tissues but retain their malignancy or the capacity to proliferate, an *immature teratoma* is formed. Immature teratomas are tumours of variable malignancy. They do not contain EC cells and their malignancy resides with stem cells of immature somatic tissues. Most frequently, the malignant stem cells in immature teratomas are neural and correspond to neuroblasts and other embryonic neural elements.

EC cells may give rise to malignant extra-embryonic cells that obliterate all other components and form pure yolk sac carcinomas or choriocarcinomas. Presumably yolk sac carcinomas and choriocarcinomas could also arise directly from activated germ cells without an intermediate EC cell-like development stage, but it seems more plausible that most such tumours arise from the extra-embryonic components of abnormal embryos formed from activated ova, or from EC cells whose differentiation is restricted to extra-embryonic structures.

Histogenetic classification of human germ cell tumours is not always possible in clinical practice. For practical reasons and to ensure easier and more meaningful clinico-pathologic correlation, several working classifications have been introduced. These are summarized in *Table 1* in parallel with corresponding animal tumours. For therapeutic and prognostic purposes it has become common practice to segregate seminomas (dysgerminomas) from all other malignant germ cell neoplasms, which are then grouped together as non-seminomatous germ cell tumours.

2.2 **Morphology of specific tumour types**

Only the most salient features of human germ cell tumours are described in this section and the reader is referred to several richly illustrated monographs (3–8) for additional details.

2.2.1 *Seminoma*

Seminoma occurs in two forms, classical and spermatocytic. Classical seminoma is composed of a uniform population of large polygonal or round cells arranged into nests surrounded by connective tissue septa (*Figure 2*). Nuclei may show considerable variation in size and shape, but mitotic figures are rare. Tumour cells have well developed cytoplasm, which often contains large amounts of glycogen and therefore appears clear in standard histological preparations. Glycogen may be easily demonstrated with special stains such as the periodic acid–Schiff reaction (PAS), with or without pre-digestion with diastase. Typically, the connective tissue septa contain infiltrates of lymphocytes and macrophages, and even granulomas with giant cell formation. In a small number of seminomas (5–10%) there are well differentiated trophoblastic giant cells. These cells are not malignant but secrete and synthesize human chorionic gonadotropin (HCG) which can be histochemically demonstrated with appropriate antibodies.

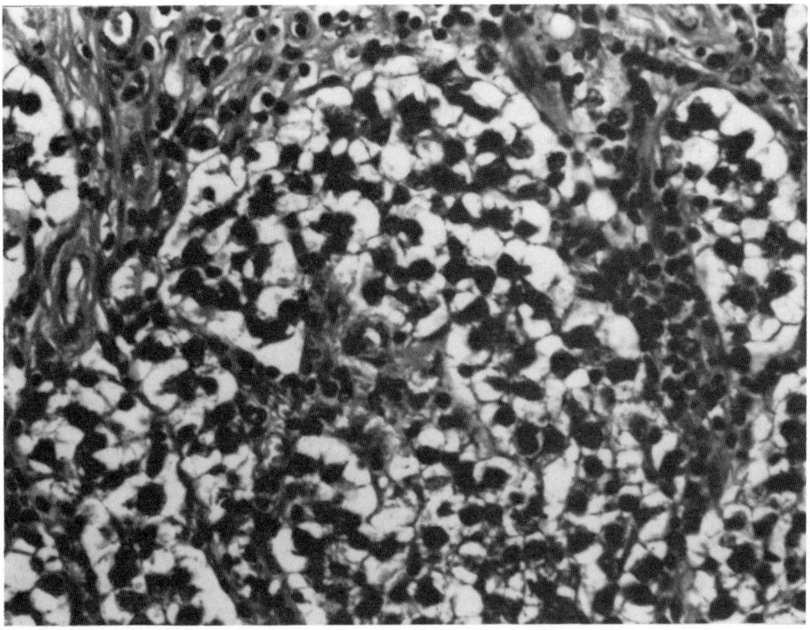

Figure 2. Classical seminoma.

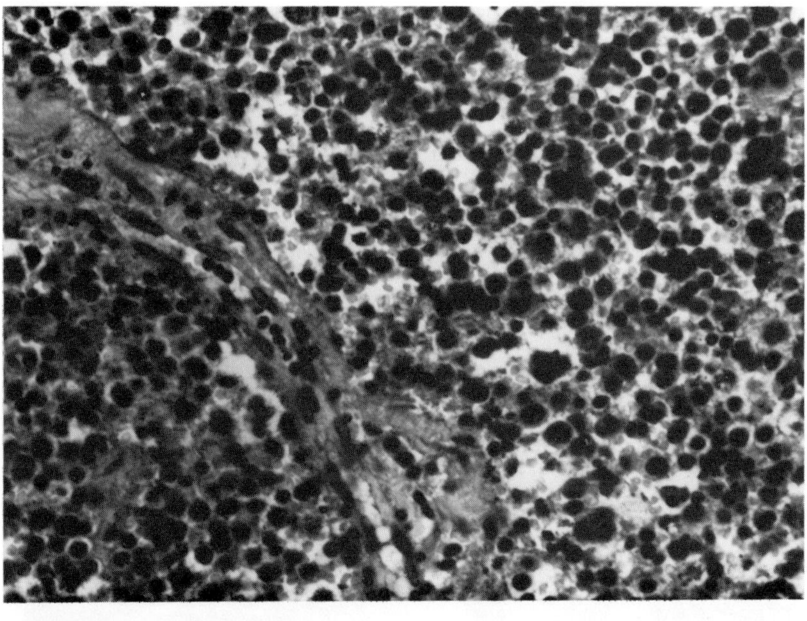

Figure 3. Spermatocytic seminoma.

In contrast to the monomorphic cell populations in classical seminomas, spermatocytic seminomas are typically composed of three distinct cell types; large, small and intermediate cells (*Figure 3*). Intermediate cells resemble in size those of classical seminomas but lack glycogen and thus do not have clear but rather eosinophilic cytoplasm. The small cells have been equated with spermatids and spermatocytes, but the evidence for spermatocytic differentiation in these tumours is circumstantial and the nature of large and small cells remains to be determined (11).

Ultrastructurally, the cells of classical seminoma have rounded or polygonal nuclei with mild indentations, finely dispersed chromatin and prominent nucleoli. Nucleolonema consists of intricately intertwined fibrils and granules or segregated fibrillar and amorphous components. The cytoplasm of tumour cells varies in complexity. Mitochondrial tubulovesicular cristae are usually grouped at one end of the cytoplasm. Various amounts of glycogen are evenly dispersed throughout the entire cytoplasm or aggregated irregularly. The Golgi apparatus is well formed and may be multiple. Various amounts of smooth endoplasmic reticulum, lysosome-like dense bodies and multivesicular bodies are seen in the cytoplasm admixed with profiles of rough endoplasmic reticulum. The plasma membrane is generally smooth, with occasional microvilli and pinocytotic invaginations. Adjacent tumour cells are interconnected with desmosomes and a few zonulae adherentes.

Enzyme histochemistry may be used to demonstrate atypically high levels of alkaline phosphatase in seminoma cells. Alkaline phosphatase in the tumour cell membranes may be demonstrated immunohistochemically with antibodies to placental alkaline phosphatase (12). There are no immunocytochemical markers unique to seminoma cells. SSEA-1 cannot be detected on seminoma cells (13) and SSEA-3, SSEA-4 and TRA-1-60 and TRA-1-81 seem to be expressed on some, but not all, tumours (Damjanov, unpublished). Approximately 7% of all seminoma contain HCG-positive giant cells (14). Seminoma cells do not contain vimentin or keratin polypeptides (15). No distinct or pathognomonic reaction patterns have been defined by lectin histochemistry (16).

2.2.2 *Embryonal carcinoma*

Pure embryonal carcinomas (*Figure 4*) are composed of polygonal cells with moderately

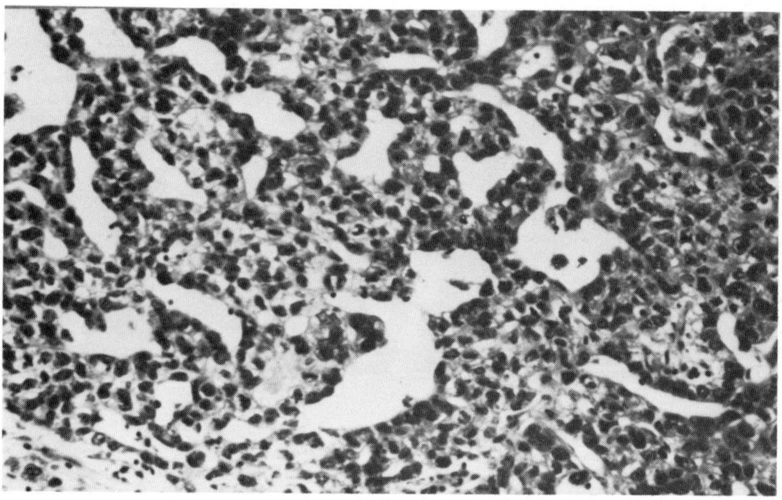

Figure 4. Embryonal carcinoma.

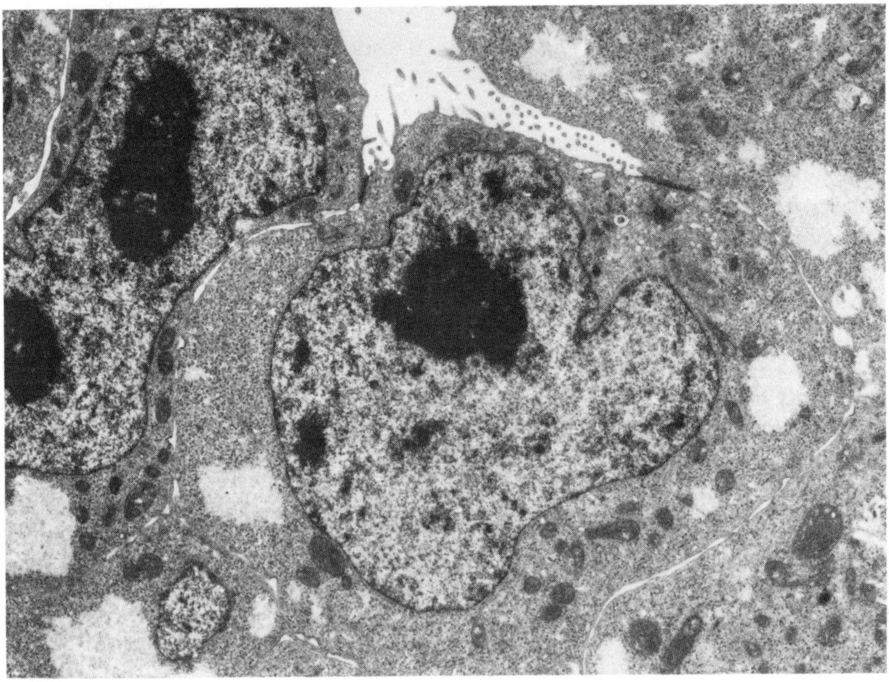

Figure 5. Electron micrograph of EC cells.

large nuclei and relatively little cytoplasm accounting for the high nucleo to cytoplasmic ratio. The cytoplasm may be eosinophilic or amphophilic, and the cell borders are indistinct. Tumour cells grow in several patterns; solid, tubular and papillary. Nuclear polymorphism and hyperchromasia are prominent and mitotic figures are frequent. Stroma is scant. Due to rapid and invasive growth, the tumour often shows foci of necrosis and haemorrhage. EC cells have been compared with primitive, nondescript embryonic cells.

Ultrastructurally, EC cells have irregularly indented nuclei (*Figure 5*). In the nucleus euchromatin predominates over irregularly clumped heterochromatin. Nucleoli are prominent, often multiple and show various irregular arrangements of their components. The cytoplasm of most cells is poorly or moderately well developed and contains predominantly free ribosomes, scattered mitochondria and a few other organelles. These include a small Golgi zone, a few profiles of rough endoplasmic reticulum and occasional lysosome-like dense bodies. Glycogen may be present but is usually inconspicuous.

EC cells are rich in alkaline phosphatase which can be demonstrated enzyme-histochemically in tissue sections. Most tumours react with some of the antibodies to human placental alkaline phosphatase, indicating that at least some of the enzyme activity is due to either the placental, or a placental-like isozyme (12). EC cells react with antibodies to SSEA-3, SSEA-4, TRA-1-60, TRA-1-81 (see also Section 4). EC cells contain the 40-, 45- and 52-kd keratin polypeptides (17).

2.2.3 *Yolk sac carcinoma*

Pure yolk sac tumours are the most common testicular germ cell tumours in infants and young children, and are the most common malignant germ cell tumour of the ovary. However, in general, yolk sac tumours are just one of the components of teratocarcinomas and account for the elevation of alpha-foetoprotein (AFP) in the serum of approximately two thirds of all patients carrying a non-seminomatous germ cell tumour in the testis and in an even higher proportion of patients with tumours in ovarian and extra-gonadal sites. Yolk sac tumours, irrespective of whether they are pure or included in teratocarcinoma, present in several patterns; reticular, composed of a meshwork of cells lining cystic, slide-like or intercommunicating irregular spaces; glanduloalveolar; papillary; solid; and myxomatous, resembling the so-called magma reticulare and the endodermal sinuses of Duval (*Figure 6*). Glomeruloid structures, called Schiller–Duval bodies, and composed of a central fibrovascular stalk lined by columnar cells projecting into a cystic space lined by hobnailed cells, resemble primitive glomeruli and are fairly characteristic of this tumour. Some tumours exhibit a polyvesicular vitelline pattern with mesothelial-like cells lining the vesicles, and tall columnar cells corresponding to the so-called Heuser's membrane (the exocoelomic membrane found during early stages of human embryogenesis) (18). Most tumours contain round hyaline globules of basement membrane-like material in intercellular spaces.

Ultrastructurally, these tumour cells differ from the undifferentiated EC cells in that they contain many more cytoplasmic organelles. Most prominently the cytoplasm of these cells contains well-developed profiles of rough endoplasmic reticulum, which often appears dilated and may contain the basement membrane-like material also found in the extracellular spaces. The tubular and papillary structures are linked by cuboidal

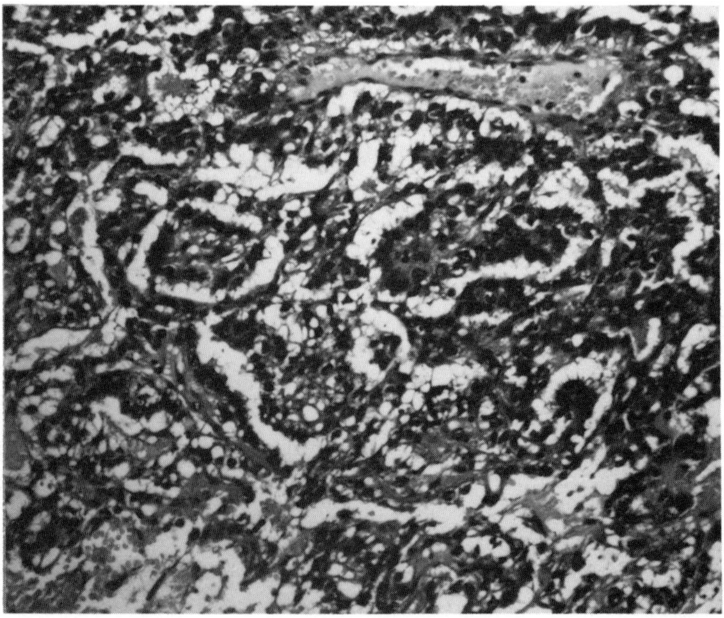

Figure 6. Yolk sac carcinoma.

cells which have bean-shaped nuclei and a well developed cytoplasm. These cells are interlocked with intense cytoplasmic processes and desmosomes. The myxomatous areas are composed of cells resembling primitive mesenchymal cells or fibroblasts. The intercellular spaces often contain large globules of amorphous basement membrane-like materials.

Immunohistochemically, AFP is the typical marker for yolk sac epithelium and is also present in the yolk sac carcinoma cells. Basement membrane-like material contains laminin, collagen type IV and probably all other basement membrane components. SSEA-3 and SSEA-4 are present in tumour cells, but SSEA-1 is lacking (13). Placental and placental-like alkaline phosphatase are also present (12). Alpha-1-antitrypsin was also demonstrated in the cytoplasm of some cells and the globular masses in the interstitial spaces (8).

2.2.4 *Choriocarcinoma*

Choriocarcinoma is a tumour composed of trophoblastic cells, and the diagnosis should not be made unless both cytotrophoblastic and syncytiotrophoblastic elements are histologically identified. Pure choriocarcinoma is an extremely rare tumour. This contrasts with the trophoblastic foci which occur in two thirds of all non-seminomatous germ cell tumours. These foci probably represent terminally differentiated, non-proliferating cells which cannot be histologically distinguished from the malignant cells forming pure choriocarcinomas. In both instances, the cytotrophoblastic cells appear as polygonal with clear, glycogen-rich cytoplasm and sharp cell outlines, and a round, to polygonal, centrally located vesicular nucleus. The syncytiotrophoblastic cells are multinucleated giant cells and have eosinophilic cytoplasm (*Figure 7*).

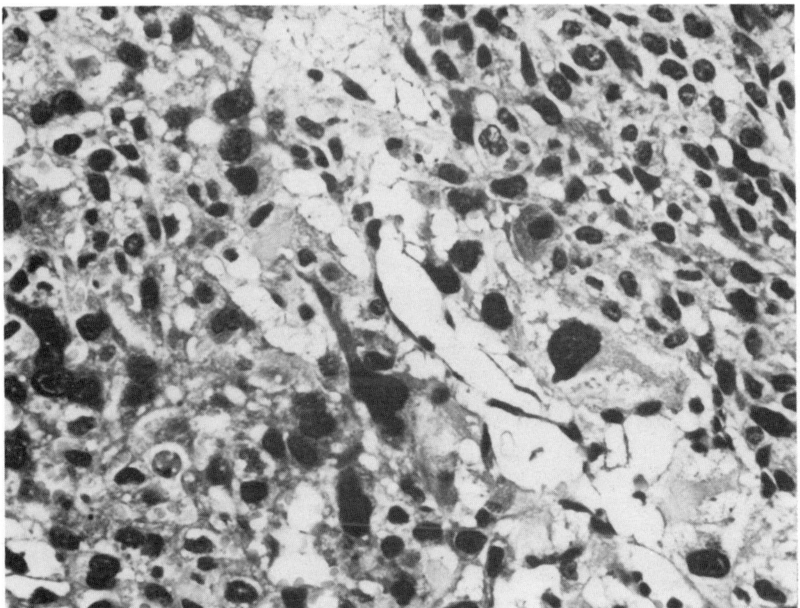

Figure 7. Choriocarcinoma.

Ultrastructurally, both cytotrophoblastic and syncytiotrophoblastic cells contain abundant cytoplasm. Cytotrophoblastic cells have fewer organelles and more glycogen, whereas syncytiotrophoblastic cells appear well equipped for protein and hormone synthesis and contain mitochondria, Golgi apparatus and rough and smooth endoplasmic reticulum.

Enzyme-histochemically, choriocarcinoma cells are rich in alkaline phosphatase. Immunohistochemically, the nature of this alkaline phosphatase has not been unequivocally determined, but it seems that it is not invariably of the placental type. HCG is the typical immunohistochemical marker for syncytiotrophoblastic cells. Antibodies to HCG also react with some mononuclear cells, which presumably correspond to either cytotrophoblastic cells or some intermediate forms in the pathway of differentiation of cytotrophoblast into trophoblast. Both the malignant trophoblastic cells and their normal equivalents in the placenta express the 40-, 45-, 52- and 54-kd keratin polypeptides. These cells also synthesize laminin (19), and thus probably all the other basement membrane components.

2.2.5 *Teratocarcinoma*

Teratocarcinomas are mixed tumours composed of EC cells and various somatic and extra-embryonic tissues (*Figure 8*). EC cells are the malignant stem cells of the tumour and are arranged in groups resembling the pattern seen in pure embryonal carcinomas. The somatic tissues present vary from one tumour to another in complexity of organization and maturity. Most simple tissues and many complex organs and organ primordia such as teeth or thyroid may be present. Extra-embryonic tissues include yolk sac and placenta-like components.

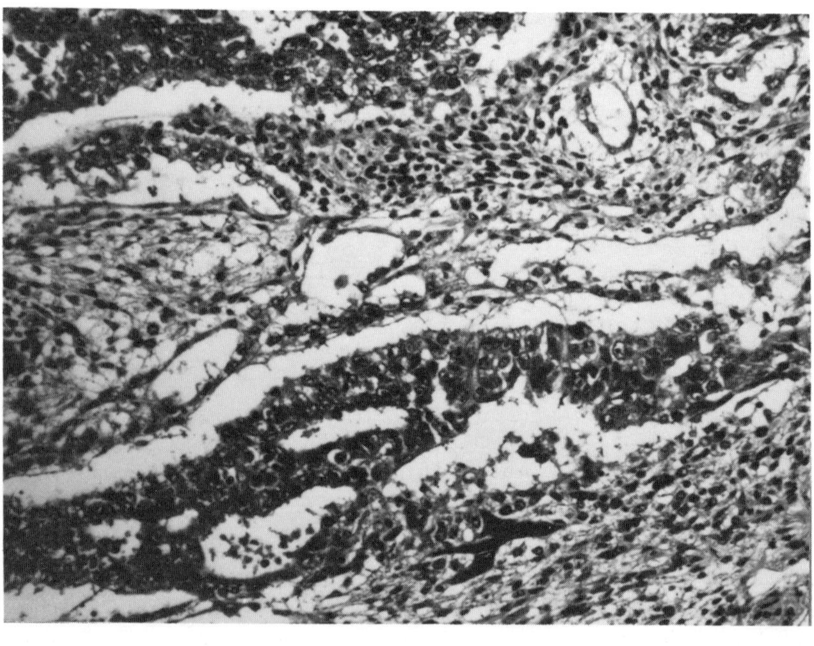

Figure 8. Teratocarcinoma.

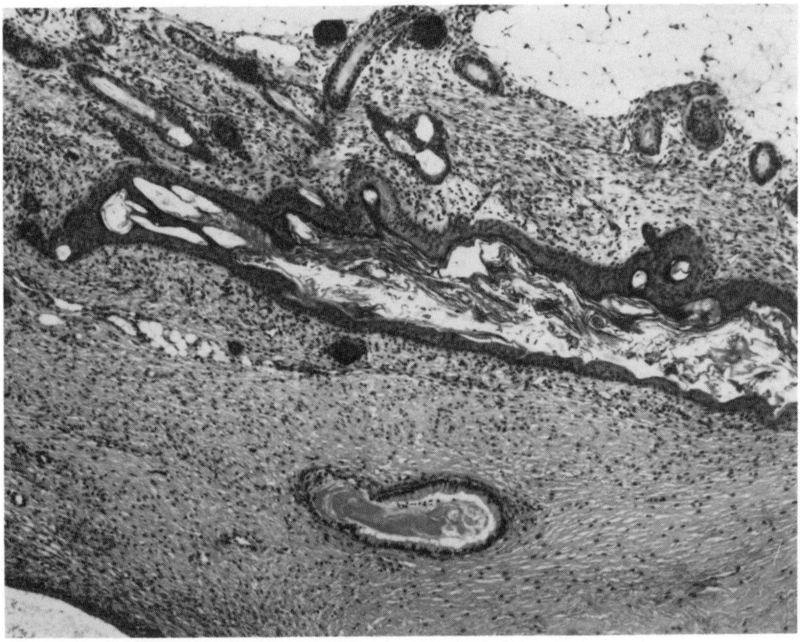

Figure 9. Mature teratoma.

2.2.6 *Teratoma*

Teratomas are benign tumours composed exclusively of non-proliferating somatic tissues. In the simplest form, teratomas present as so-called dermoid cysts and are composed of skin and appendages (*Figure 9*). This is the most common germ cell tumour of the ovary. In the testis, teratomas are less common than teratocarcinomas and the diagnosis of benign teratoma of the testis should be made cautiously only after examination of numerous sections to exclude the oversight of a small focus of malignant stem cells. Serologic tests for AFP and HCG are also useful in this respect and any tumour presenting with elevation of these two serum markers should be considered as teratocarcinoma rather than teratoma even if the pathologist finds no malignant components on histological examination of the resected tissues.

2.2.7 *Immature teratomas*

Immature somatic tissues found in most teratocarcinomas carry the potential for malignant growth. However, it is uncommon for the immature cells forming foetal tissues in teratocarcinomas to proliferate as rapidly as the EC cells, and in most cases EC cells are the only life-threatening malignant cell population in these tumours. However, there are solid teratoid tumours, especially in the ovaries of girls and young women, which are devoid of EC cells but still act as malignant neoplasms (7). In these immature teratomas, the malignant stem cell population consists of neuroectodermal stem cells (*Figure 10*). Although these cells are committed and destined to differentiate into neural and glial cells, they have a tendency for uncontrolled growth and proliferation, may metastasize and, for all practical reasons, behave like malignant cells. The extent of

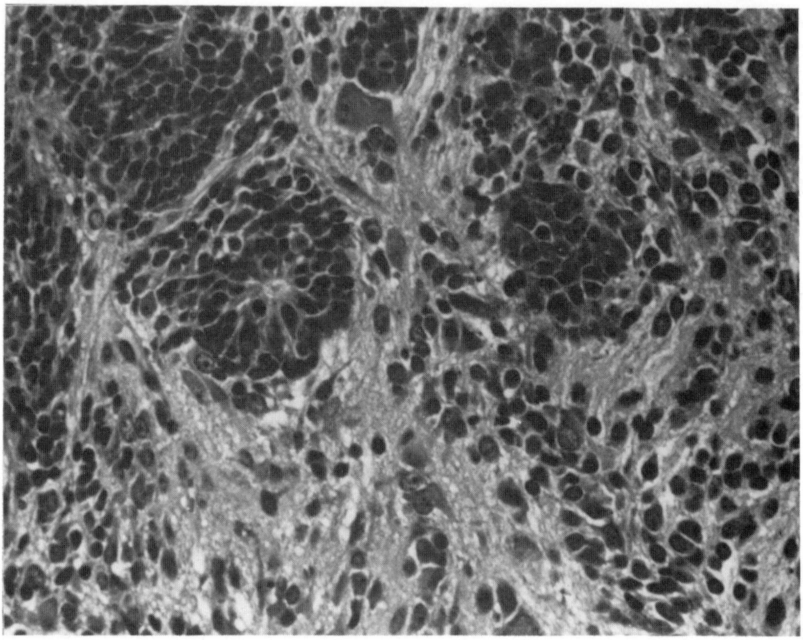

Figure 10. Immature (neural) teratoma.

malignancy of neuroectodermal stem cells varies from one case to another and probably reflects the extent to which these cells retain their potential to differentiate relative to the extent of their acquired malignant phenotype.

2.2.8 *Malignant alteration of benign teratomas*

Benign teratomas which have not been surgically removed may undergo malignant transformation that does not differ from equivalent changes taking place in similar somatic tissues in eutopic locations. Thus, skin in a dermoid of the ovary may give rise to squamous cell carcinomas, fibrosarcomas may arise in the connective tissue component of benign teratomas and the glandular elements may be the site of origin of adenocarcinomas. Squamous cell carcinomas arising in cystic teratomas (dermoids) of the ovary are the most common secondary malignancy arising in teratomas (7). These tumours have the same biological characteristics as squamous cell carcinomas arising in the skin or lung and lower female genital tract.

2.2.9 *Morphology of tumours after chemotherapy*

Following the resection of primary germ cell tumours, patients with residual metastatic lesions, which are either not clinically obvious or unresectable due to technical problems, receive chemotherapy. After treatment according to standardized chemotherapeutic protocols, the residual tumour is usually eventually resected. In most instances these distant metastases contain no malignant cells, which have presumably been eradicated by the treatment. The residual tumour masses from metastatic teratocarcinoma consist of fibrotic connective tissue and usually well differentiated glandular structures. Foci of metastatic embryonal carcinoma or yolk sac carcinoma show only fibrosis and residual necrotic areas. Tumours that are resistant to treatment will usually contain not only fibrosis and necrosis, but also viable tumour cells, which may be used for *in vitro* explantation.

2.3 **Handling of tissue for morphological examination**

For any morphologic study it is essential that the tissue be properly prepared and preserved. Routine pathological diagnosis is usually made on histological slides prepared from formalin-fixed, paraffin-embedded tissue. Some additional histochemical studies can be performed on the same tissue, but since most enzymes do not survive standard histological preparatory techniques and many immunohistochemically detectable antigens are destroyed by these procedures, it is recommended that all histochemical examinations be performed on cryostat sections. It is therefore advantageous to save freshly frozen tissue for special studies if they are indicated or desired. Similarly, electron microscopic studies can be performed on formalin-fixed material, and even on tissue already embedded in paraffin, but optimal results are obtained with properly fixed and properly handled tissue earmarked for electron microscopic examination. Thus, ideally, any surgically resected tumour or experimentally induced xenograft (Section 3.4) to be studied by special techniques should not only be fixed for light microscopy, but should also be prepared for electron microscopy and freshly frozen for enzyme histochemical or immunohistochemical studies.

2.3.1 *Fixation*

Tissue surgically removed from the patient or excised from a nude mouse should be sampled as soon after removal as possible, ideally within several minutes. In clinical practice this may not be practicable and most tumours are obtained 15 min to several hours after excision. Although such tumours may not seem suitable for electron microscopy, tissue obtained even 24 h after removal may be preserved well enough for adequate histological and even histochemical studies.

For light microscopy, place several tissue slices not thicker than 0.5 cm in the same bottle, immerse in 0.1 M phosphate-buffered (pH 7.3) 3.8% formaldehyde and fix overnight. The container should contain 10–20 times more fixative than tissue. Formaldehyde solution can be made fresh from paraformaldehyde powder, or by diluting 1:10 the standard saturated solution of formaldehyde which is commercially distributed as formalin (formalin is 38% formaldehyde). Tissue may be stored in formalin indefinitely provided that the bottle is tightly stoppered or, even better, sealed off with wax to prevent evaporation of the fixative.

For electron microscopy, cut the tissue into cubes approximately 1–2 mm in diameter. Ideally, place a slice of tissue into a drop of fixative as soon as possible and cut the tissue under the fluid with a sharp razor blade. Immersion in the fixative is important to prevent drying out of tissue and the use of a sharp razor blade is important to avoid introducing crush artifacts. Handle the tissue gently using forceps and take care not to compress or squeeze the tissue. If a portion of tissue is accidently squeezed, discard that part and only transfer properly handled cubes to the container with fixative. Transfer the small cubes either by touching them lightly with the forceps or by pushing them from the cut surface together with the drop of fixative. Wooden sticks can also be used, rolling them over the drop of fixative to let the tissue attach to the wood. The fixative recommended for routine electron microscopy contains 2.5–4% glutaraldehyde in a neutral buffer such as 0.1 M phosphate or cacodylate buffer. Ideally, the fixation should be performed using chilled fixative on ice (4°C). Small pieces of tissue are adequately fixed in 3–5 h, but the tissue may be kept in the fixative and chilled to 4°C for longer periods. Freshly prepared, buffered paraformaldehyde is the fixative of choice for ultrastructural immunohistochemistry and the recommended time of fixation is 3 h at 4°C.

For enzyme histochemical and immunohistochemical studies freeze freshly-obtained tumour tissue. Best results are obtained by snap freezing a fragment of tissue, measuring approximately 2 × 0.5 × 1 cm, in methylbutane pre-cooled in liquid nitrogen to −70°C. Mount the tissue on cardboard and cover it with cryogel such as OCT (Lab-Tek Products, Naperville, IL). Pick up the specimen with a pair of forceps and immerse into the methylbutane for 20–30 s. The cryogel becomes white and firm and the tissue becomes pale. The specimen can either be mounted on a chuck for sectioning in the cryostat, or stored at −70°C. For storage the entire specimen must be covered with cryogel. To prevent drying out wrap the cryogel-covered tissues in aluminum foil and store them in tightly closed containers. We also add some ice, scraping it from the sides of the freezer, to assure humidity inside the storage chamber.

Enzyme histochemical and immunohistochemical techniques to be applied to cryostat sections can be chosen at will and the reader is referred to one of the standard texts dealing with these techniques (20–22). A few cautionary notes are added.

(i) Some enzyme histochemical techniques work only on *unfixed* tissue. Others work on tissues fixed with some, but not with other, fixatives. It is thus important to establish which preparatory technique should be used prior to incubation.

(ii) Immunohistochemical techniques may be applied to unfixed or fixed tissues but some antigens are destroyed at different rates by different fixatives. Thus the most suitable fixation protocol must be determined empirically by trying out several fixatives and varying the length of fixation.

(iii) The optimal dilution of the primary and secondary antibody for indirect immunochemistry should be determined for each antibody by a so-called 'checkerboard titration' by varying the dilution of both antibodies in a pre-determined manner. Most commercial antibodies that are labelled with a fluorochrome work at dilutions of 1:50 to 1:100. The concentration of the primary antibody may be titrated in a range from 1:10 to 1:500, but some antibodies give satisfactory results even at dilutions of 1:10 000.

(iv) Chromogens visible by light microscopy have some advantages, whereas antibodies labelled with fluorescent dyes have others. Slides stained in immunoperoxidase or immunoalkaline phosphatase may be observed by light microscopy and provide a permanent record. On the other hand, these slides do not always exhibit enough contrast, or chromogen deposits may obscure details so that photodocumentation may be suboptimal. Fluorescence microscopy slides are not permanent; they fade under u.v. light, and should be stored cold. On the other hand, this technique is more sensitive for screening of unknown antibodies,

antibodies of low affinity and those that bind selectively to intracytoplasmic organelles. Cellular details, e.g. cytoplasmic filaments, can be visualized much better with fluorescent techniques. Double staining may be more conveniently performed with fluorescein isothiocyanate (FITC) and rhodamine isothiocyanate (RITC) or Texas red than with any chromogen adapted for light microscopic immunohistochemistry. Finally, photographic documentation of the results, especially if one deals wih weakly reactive antibodies, is much easier on immunofluorescence microscopy than by any other approach.

(v) One should not forget that immunohistochemistry is an empirical discipline and that proper positive and negative controls must be included on every occasion.

3. DERIVATION OF CELL LINES

3.1 Summary of existing lines

3.1.1 *Xenograft lines*

Tumour-derived cells may be cultured by serially passaging xenografts in nude (athymic, *nu/nu*) mice (see Section 3.4). Extant xenograft lines derived from human germ cell tumours are listed in *Table 2*. Xenografts resemble the tumours from which they were derived, both morphologically and functionally, although the level of morphological differentiation is often reduced in the xenografts compared with the original tumour. Most of the xenografts with a yolk sac component produce AFP in the xenogeneic host. The line HX36, which has a choriocarcinoma component, produces HCG. Pure seminoma and germ cell tumour-derived choriocarcinoma lines are not currently available. HX36 was derived from a combined tumour containing a teratoma component, but somatic differentiation was lost in the xenograft. Somatic differentiation does occur, however, in xenografts from some established cell lines *in vitro* (*Table 3*).

Table 2. Re-transplantable human teratoma xenografts.

Line	*Reference*	*Tumour explant histology*	*Xenograft host*	*Xenograft histology*
HYST	23	Y	NM	Y
OE	24,25	Y	NM	Y
TE	24,25	Y	NM	Y
TT-I-JCK	24,25	Y	NM	Y
EST-1	26,27	Y	NM	Y
EST-2	26,27	Y	NM	Y
ECCS	28	EC	NM	EC
HX36	29	EC, C	ISM	EC, C
HX39	30,31	EC	ISM/NM	EC
HX53	30,31	S, Y	ISM	'S', Y
HX57	30,31	Y	ISM/NM	Y
HX67	30	S, EC, T, Y	ISM	S, EC, Y
HX84	30,31	EC	ISM/NM	EC
HX111	30,31	EC	ISM	EC
HX112	30,31	Y	ISM	Y

Abbreviations: C = choriocarcinoma; EC = embryonal carcinoma; ISM = immunosuppressed mouse; NM = nude mouse; S = seminoma; T = teratoma; Y = yolk sac tumour.

Table 3. Human teratoma cell lines established *in vitro*.

Cell line	*References*	*Tumour explant histology*	*Chromosomes*	*Xenograft*		*Establishment conditions*
			Mode (range); Specific abnormality	*Host*	*Histology*	
Testicular						
TERA-1	32,33	EC, T	58(43–84) XO; m 1			Fragments trypsin McCoy's 5A; 15% FCS; collagen-coated flasks
TERA-2	32–34	EC, T	61XY(55–67); m 1; i(12p)?	NM	EC, T	
Susa	35	EC, T	58 including Y			Fine fragments DMEM; 10% FCS and 10% CS; STO feeders (mitomycin C or irradiated)
NEC-8	36	EC	85–88(70–183)	H	EC	Collagenase; RPMI-1640; 20% FCS
833KE	33,37–39	S, EC, T, C	57(43–62) X and Y; m 1, i(12p)?	NM	EC	Coverslip method; RPMI-1640; 15% FCS; 10% TPB
577MR	33,39	EC, T	55(42–105) X and Y; m 1			Fragments; RPMI-1640; 15% FCS; 10% TPB; explanted in a small amount of medium or explants under coverslips, or explants/tumour cells on feeder of mitomycin C-treated HLF
577ML	33,39	EC, T	55(46–94) X and Y; m 1			
577MF	33,39,40	EC, T	50(39–103) X and Y; m 1	NM	Undiff. Ca	
1075L	44	EC, C				
1156Q	33,39,40	S, EC, C	54(45–62) X and Y; m 1			
1218E	33,39,40	S, EC	106(83–117) X and Y; m 1			
1242B	33,39	EC	83(48–105) X and Y; m 1			
12550	33,39	S, EC, T	51(33–102) X and Y; m 1			
1411H	46	EC, T, Y		NM	EC, Y	
14111HRQmet	46	EC, Y		NM	EC, Y	
1428A	33	EC, Y	61(53–102) X and Y; m 1			
1777N-Pr	45	EC, T				
1777N-RP	45	EC				
2044L	33,39	EC	53(47–108) X and Y; m 1			
2061H	33,39	EC, T	88(56–107) X and Y; m 1			
2102Ep	33,39–41	EC, T, Y	57(46–140) X and Y; m 1	NM	EC	
2102ER	33,39	EC, T	54(44–60) X and Y; m 1			

LICR-LON HT-1	42	EC, T, C				Minced DMEM; 10% FCS; gelatin-coated flasks; initial growth only in presence of non-malignant cells from same tumour
LICR-LON HT-3	42	EC, Y				
LICR-LON HT-5	42	S, EC, T, Y				
LICR-LON HT-7	42	EC, Y				
LICR-LON HT39/7	42,43	EC	46XY(42–62)	ISM	EC	Via nude mouse xenograft
ITO	47	S, T, Y	45	H; NM	EC	
NEC 14	47	EC, Y	56 (wide range)	H	EC	
NEC 15	47	EC, C	52 (wide range)	H	EC	
Scha	Unpublished	S, EC	62 X and Y, m 1; 'i(12p)'			Minced; RMPI-1640; 15% FCS
Huttke	Evans, personal communication	S, EC, T				
Ovarian						
PA-1	48,49	IT, T	46, XX, t(15;20)	NM	IT, T	Ascites in MEM; 10% FCS
IMa	50	C	56, X(51–59)	H	C	Minced; RPMI-1640; 20% FCS
YK	51	IT		NM	EC	Minced; trypsin; RPMI-1640; 10% FCS
HUOT	52	IT,T	52–56(43–256) three markers	NM	Anaplastic Ca	Minced; pronase; Ham's F-12; 15% FCS
Extra-gonadal						
U-1161	53	S, IT, T	78,XXYY(50–160), 14 markers, including m 1			Fragments MEM; 10% FCS
1618K	54	EC		NM	EC, T	Mitomycin C-treated HLF feeder
HaZ-1	55	T	46,XY,t(6;6;11)			Minced; collagenase, RPMI-1640; 20% FCS
HaZ-2	55	IT	46,XY,t(6;6;11)			
HaZ-3	55	IT	46,XY,t(6;6;11)			
HOGT	56	T	46,XY	NM	T	Minced; pronase; Ham's F-12; 20% FCS; cloned in agar

Abbreviations: C = choriocarcinoma; Ca = carcinoma; CS = calf serum; EC = embryonal carcinoma; FCS = foetal calf serum; H = hamster; HLF = human lung fibroblasts; ISM = immunosuppressed mouse; IT = immature teratoma; m = marker; NM = nude mouse; S = seminoma; T = teratoma; TPB = tryptose phosphate broth, Y = yolk sac carcinoma.

3.1.2 *Cell lines in vitro*

Human teratoma cell lines established *in vitro* are listed in *Table 3* (only references to the establishment and initial characterization of the cell lines and subclones are provided). Most of the lines were derived from primary or metastatic testicular cancers. These tumours, probably with the exception of infantile orchidoblastoma (yolk sac carcinoma), are virtually always aneuploid and often have chromosome numbers in the hypotriploid range. Many of the cell lines have non-random abnormalities in chromosome 1. However, the i(12p) isochromosome, found in short-term tissue culture of primary and metastatic testicular germ cell tumours (57,58), is considered specific for these tumours, but has not been clearly demonstrated in the established cell lines. A similar marker identified as del(12)(q13) in TERA-2 (34) might be an i(12p), since the two are difficult to distinguish. We have also identified a similar marker chromosome in the unpublished cell line Scha, but we believe that in this case the marker is not an i(12p) chromosome, but a translocation of 12p to another chromosome fragment, possibly a 22q.

To date only four cell lines have been established from ovarian germ cell tumours. The cell line published by Kimoto *et al.* (59) was probably not derived from a germ cell tumour; the histology and the age of the patient are more compatible with common epithelial cancer. Malignant germ cell tumours of the ovary, like ovarian dermoid cysts, are usually diploid. Immature teratomas with a higher grade of malignancy may be slightly aneuploid (60). Specific chromosomal markers have not been established for ovarian germ cell tumours.

Very few cell lines have been derived from extra-gonadal germ cell tumours. In both males and females extra-gonadal germ cell tumours from different primary sites are usually euploid, most often diploid (61).

A comparison of the histology of original tumours and xenografts derived from cell lines *in vitro* shows that many cultured cells have lost their capacity for differentiation. Most of the grafted cells develop into undifferentiated tumours composed of cells with an EC cell morphology. Exceptions include TERA-2, PA-1, 1618K and HOGT, which show some somatic differentiation. The only clonal cell lines that so far show convincing somatic differentiation *in vitro* were derived from TERA-2 and are discussed in Section 4. The capacity for spontaneous differentiation of cell lines *in vitro* is predictable from the histology of the original tumour. Compared with the original tumour a cell line will not show new differentiation lineages; at best the same range of differentiation is seen, but in most instances less differentiation is evident. The differentiation in metastases relates in the same way to the differentiation in primary tumours (62). Therefore, for the study of somatic differentiation, cell lines should be established from tumours with a teratoma component. Cell lines derived from tumours with extra-embryonic differentiation can be expected to show extra-embryonic differentiation *in vitro*.

3.2 **Methods for obtaining tumours**

3.2.1 *Specimens*

Ideally obtain tumour specimens for tissue culture from patients before the start of chemo- or radiotherapy regimes. Specimens should be submitted sterile and the period between surgical removal and preparation for tissue culture should be as short as possible. To prevent dehydration place the specimen in a humidified container. Avoid rapid changes in temperature. Small (incisional) biopsy specimens are best kept submerged in pre-warmed complete tissue culture medium (see Section 3.3).

3.2.2 *Sampling of tumours*

Handle the tumour samples for tissue culture in a laminar flow hood, preferably in collaboration with the pathologist who is making the diagnosis. Make frozen sections during sampling to check that tumour tissue rather than connective tissue is being taken. In addition analysis of frozen sections allows determination of whether the tumour has been adequately sampled from viable areas and from the different histological components.

3.2.3 *Storage of cells and tissue*

Guided by the information gained from the examination of frozen sections, remove one or more blocks of $1-2$ cm^3 tissue from the tumour. Disperse about half of the tissue, as described in Section 3.3, for tissue culture. Store the remaining tissue in liquid nitrogen, after making imprints on glass slides for cytology, immunocytochemistry and DNA staining as required. If, before freezing, the tissue is divided into small pieces ($1-2$ mm^3) and immersed in a freezing mixture of 90% foetal bovine serum (FBS) plus 10% dimethyl sulphoxide (DMSO), many of the cells should remain viable.

3.2.4 *Patient data*

Germ cell tumours are frequently composed of many different entities and occur in a variety of clinical settings. For example, yolk sac tumour may be the single component in an infantile orchidoblastoma, or in an ovarian germ cell tumour, but rarely in a testicular germ cell tumour of an adult male. It is therefore important to obtain as much information about the patient and the tumour as possible. An example of a checklist is given in *Table 4*. Cell lines derived from carefully documented tumours can then be studied in the proper epidemiological and etiological context.

Table 4. Desirable data on patients and tumours.

Patient
Identification data (should not interfere with privacy)
Sex
Age
Race
Blood types
Clinical history
Other diseases
Cancer among relatives
Tumour
Site (primary; metastasis from)
Histological classification
Stage
Previous treatments
Serum tumour markers (pre- and post-treatment)
Follow-up

3.3 **Methods for establishing cell lines *in vitro***

3.3.1 *Initiation of tumour cell lines*

The following techniques have been applied for the initiation of germ cell tumour cell lines: spillout of tumour cells; explants; explants under coverslips; and proteolytic enzymatic digestion with collagenase, trypsin or pronase coupled with DNase digestion to remove DNA released by dead cells. Collagenase is the least traumatic proteolytic enzyme; it gives a higher yield and viability than mechanical disaggregation, but the contamination with stromal cells is heavier. Rigorous digestion to obtain single cells is neither necessary nor desirable, when initiating cell lines. For short-term tissue culture, for example for cytogenetic studies, the collagenase digestion technique is recommended. To maximize the chances of obtaining a permanent cell line tumour specimens should be treated using all, or a combination of, the following methods.

(i) *Spillout method*

(1) Place the tumour tissue in a sterile Petri dish, cover it with complete growth medium (Section 3.3.2) and dissect out non-tumour tissue and necrotic areas.
(2) Wash the specimen with medium in a second dish, and transfer to a third dish containing complete growth medium.
(3) Mince the tissue with crossed, sharp surgical blades, or with sharp, curved scissors to obtain tissue fragments about 1–2 mm in diameter.
(4) Decant the medium containing the released cells into a centrifuge tube and wash the cells at least three times by low-speed centrifugaton (500 r.p.m., 5 min).
(5) Resuspend the cells in medium and from a sample estimate cell viability by trypan blue exclusion. Usually not more than 20% of cells are viable.
(6) Plate the cells onto tissue-culturing dishes as desired (Section 3.3.3).

The yield and the viability is low with this method, but stromal contamination is limited. Further disruption of the tissue enhances contamination with non-tumour cells.

(ii) *Explant method.* Some of the tissue fragments obtained after mincing [step (3) above] can be used as explants. One problem encountered with this method is to ensure that the tissue fragments become attached to the culture surface. This can be achieved either by using only a small amount of medium in the tissue culture dish, or by placing coverslips over the explants to prevent them from floating. To do this place the tissue fragments in a small amount of medium between two drops of sterile silicone grease (Dow Corning Corp.) in a tissue culture Petri dish. Press a sterile coverslip gently into the silicone grease, thereby immobilizing the tissue fragments. The number of coverslips that can be used in one dish is limited according to the size of the Petri dish. Change the medium as necessary by aspiration. Coverslips on which pure tumour colonies have grown can be handled separately (see Section 3.3.7).

(iii) *Enzymatic disaggregation.* The remaining minced tissue [step (3) above] contains a large number of viable tumour cells. These are released by collagenase/DNase digestion as follows.

(1) Incubate the minced tissue in a solution of 0.8% (w/v) collagenase (Collagenase II, Worthington Diagnostic Systems Inc.) and 0.002% (w/v) DNase I (Sigma

Chemical Co.) in complete culture medium (supplemented with 5–10% FBS), at 37°C in a humidified atmosphere of 5% CO_2 in air.

(2) Incubate for 2–3 h. Usually the tissue will be digested after this time. Longer incubations, which may be necessary in the case of firm tissue, will not harm the tumour cells.

(3) Disaggregate the digested tissue by pipetting and wash the cells three times with pre-warmed medium by low-speed centrifugation.

(4) From a sample estimate cell viability by trypan blue excision and plate the cells as desired.

(iv) *Handling initial cultures.* Manipulate the explant cultures as little as possible and change the medium only partially, when necessary, in order to allow the tissue to condition the medium. Cultures initiated from minced or collagenase suspensions usually contain debris and floating dead cells which should be removed soon after attachment of the viable cells. In this case initially incubate the cultures overnight, completely change the medium on the following day and transfer the non-attached material to new tissue culture dishes or flasks. This step allows salvage of tumour cells growing in suspension. In subsequent medium changes, the medium is only partially replaced to allow conditioning of the medium by the tumour cells. Patience is very important, as cultures may appear dormant for long periods and yet give rise to cell lines.

3.3.2 *Tissue culture media and additives*

The following commercially available media have been used in the derivation of germ cell tumour cell lines; McCoy's 5A, Dulbecco's modified Eagle's medium (DMEM). Ham's F12, Eagle's minimal essential medium (MEM) and RPMI-1640 (*Table 3*). Supplement these media with 10–20% heat-inactivated FBS, 2 mM L-glutamine, and antibiotics (100 units/ml penicillin and 100 μg/ml streptomycin, or 50 μg/ml gentamycin). When dealing with contaminated specimens also add 2.5 μg/ml of fungizone (amphotericin B; Flow Labs, Gibco) to control mould. Bronson *et al.* (37) recommend adding 10% tryptose phosphate broth (Difco) to the growth medium. Incubate the medium at 37°C in a humidified atmosphere of 5–10% CO_2 in air. It is probably not critical as to which basal medium is used for the derivation of lines.

3.3.3 *Surface coatings and feeder layers*

Tissue culture is performed in commercially available, disposable plastic flasks or Petri dishes. Attachment of tumour cells can sometimes be improved by coating with gelatin or collagen. Alternatively, covering the culture surface with a layer of feeder cells not only offers an adhesive substratum, but may also encourage growth of the tumour cells. Most importantly, the differentiation capacity of many mouse EC cells is better preserved when they are maintained on feeder layers. However, as is evident from the cell lines TERA-2 and PA-1, culturing of human EC cells without feeder cells does not necessarily lead to loss of differentiation capacity. The following feeder cells have been successfully applied in the derivation of germ cell tumor cell lines or subclones; STO-transformed mouse cells (63), 10T1/2 cells (64), human diploid fibroblasts (44) and BHK-21 cells (42). The cell lines LICR-LON HT-1, H-5 and H-7 need fibroblasts derived from the original tumour as feeder cells (42).

(i) *Gelatin coating.* For gelatin coating, incubate the culture surface with a 0.1% (w/v) aqueous solution of gelatin (swine skin, type 1 gelatin; Sigma Chemical Co.) for 2 h at 5°C, and air dry at 40°C for 24 h.

(ii) *Collagen coating.*

(1) Place 5–10 rat tails in 95% alcohol for 10 min. Further manipulation is done under sterile conditions and with sterile materials.
(2) Break the tails, and dissect and disrupt the tendons in distilled water using forceps.
(3) Incubate the released collagen fibres in a closed bottle in 150 ml of 0.1% (v/v) acetic acid for 48 h at 4°C. The collagen fibres will swell up.
(4) Centrifuge for 1 h at 2300 r.p.m. at 4°C and dialyse the supernatant against distilled water for 24–48 h until the collagen solution becomes syrupy.
(5) To coat culture dishes, wet the surface with this solution (1–2 drops per cm^2) and precipitate the collagen on the culture surface by exposure to ammonia vapour.
(6) Wash the coated surface several times with distilled water and finally with medium. If necessary sterilize the coated plates by exposure to u.v. light.

(iii) *Preparation of feeders.* Grow feeder cells of choice to subconfluence and treat with culture medium containing 2 μg/ml of mitomycin C (Sigma Chemical Co.) for 2–3 h. Alternatively expose the cell layers to X-ray or γ-ray irradiation (6000–10 000 rads) to stop growth. Harvest the feeder cells and re-seed on fresh plates at a density of about 5×10^4 cells/cm^2 to obtain a confluent monolayer.

3.3.4 *Growth in agar*

For human germ cell tumour lines the efficiency of colony formation in agar is low and, for the derivation of new lines, growth in agar is generally not the most suitable method. However, Cotte *et al.* (42) have successfully used the following modification of the method of MacPherson (65).

(i) Coat sterile plastic tubes (25 cm^3) with a thin layer of 0.2% (w/v) gelatin to provide good adhesion between agar gel and the walls of the tube.
(ii) Add 5 ml of 0.3% Difco agar in DMEM containing 10% FBS and 10^5 LICR.P1.7 cells (a line developed from a human mid-term placenta) as a feeder layer, and incubate at 37°C for 4 h.
(iii) Add 10 ml of 0.5% agar in DMEM containing 10% FBS and incubate for a further 5 days at 37°C to permit diffusion of the feeder cell products into the top layer of agar.
(iv) Gently layer 5 ml of 0.3% agar in DMEM containing 10% FBS and tumour cells onto the top of the agar layer.

3.3.5 *Inhibition of fibroblasts*

Almost any attempt to derive tumour cell lines is complicated by the growth of normal host cells. Most often these cells, loosely referred to as fibroblasts, overgrow and finally choke the growth of tumour cells, which may have initially looked promising. Unfortunately, there are no sure ways to avoid this. Methods designed to separate fibroblasts from tumour cells in the original tumour-derived cell suspension have the disadvantage of extra manipulations and delay the initiation of tissue culture. Moreover, the preparation of single cell suspensions, on which these techniques depend, is harmful to the tumour cells as growth of tumour cells in primary cultures commonly starts from small clumps, not from single tumour cells. However a number of techniques may be tried in order to deal with fibroblast outgrowth. Collagenase (added to a concentration of 0.5 μg/ml growth medium) can be used to suppress growth of contaminating fibroblasts. Alternatively, loosely attached tumour cell colonies can be selectively harvested under a microscope by judicious trypsinization using a micropipette. The released colonies of cells can be transferred individually to the wells of a 24-well tissue culture plate. Isolated colonies of cells which possess a distinct cell morphology can also be rescued by scraping off the surrounding fibroblasts with a rubber policeman. A variant of these approaches is provided by the coverslip method (Section 3.3.1) since it allows the selection of coverslips with pure tumour cell colonies, free from fibroblasts, from the primary culture. Also, theoretically, when fibroblasts start to outgrow the tumour cells, temporary culture in semi-solid media should suppress the growth of fibroblasts which are anchorage dependent, but this method has not been used in the derivation of germ cell tumour cell lines. Finally it is noteworthy that Cotte *et al.* (42) rescued the EC cells from overgrowth by fibroblasts by regular subculturing. However, three of the EC cell lines that they derived continued to need tumour-derived fibroblasts as autologous feeder cells. Apparently it is not always necessary, or desirable, to get rid of the fibroblasts.

3.3.6 *Cloning*

The only reliable way to derive clonal lines from tumour cell lines is by picking single cells. The clones NT2/B9, NT2/D1 and NT2/D3, for example, were isolated in this way from NTERA-2, a xenograft-derived subline of TERA-2 (34). The cloning procedure is given below.

(i) Pick single cells with a micropipette from a suspension of trypsinized cells under a dissecting microscope, and transfer them individually in small drops of medium to the wells of 24-well tissue culture plates.

(ii) Confirm the presence of only a single cell in each well by microscopic examination and add sufficient irradiated feeder cells to give a confluent monolayer.

(iii) Expand the colonies slowly and maintain a high cell density splitting the cells at a ratio of between 1:2 and 1:4.

(iv) Harvest the cells by scraping and, if necessary, add feeder cell in the earliest subcultures, even when the original tumour cell line was feeder independent.

3.3.7 *Distinguishing tumour cells from non-tumour cells*

Routine microscopic examination of growing cultures is essential for distinguishing tumour cells. EC cells and EC colonies have a distinct epithelioid morphology in tissue culture (see Section 4), which is not mimicked by normal host cells. In comparison, tumour cells derived from non-EC elements in germ cell tumours may sometimes have a fibroblastoid appearance and may therefore be confused with host-derived cells. In the case of testicular germ cell tumours, measurement of DNA content and karyotyping are the most reliable indicators of the nature of the cells. Normal host cells have a euploid DNA content and a 46 XY karyotype which is stable in culture. Tumour cells are aneuploid with highly abnormal karyotypes, often with abnormalities in chromosome 1 (*Table 3*). However, extra-gonadal (61) and ovarian germ cell tumours (60) are commonly euploid or near-euploid, and may have a normal karyotype (*Table 3*). For these tumours the phenotypic markers described in Sections 2 and 4 should be used to establish the nature of cells in culture. However, even the extra-gonadal-derived cell lines Haz-1, Haz-2 and Haz-3, although euploid, have a balanced translocation indicating their tumour cell nature (55). By measuring DNA content, cells in culture can be compared with cells in the original tumour, even when only paraffin-embedded tissue is available. A brief description of the methods used for measuring DNA content in fresh or frozen tumour tissue or cells, and in paraffin-embedded tissue, is given below but the indicated references should be consulted for details.

(i) *Measuring DNA content in fresh or frozen tumour tissue or cells (references 66–69).* Isolate nuclei using a detergent–trypsin method and stain with propidium iodide (Calbiochem, San Diego). Use chicken and trout red blood cells as internal reference standards. Measure DNA using a flow cytofluorimeter.

(ii) *Measuring DNA content in paraffin-embedded tissue (70).* Cut 30 μm sections and deparaffinate. Digest the tissue sections using 0.5% (w/v) pepsin (Sigma Chemical Co.). Stain isolated nuclei with 4,6-diamidino-2-phenylindole dihydrochloride (DAPI) (Boehringer Mannheim), and measure the DNA content using a flow cytofluorimeter.

(iii) *Karyotyping.* Karyotyping of solid tumours is a specialist technique and should be performed in collaboration with an experienced cytogeneticist. Healthy, fast growing, subconfluent cultures are required. However, efforts should be made to obtain karyotypes as soon as possible after establishment of the tumour in culture, since changes may occur following extended growth *in vitro*. If possible the normal karyotype of the patient should be established. This is most simply achieved by karyotyping cultures of peripheral lymphocytes, stimulated with phytohaemagglutinin.

3.3.8 *Derivation of non-tumour cells*

For comparative studies it is useful to have other host cells in addition to tumour cells. Usually, in the first stage of the establishment of tumour cell lines, contaminating fibroblasts are present in the cultures. Store samples of these fibroblasts frozen in liquid nitrogen.

Additionally, Epstein-Barr virus (EBV)-transformed B-cells are very useful as comparative material. These may develop spontaneously in cultures, probably more often when using the coverslip method (44). These lymphoid cells grow in suspension and can be harvested easily and stored in liquid nitrogen. Unlike fibroblasts, B-cells are, themselves, transformed cells and have the advantage of unlimited growth in culture.

3.3.9 *Maintenance culture*

In most of our studies we have cultured cells in the high glucose formulation of DMEM (purchased from any of the major suppliers of tissue culture materials) and supplemented with 10% (v/v) FBS. Others have used different media (e.g. RPMI-1640) with equal

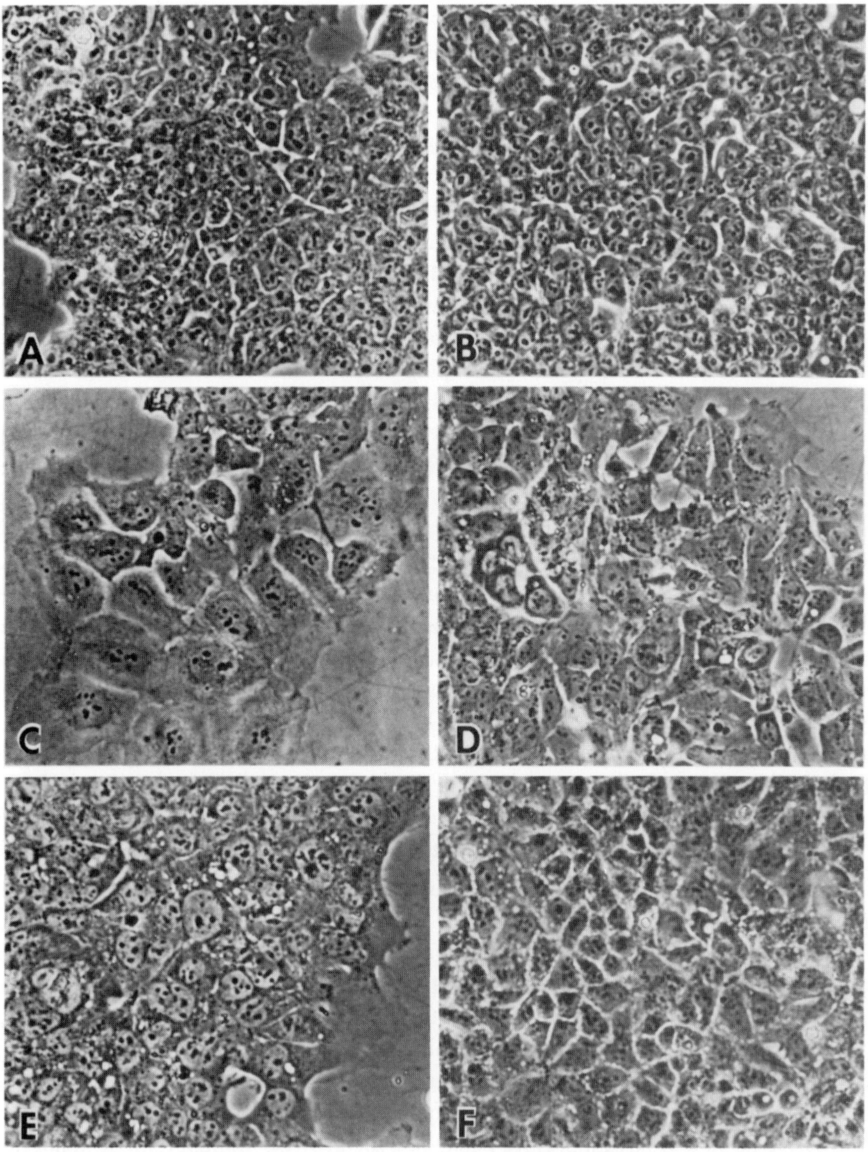

Figure 11. Phase contrast photomicrographs of human teratocarcinoma and related cells growing *in vitro*: **A** and **B**, N2102Ep cl.2A6 cells (41) and NTERA-2 cl.D1 cells (34), respectively, both growing at high density and showing a typical EC morphology. **C** and **D**, N2102Ep cl.2A6 cells at low density and NTERA-2 cl. D1 cells grown for 7 days in 10 μM retinoic acid, respectively; note their distinct morphology compared with their parental EC cells. **E**, BeWo, gestational choriocarcinoma cells (71); **F**, 1411H, yolk sac carcinoma cells (46).

success. However, comparative studies have not been made and it is worth remembering that, especially in the case of germ cell tumour cell lines, which are particularly sensitive to culture conditions, cells may not behave identically in different media. *Figure 11* shows the morphology of cells, by phase contrast microscopy, in cultures of the two human EC cell lines 2102Ep (clonal subline N2102Ep cl.2A6; ref. 41) and TERA-2 (clonal subline NT2/D1; ref. 34) as well as the yolk sac line 1411H (46) and the choriocarcinoma line BeWo (71). Note the uniform morphology of small round cells with little cytoplasm and nuclei containing few prominent nucleoli in the EC cultures, in contrast to the cells in the 1411H and BeWo cultures. We have found that to retain this typical phenotype of the EC cell lines it is necessary to maintain cultures at high cell densities. At low cell densities, morphological changes and some differentiation may occur (see Section 4.3). In the case of 2102Ep, cultures should be harvested using trypsin EDTA [0.25% w/v trypsin, 2 mM EDTA in Dulbecco's phosphate-buffered saline (PBS), lacking calcium and magnesium] and the cells re-seeded at 5×10^6 cells per 75 cm^2 flask. However, trypsinization is undesirable for TERA-2 cells and cultures should be subcultured by physically scraping cells from the substrate, and confluent cultures should be diluted in a ratio of 1:3 – 1:4 to ensure high cell densities. For scraping cultures the following method is simple and convenient. Cover the cell layer with about 5 – 10 ml medium (for a 75 cm^2 flask) and tip into the flask about 20 acid-washed sterile glass beads (3 mm diameter). Roll the beads in the flask just vigorously enough to detach the cells in clumps. Provided some medium is present and violent agitation is avoided, the cells are not damaged by this procedure. Similar approaches should be adopted for other newly isolated cell lines to minimize the chance of losing any pluripotent EC cells through differentiation. We have not used feeder layers for either 2102Ep or TERA-2 EC cells, except when cloning the cells. However, certain mouse EC lines are best prevented from differentiation by culturing on fibroblast feeder layers. If human EC cell lines have been isolated on feeder layers care must be taken in adapting them to grow without feeder layers to ensure that this does not result in loss of the EC phenotype.

3.4 Growth of tumours in xenogeneic hosts

The use of human germ cell tumor xenografts grown in immunodeficient animal hosts are important for both the establishment of xenograft lines and for the characterization of *in vitro* cell lines. Nude mice, homozygous for the recessive *nu* mutation, are congenitally athymic and are consequently deficient in many aspects of their cellular immune system. Accordingly they often fail to reject tissue even from other species and are commonly used as hosts for growing human tumours. However, because they are also very susceptible to infection they must be kept under very clean, though not necessarily sterile, conditions and so are expensive both to purchase and to maintain. Alternative hosts that have therefore been used for growing human tumour xenografts include mice that have been neonatally thymectomized and subsequently treated with cytosine arabinoside and irradiation (72), mice treated with cyclosporin A (73), and the cheek pouch of hamsters. We have no experience of the latter techniques and refer the reader to the appropriate references for practical details. For additional information on the use of nude mice consult the review on this subject by Giovanella and Fogh (74).

3.4.1 *Derivation of xenograft lines*

Xenografting of human germ cell tumours is carried out to provide *in vivo* models of tumours in human patients and also to serve as a source of material for subsequent culture *in vitro*. However, in the experience of Giovanella and Fogh (74), nude mouse-grown human tumours do not provide advantages over original tumours as source material for tissue culture. If a cell line is grown *in vitro* from a xenograft, the origin of the growing cells must be established by isozyme or chromosome analysis; murine host cells may be transformed and grow as continuous lines, ultimately completely replacing the human tumour cells.

Initiate xenografts by implanting small tumour fragments (1 − 2 mm^3) subcutaneously using a trocard. If the tumour fails to grow subcutaneously the kidney capsule may provide an alternative and more suitable site. However, the surgery involved (see Chapter 1) is more time consuming and may introduce a greater danger of infection in the immunosuppressed host. Accordingly we recommend first trying the subcutaneous site, where tumour growth is also easier to follow. Following implantation monitor the tumour growth at weekly intervals. Kill the host and remove the tumour when it reaches 1 − 2 cm in diameter. Transfer pieces of the tumour (~1 − 2 mm^3) to a new host. Note that if the tumours are allowed to become larger than the recommended size much tissue is lost to necrosis.

3.4.2 *Xenografting of human germ cell tumour lines established in vitro*

Most experimenters try to produce xenograft tumours from *in vitro* cell lines, both to test the tumorigenicity of the cell line and to characterize its morphological and functional differentiation *in vivo* in terms of the standard histology of human germ cell tumours. However, many tumour lines do not produce tumours in immunodeprived hosts. The reasons for this are unclear, but cell lines that do produce xenograft tumours may usually be serially passaged with relative ease.

Xenograft tumours from cell lines have been grown in the same range of hosts as primary xenograft lines. Inject cells from *in vitro* cultures in a small volume of fluid (e.g. PBS or medium), or insert a cell pellet using a trocard. If injection of suspended cells is unsuccessful, the implantation of gelatin sponges containing large colonies of tumour cells may provide a useful alternative way of initiating a tumour (43). Monitor tumour growth as for primary xenografts.

Tumours produced in nude mice from human EC cell lines established *in vitro* usually have the typical morphology of pure embryonal carcinoma. NTERA-2 xenografts are, however, unique in so far as they consist not only of EC cells, but also contain various somatic tissues such as neural, muscular and glandular structures (34).

4. THE PROPERTIES OF ESTABLISHED LINES

Many cell types may be present in a specimen of a germ cell tumour and be represented in cell lines grown out from that tumour. In newly established cell lines, therefore, it is important to document the cell types present and, if possible, to establish clonal sublines representative of single cell types. In the mouse, EC cells are the stem cells capable of differentiating into the various cell types found, and it seems likely that the same is true of human germ cell tumours. Thus most attention is paid to this cell type. Other cell types from the germ cell tumours provide important sources for comparative data, and may indeed be important when studying clinical aspects of this disease, but investigation of the processes whereby a pluripotent stem cell differentiates into multiple cell types provides a unique goal for studies of germ cell tumours. Unfortunately the very fact of differentiation means that these stem cells will tend to be lost unless ways to control their differentiation are achieved. The usual result is the selection of EC cells that have lost their ability to differentiate. Indeed of all the probable human EC cell lines that have been isolated, few have yet been shown to possess a marked capacity for differentiation into a defined somatic cell type (e.g. the differentiation of TERA-2 EC cells to neurons). Of the others, only limited differentiation to extra-embryonic cell types, or, under certain conditions, phenotypic changes that have not been correlated with a known embryonic or extra-embryonic cell type, have been found. In this section we discuss methods for defining human EC cells in culture, and distinguishing them from other cells found in germ cell tumours, by using as examples the properties of four established cell lines. The morphology of these cells *in vitro* is illustrated in *Figure 11*. Two of these lines, 2102Ep and TERA-2 are composed of human EC cells. Cloned lines from 2102Ep form pure EC tumours in nude mice and, under certain culture conditions, differentiate to a limited extent into trophoblastic giant cells. Cultures of TERA-2, on the other hand, often contain multiple morphologically distinct cell types and from this line we have isolated cloned sublines (e.g. NTERA-2 cl D1; abbreviated NT2/D1) which can be maintained as nearly pure EC cell cultures but which differentiate extensively, either spontaneously during tumour formation in nude mice, or after exposure to RA in culture. The third line we shall use for comparision, 1411H, exhibits a marked difference in phenotype in culture and forms yolk sac tumours in nude mice. Although clones of 1411H have been reported to form mixed EC and yolk sac tumours, it seems likely that the 'EC' elements of 1411H tumours are different from the EC tumours produced by 2102Ep or the EC elements in tumours of TERA-2; this reflects the problems associated with morphological studies and points to the necessity for developing more objective criteria for classifying cell types. The final cell type included for comparison is BeWo, a gestational choriocarcinoma cell line (71). Strictly, BeWo was not derived from a germ cell tumour, but it is representative of a cell type common in human germ cell tumours. When pure choriocarcinoma lines of this origin are not available, lines such as BeWo can provide useful data about what may be expected of these cells.

4.1 Markers for human EC cells in culture

4.1.1 *Cell surface antigens*

Cell surface antigens provide convenient markers of cell type since their expression can be monitored rapidly and quantitatively using small numbers of cells. Also, using fluorescence-activated cell sorting (FACS), these antigens may be used to monitor the proportions of subsets of cells in mixed cultures and even to purify viable cells of particular types for further study in isolation (see also Section 4.4). The reagents used for this work are now, almost exclusively, monoclonal antibodies. We shall concentrate only on the antibodies we have defined already, and the techniques in which we have used them. *Table 5* lists several antibodies that we have used to recognize antigens characteristically expressed by human EC cells. Several others have also been investigated by other researchers. As discussed in Section 1, histological studies of human

Table 5. Monoclonal antibodies detecting cell surface antigenic markers of human EC cells.

Antibody			*Antigen*	*Expression by*				*References*
Name	*Class*	*Species*		*2102Ep*	*NT2/D1*	*1411H*	*BeWo*	
MC631	IgM	Rat	SSEA-3	+	+	–	–	75,76
MC813-70	IgG_3	Mouse	SSEA-4	+	+	–	–	77
TRA-1-60	IgM	Mouse	TRA-1-60	+	+	–	–	78
TRA-1-81	IgM	Mouse	TRA-1-81	+	+	–	–	78

germ cell tumours have indicated that cells recognized as EC cells often express the four antigens defined by the monoclonal antibodies listed in *Table 5*. Similarly, cells thought, on various grounds, to be EC cells in several human germ cell tumour-derived lines also express these antigens. Thus N2102Ep cl.2A6 and NT2/D1 EC cells express all four antigens. However, exceptions to the rule are being recognized, although it is not yet clear whether such exceptions represent rare variants (mutants?) with no other distinguishing features from the 'standard' EC cell, or whether they will eventually permit us to define distinct subsets of EC cells with different biological properties. As we have noted already, the histological examination of 1411H tumours reveals the presence of EC elements. As 1411H cells lack all the four antigens listed in *Table 5*, exhibit a quite different morphology *in vitro* and exhibit features of yolk sac carcinoma cells, we believe that these cells represent a different class from those represented by 2102Ep and NT2/D1 EC cells. However, we and others have identified in the TERA-2 line, cells that in many respects including morphology, capacity for differentiation and expression of the TRA-1-60 and TRA-1-81 antigens are indistinguishable from NT2/D1 EC cells; nevertheless they do not express SSEA-3 and SSEA-4. This emphasizes that, in our current state of knowledge, it is unwise to draw conclusions concerning the phenotype of cells from the expression of one parameter alone.

Expression of the EC antigens described here, or other surface antigens discussed in Section 4.4, may be readily determined by indirect immunobinding assays, using a second antibody labelled with either ^{125}I (radioimmunoassays), or with fluorescein or rhodamine (immunofluorescence assays). The following protocols are used in our laboratories.

(i) *Radioimmunoassay (RIA)*

(1) Harvest the cells using trypsin EDTA and prepare a single cell suspension containing 2×10^6 cells/ml in medium buffered at pH 7.2 with 10 mM Hepes and containing 5–10% FBS.

(2) In the wells of a round-bottomed 96-well plate (e.g. Costar, Falcon) mix 50 μl of cell suspension and 50 μl of first antibody (MC631, etc.) diluted as required, and then incubate at 4°C, preferably shaking continuously on a plate shaker (e.g. Dynatech Labs). While incubating, the plates should be sealed to prevent spillage from one well to the next. Pre-cut self-adhesive acetate sheets (e.g. Dynatech, Flow Labs) are convenient for this purpose.

(3) Following incubation, pellet the cells by centrifuging the plate at 400 *g* for 2–3 min and aspirate the supernatant. Wash the cells by adding 100 μl of ice-cold wash buffer (PBS containing 5% FBS and 0.1% sodium azide) to each well; resuspend the cells by shaking and pellet them again by centrifuging.

(4) Aspirate the supernatant and wash twice more.

(5) Add 50 μl of wash buffer containing 50 000 c.p.m. of ^{125}I-labelled, affinity-purified goat anti-Ig (specific for species and class of Ig as appropriate to the first antibody), and again incubate the plate at 4°C for 1 h with shaking. Radiolabel suitable commercially available (e.g. Cooper Biomedical) affinity-purified second antibodies to a specific activity of about 50 000 c.p.m./μg protein using the chloramine-T method (79).

(6) After the second incubation wash the cells three times as before and finally suspend in 50 μl of wash buffer. Swab this suspension from each well with a cotton Q-tip and count in a gamma-counter.

Each test should be carried out at least in triplicate, and a prior titration of the first antibody will be required to determine an appropriate dilution (usually the greatest dilution that gives maximal binding on a standard target cell). To estimate non-specific background binding, a test with a negative-control antibody (i.e. one known not to bind to the cells) must be included. We usually use the antibody produced by the myeloma P3X63Ag8 (80) since derivatives of these cells are the usual fusion partners for most hybridomas that have been made.

(ii) *Immunofluorescence assay.* This assay is carried out as the RIA, with the following differences. Firstly, accurate determination of the cell concentration is not as important since the final assay is made on a single cell basis. We have found it convenient, though not essential, to use more cells than in the RIA; typically we use a suspension of 10^7/ml rather than 2×10^6/ml. Secondly, in place of an ^{125}I-labelled second antibody, a FITC- or RITC-conjugated second antibody is used. Good commercial preparations are generally available (e.g. Cooper Biomedical). For most purposes it is sufficient to use the cheaper conjugates made using only partially purified Ig. However, the more expensive affinity-purified antibodies and $F(ab')_2$ fragments may give lower backgrounds. Each preparation must be titred before use.

Follow the procedure for RIA substituting fluorochrome-conjugated antibody for the [^{125}I]antibody. After the second incubation (step 6), the washed cells can be either mounted on a glass slide in 50% glycerol:50% PBS and coverslipped for microscopic assessment of fluorescence, or suspended to about 10^6/ml in wash buffer for flow cytofluorometry. *Figure 12* illustrates the analysis of SSEA-3 expression by NT2/D1 cells using flow cytofluorimetry. For a detailed treatment of flow cytofluorimetry, refer to ref. 81 and ref. 82.

4.1.2 *Intermediate filaments*

In addition to cell surface antigens, internal cytoplasmic proteins may serve as useful markers of cell type. Amongst such markers are the intermediate filament proteins, which comprise part of the cytoskeleton of cells and show marked specificity for different cell types. Monoclonal antibodies to each of the main classes of intermediate filament proteins [cytokeratins, expressed by epithelial cells; vimentin, typically expressed by mesenchymal cells; neurofilaments, expressed by neurons; glial fibrillary acidic protein (GFAP), expressed by gial cells; and desmin, expressed by muscle cells], are now commecially available (e.g. Labsystems Inc.; Miles Scientific; The Amersham Corporation). We have found that human EC cells in culture, as typified by 2102Ep and TERA-2 EC cells, and *in situ* in tumours, express the 40-, 45- and 52-kd cytokeratin proteins characteristic of simple epithelia (17). These cells do not express vimentin, although a few vimentin-positive cells, thought to represent spontaneous differentiated derivatives, are always found in cultures of the cloned 2102Ep and TERA-2 lines. By contrast murine EC cells typically express vimentin rather than cytokeratin. The 1411H and BeWo cells express a similar spectrum of cytokeratin proteins to the human EC

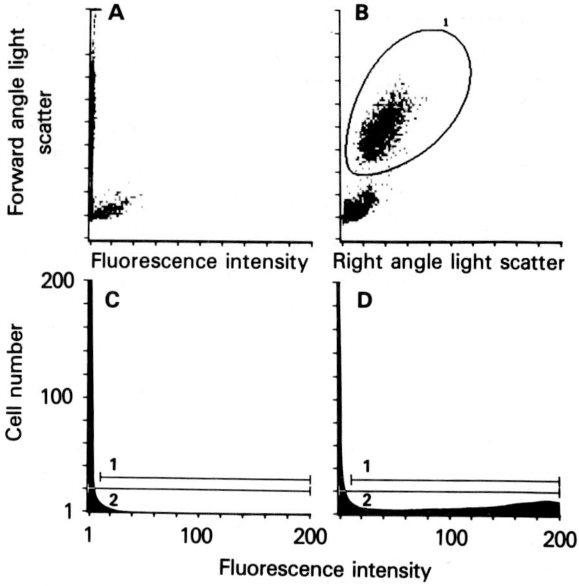

Figure 12 Flow cytofluorometric analysis of SSEA-3 expression by NT2/D1 EC cells, using an Ortho Instruments Cytofluorograf. **Panels A** and **B** show the raw data for three parameters (forward-angle light scatter versus relative fluorescence intensity, **panel A**; and forward-angle light scatter versus right-angle light scatter, **panel B**) collected from cells that had been incubated with the negative control antibody P3X63Ag8 followed by a FITC-conjugated goat anti-IgM antibody; the intensity of each spot increases with the number of cells exhibiting the coordinates of that spot. Forward-angle light scatter is a function of cell size, whereas right-angle scatter depends upon the internal structure of the cells. In **panel B** two main populations of cells are evident. Comparison with **panel A** shows that the lower population exhibits significant fluorescence; this population is composed of dead cells and other debris. On the other hand, the upper population, in the eliptical region marked 1, shows little or no fluorescence; this population consists of single live cells and data collected for these cells alone are shown in the form of a histogram in **panel C. Panel D** shows a similar histogram for cells incubated with antibody to SSEA-3. A comparison of these two histrograms permits determination of the level of expression of SSEA-3. The data are best analyzed by determining the percentage of cells falling within certain regions of the histogram, and also the mean fluorescence intensity of the cells within those regions. We typically use two regions: region 1 on the histogram includes all the cells fluorescing with an intensity greater than an arbitrary threshold, set to exclude most (95–99%) of the cells incubated with the negative control antibody; in this example only 2% of such cells fluoresce in region 1, and these are considered to represent non-specific background staining. On the other hand, after incubation with anti-SSEA 3, 87% of the cells fall in region 1, and are hence scored SSEA-3-positive. Region 2 is drawn to include all the cells in the histogram and is used to obtain the mean fluorescence intensity of the whole histogram; in this example the mean fluorescence, in arbitrary units, after reaction with anti-SSEA-3 is 146, compared with a background fluorescence of 3 after reaction with the negative-control antibody. Note, however, that the value of 146 is an underestimate since many of the cells are fluorescing off scale but are scored by the cell sorter as fluorescing with an intensity of 200. The program on the Ortho Cytofluorgraf permits collection of similar data from several additional regions drawn over any part of the histogram as required. Flow cytofluorometers produced by other manufacturers permit similar types of data analysis.

cells, except that BeWo, like normal trophoblast, also expresses the 54-kd cytokeratin protein. Neither 1411H nor BeWo expresses vimentin (83).

The expression of these intermediate filaments may be conveniently studied by indirect immunofluorescence.

(i) Seed the cells to be tested on coverslips and culture for 2–3 days until a subconfluent monolayer is formed; if the cells are too dense a high background

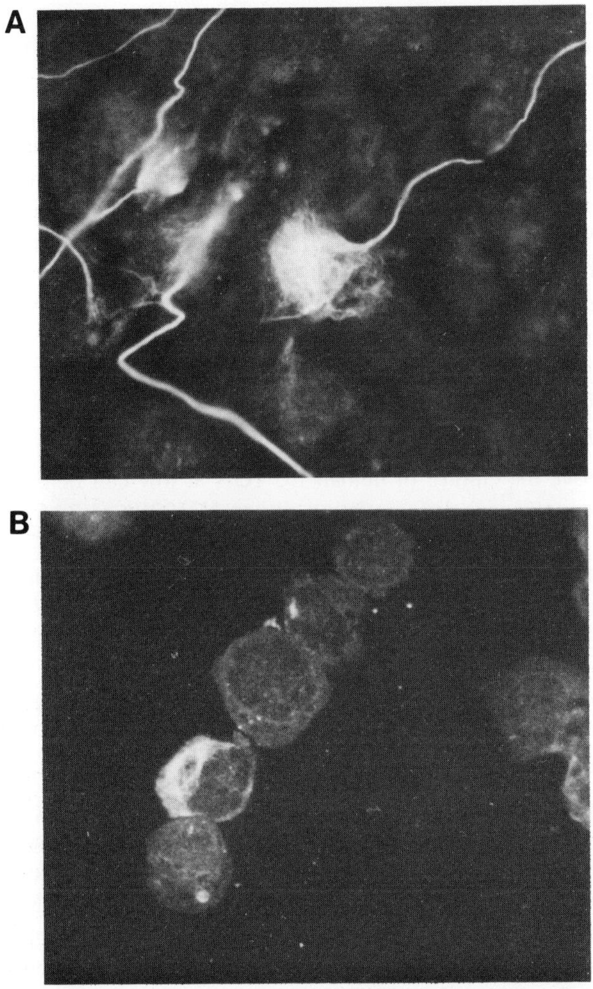

Figure 13. Detection of neurofilament-positive cells in cultures of NT2/D1 cells induced to differentiate by 10 μM retinoic acid. (**A**) Detection by staining monolayers of cells fixed *in situ* on the coverslips on which they were growing. (**B**) Detection by staining of cytospin preparations. In each case acetone-fixed cells were stained with the monoclonal antibody 7H11, which detects the 200-kd neurofilament protein, followed by FITC-conjugated goat anti-mouse Ig.

fluorescence may result whereas, as discussed, growth of human EC cells at a low density often results in their differentiation to some extent.

(ii) Wash the coverslips carrying the attached cells three times with PBS, and then three times with acetone, cooled to -20°C.

(iii) Leave the cells immersed in the final acetone wash for at least 5 min, then remove and allow to air dry.

An alternative procedure is to harvest the cells wtih trypsin EDTA and centrifuge the cells onto a slide using a cytospin centrifuge. The exact procedure depends upon the apparatus available, but typically 0.1 ml of a cell suspension containing $10^5 - 10^6$

cells/ml is loaded in the centrifuge and the cells are spun onto a glass slide while the liquid is removed by an absorbent pad. Once the cells are attached to the slide, fix in −20°C acetone as above. The advantage of growing the cells on coverslips is that the attached-cell morphology is retained; the advantage of the cytospin procedure is that cells are better separated and it is easier to count the numbers of positive and negative cells more precisely. These complementary advantages become a more important consideration when the cell cultures are markedly heterogeneous as when studying differentiation (see Section 4.2). Once acetone-fixed preparations of the cells are obtained, either on coverslips or slides, they are processed for indirect immunohistochemistry as discussed in Section 2. Examples of neurofilament-positive cells in NT2/D1 cultures, stained by these two techniques, are shown in *Figure 13*.

4.2 **Methods for inducing differentiation**

4.2.1 *Low density plating*

As mentioned above the culture of human EC cells at low cell densities often tends to promote their spontaneous differentiation. When N2102Ep c1.2A6 cells are seeded at 5×10^6 cells per 75 cm^2 flask, the majority of cells retain an EC phenotype, but when seeded at 10^5 cells per 75 cm^2 flask many cells exhibit a distinct morphology which is larger and flatter than the typical EC morphology (see *Figure 11*). Other changes in phenotype are also evident (see Section 4.3.2) and a few of the cells appear to exhibit features of trophoblastic giant cells. Similar changes occur when NT2/D1 EC cells are seeded at low density, although no trophoblastic differentiation has been detected.

4.2.2 *Chemical inducers*

Whereas all-*trans*-retinoic acid is a potent inducer of differentiation for mouse EC cells (see Chapter 2), so far the only human EC cells that have been reported to respond to this inducer are the TERA-2 EC cells (*Figure 11*) (84). Nevertheless, it should be tested on any newly derived lines. We purchase all-*trans*-RA from Kodak or Sigma and store the powder in the dark under nitrogen at −70°C; it is light-sensitive and should be handled under subdued lighting. From the powder, prepare a 10 mM stock solution by dissolving 3 mg/ml in DMSO and store under nitrogen below 0°C. For inducing the differentiation of NT2/D1 EC cells, dilute the stock solution to 10 μm or 1 μm RA in fresh medium, and seed the cells in this, at 10^6 cells per 75 cm^2 flask. These concentrations of RA are sufficient to induce almost all the cells to differentiate; at lower concentrations of RNA, significant numbers of undifferentiated EC cells persist and eventually overgrow the differentiated cells. This requirement for high concentrations of RA contrasts with the lower concentrations effective in the mouse; however, the optimal concentration for any new human lines should be individually determined.

In the case of NT2/D1 cells several other agents, including bromodeoxyuridine (BUdR, 1−3 μM), hexamethylene bisacetamide (5−10 mM), and dimethylacetamide (5−10 mM) also induce differentiation, but they are less effective than RA. Also they require a careful balance between the concentrations that kill all the cells and the concentrations that fail to induce any significant differentiation; 2- to 3-fold higher concentrations than those indicated generally kill all NT2/D1 cells so treated, whereas 2-

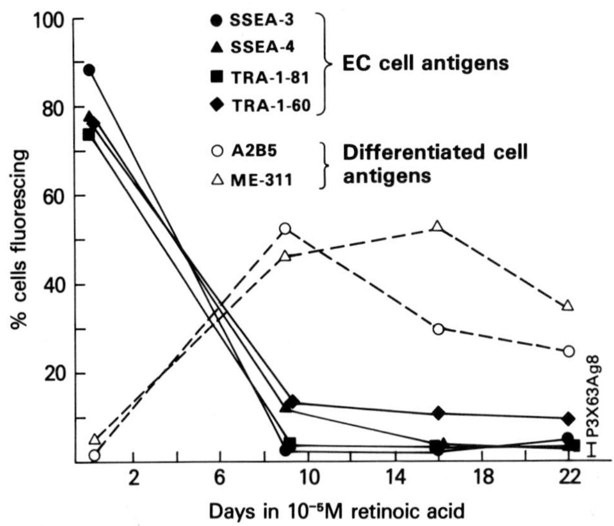

Figure 14. The disappearance of EC marker antigens (SSEA-3, SSEA-4, TR-1-60 and TR-1-81) and the appearance of other antigens (A2B5 and ME311) during the differentiation of NT2/D1 EC cells, induced with retinoic acid. Note that whereas almost no SSEA-3, SSEA-4 or TRA-1-81-positive cells are detected after 2 weeks in retinoic acid, a small but significant population of apparently differentiated cells expressing TRA-1-60 persists.

to 3-fold lower concentrations are ineffective. In addition to these agents, high concentrations of DMSO (>1% or 128 mM) also induce some differentiation although 0.1% DMSO, the concentration in medium containing 10 μM RA prepared as described above, is ineffective.

4.3 Markers for differentiation

4.3.1 *Surface antigens*

Surface antigen markers again provide the most versatile tools for studying differentiation since, as well as convenient tools for analysis, they can also be used in immunological techniques to manipulate the composition of differentiating cultures by isolating or deleting specific subsets of cells. The analytical techniques are the same as discussed in Section 4.2. Surface antigen markers fall into two categories. Firstly, the disappearance of antigens typically expressed by the EC stem cells can be followed. Thus, for example, when NT2/D1 cells differentiate after exposure to RA, cells expressing the EC-specific antigens listed in *Table 5* disappear (*Figure 14*). Two to three weeks after exposure to RA, almost no cells expressing SSEA-3 or SSEA-4, or TRA-1-81 are detectable. On the other hand, a few cells expressing TRA-1-60 continue to be detected, but these probably represent differentiated cells that express this antigen in addition to the EC cells. The second group of antigens are those expressed by the differentiated cells but not by the EC cells. Examples of changes in expression of such antigens are shown in *Figure 14*. SSEA-1 detected by monoclonal antibody 480 (85) is not expressed by human EC cells, although it is expressed by mouse EC cells and does appear transiently on some NT2/D1 cells after differentiation (not shown in *Figure 14*). A2B5 is another monoclonal antibody that detects an antigen found on various cells of the nervous sytem (86); the A2B5 antigen is not expressed by human EC cells but does appear on a subset of the differentiated derivatives of NT2/D1, including the neurons. Another antigen, ME311, defined by a monoclonal antibody to melanoma cells (87), is expressed on an overlapping subset of differentiated cells, but again not by the stem cells. Although these antigens are currently providing useful tools for following the differentiation of NT2/D1 EC cells, they clearly are not an exhaustive list. Others may come from considering specific antigens known to be expressed by cell types that are seen in the differentiated cultures (e.g. neurons); still others may come from producing new monoclonal antibodies by immunizing with differentiated cultures and selecting antibodies that do not recognize the stem cells. In either case different antibodies may be useful for analyzing the differentiation of other stem cell lines. Note that whereas the appearance of major histocompatibility complex (MHC) antigens has proved useful for studying the differentiation of murine EC cells, which do not express these antigens, the corresponding human MHC antigens (HLA) have not proved particularly useful for analyzing differentiation of human EC cells since these stem cells themselves often express low levels of both the class I antigens and β_2-microglobulin (β_2m). Further, the level of HLA and β_2m expression by human EC cells is often variable and unpredictable while it is also inducible by interferon without evidence of differentiation (authors' unpublished observations).

4.3.2 *Fluorescence-activated cell sorting*

The analysis of expression of cell surface antigens by flow cytofluorimetry is readily scaled up and adapted for isolating antigen-positive and -negative cells from a heterogeneous population. A cell culture can be prepared for sorting as follows.

(i) Prepare a sterile, single cell suspension using trypsin EDTA, wash the cells with fresh medium, and transfer 10^7 cells to a 15 ml centrifuge tube.

(ii) Pellet the cells, aspirate the medium and resuspend in 1 ml of the specific monoclonal antibody, at a dilution determined by analytical titration. The diluted antibody should be pre-sterilized by filtration through a 0.2 μm Millex filter (Millipore).

(iii) Incubate for 1 h at 4°C with gentle shaking, and wash the cells three times with fresh medium. Keep the cells cold and omit the azide used in the analytical assay wash buffer. However, we have found that if the monoclonal antibody preparation used has 0.02% azide added for storage purposes, the brief 1 h exposure during the first incubation does not usually harm the cells provided that the azide is removed during the first set of washings.

(iv) Resuspend the cells in 1 ml of filter-sterilized fluorescein- or rhodamine-conjugated second antibody, again at a dilution ascertained by analytical titration, and incubate at 4°C for a further hour.

(v) Wash the cells three times and resuspend in PBS containing 5% FBS, at a concentration of about 5×10^6 cells/ml.

In addition to the cell suspension labelled with specific antibody, a small sample of cells (e.g. 10^6) should be reacted with a negative control antibody, in order to set the fluorescence limits for distinguishing antigen-positive and -negative cells. The exact sorting protocol will depend on the cell sorter available and the nature of the experi-

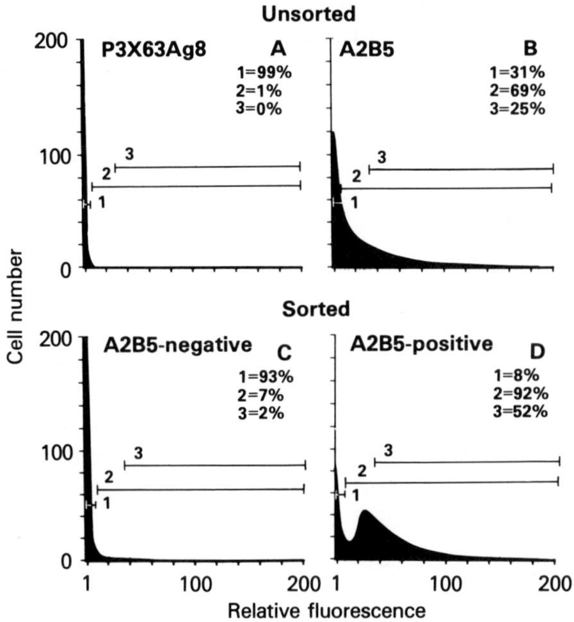

Figure 15. An example of fluorescence-activated cell sorting of retinoic acid-induced, differentiated NT2/D1 cells, into A2B5-positive and -negative populations. **Panels A** and **B** show the fluorescence histograms of unsorted cells following reaction with P3X63Ag8 negative-control antibody and A2B5 monoclonal antibody, respectively. Regions 1 and 2 are set to encompass the regions of the histograms that will represent antigen-negative and antigen-positive cells, respectively. A third region is set to encompass the 25% most strongly A2B5-positive cells, and only cells from regions 1 (A2B5-negative) and region 3 (A2B5-positive) were selected during sorting. **Panels C** and **D** show the fluorescence histograms of the two subpopulations, re-analysed when the sort was completed. Note that although theoretically 100% of the A2B5-positive cells should be expected in region 3, bleaching of the fluorochrome results in a shift of the histogram towards the negative side and in the example only 52% of the A2B5-positive cells remained in region 3. Nevertheless the sorted A2B5-positive and -negative populations were clearly different from one another; when in the experiment illustrated here these cells were re-seeded, approximately 10 times more neurons were found in the A2B5-positive cultures than in the A2B5-negative cultures, consistent with our observation that NT2/D1-derived neurons express the A2B5 antigen.

ment. An example set of results is shown in *Figure 15*. Cell recoveries depend on factors including how fast the sort is made, with what precision the sorter is set to select a particular cell type and avoid other cells, what percentage of cells fall within the selected populations, the viability of the cell preparation and the success at obtaining a single cell suspension for sorting. However, generally 10^7 cells can be handled in 1 h of sorting time and 50% of the theoretical yield should be attainable. The sorted cells are collected directly into tubes containing FBS; this will be rapidly diluted by the running buffer of the sorter (PBS). Once the sort is completed, pellet the isolated cells by centrifugation, resuspend in fresh medium containing antibiotics, and reseed as desired. At this time take a small sample of the sorted cells from each population and re-analyse to check that the sorter indeed selected the cell types for which it was set (see *Figure 15*).

4.3.3 *Intermediate filaments*

As described in Section 4.2, the various intermediate filaments are readily detected by immunofluorescence using commercially available monoclonal antibodies, and may give clues as to the types of cells present in differentiated cultures of teratocarcinomas. In particular, the neurofilaments GFAP and desmin can provide indications of neuronal, glial or muscle differentiation, respectively. Neurofilament-positive neurons are readily detectable among the differentiated derivatives of NT2/D1 cells (*Figure 13*). GFAP- and desmin-positive cells have not been detected in these cultures, but they do occur in human germ cell tumours and so should be monitored when the differentiation of new EC lines is investigated.

4.3.4 *Secreted products – HCG, AFP, fibronectin and laminin*

(i) *HCG and AFP.* HCG and AFP are useful markers for monitoring the differentiation of human EC cells into trophectodermal/choriocarcinoma cell types and yolk sac cell types, respectively. As discussed in Section 2, they may be detected by immunofluorescence assays on fixed cells. Alternatively we have used the following radioimmunoassay to detect these secreted proteins in culture supernatants.

(1) In the wells of a 96-well round-bottomed microtiter plate, mix 50 μl of supernatant sample and 50 μl of PBS + 1% (w/v) bovine serum albumin (BSA) containing 50 000 c.p.m. of ^{125}I-labelled HCG (or AFP) iodination grade (Calbiochem), iodinated to approximately 50 000 c.p.m./ng by the chloramine T method (79).
(2) Add 50 μl of rabbit anti-HCG (or AFP) serum (Cooper Biomedical) diluted in PBS + 1% (w/v) BSA and incubate at room temperature for 1 h. The antibody should be diluted to give 50% maximum binding of the labelled HCG (or AFP) in the absence of competing cold HCG (or AFP), as determined by a prior experiment; an appropriate dilution is typically in the range of 1:10 000–1:20 000
(3) After 1 h, add 25 μl of a 10% suspension of heat-inactivated *Staphylococcus aureus* (Cowan I strain), commercially available from BRL as 'Immunoprecipitin'. The bacteria, which are stored frozen, should be washed three times immediately before use and resuspended in 1% (w/v) BSA in PBS.
(4) Incubate for a further hour with shaking, then pellet the bacteria, which bind the immunoglobulin by virtue of their content of protein A, by centrifuging the plate at 700 *g* for 10 min.
(5) Aspirate the superantant and wash the bacteria twice by suspending in 0.15 M NaCl, 0.01 M Tris-HCl pH 8, 0.25% Nonidet P-40 (NP-40).
(6) After the final wash, pellet the bacteria again, swab them from the well with a cotton Q-tip and count the bound [^{125}I]HCG (or AFP) in a gamma counter.

Each assay should be done in triplicate and for each experiment a calibration curve prepared using a standard preparation of HCG and AFP. Increasing amounts of cold HCG (or AFP) in the sample cause a reduction in the amount of [^{125}I]HCG (or AFP) bound. As described the range of this assay is about 5–200 ng HCG (or AFP)/ml. It can be made more sensitive by making a 10-fold concentration of the culture supernatant using a Minicon B15 concentrator (Amicon Corporation). Commercial kits are also available for the assay of both HCG and AFP.

(ii) *Fibronectin and laminin.* We have found that human EC cells synthesize very little fibronectin, but that some of their differentiated derivatives do. Although surface-associated fibronectin can sometimes be detected by immunofluorescence of cell monolayers on coverslips, this protein is sometimes synthesized and secreted without forming a detectable extracellular matrix. This is, indeed, true of human teratocarcinomas and the synthesis of fibronectin by these cells is best determined by metabolic labelling of secreted proteins with [^{35}S]methionine and analysing these by immunoprecipitation and SDS-electrophoresis. We have used the following procedure to study changes in the synthesis of fibronectin by 2102Ep EC cells as they differentiate at low cell density. The same method is applicable to the analysis of laminin secretion.

(1) Seed cultures at the desired cell density the day before beginning to metabolically label secreted proteins. Wash the medium from the cells and replace it with methionine-free medium (4 ml per 75 cm^2 flask) containing 2% dialysed FBS, 50 μg/ml β-aminopropionitrile fumarate and 75 μCi/ml [^{35}S]methionine (specific activity >400 Ci/mmol).

(2) Culture the cells in this medium for 12–18 h (less if they begin to die), recover the medium and add phenylmethylsulphonyl fluoride to 2 mM, EDTA to 10 mM and cold methionine to 1 mM.

(3) Remove cell debris by centrifugation and determine the incorporation of [^{35}S]-methionine by counting an aliquot of the medium following precipitation with 10% trichloroacetic acid (TCA).

(4) To a volume of labelled culture supernatant containing 10^6 TCA-precipitable counts, add 1–3 μl of ascites containing anti-human fibronectin monoclonal antibody (MAS 037c: Sera Labs) (or anti-laminin activity) and 30 μl of rabbit anti-mouse immunoglobulin serum. (These volumes are only approximate and should be determined for each set of reagents by titration in order to achieve optimal precipitation.)

(5) Incubate overnight at 4°C and pellet the immune precipitate by centrifuging in a microfuge (about 10 000 *g* for 1–2 min is generally sufficient).

(6) Wash the precipitates five times, or until no counts are detected in the washings, in 10 mM Tris-HCl pH 7.5, 0.5 M NaCl, 0.5% NP-40, dispersing it each time.

(7) Reduce and disperse the precipitate in 50 μl of electrophoresis sample buffer (0.1 M Tris-HCl, pH 6.8, 2 μM EDTA, 15% glycerol, 2% SDS and 0.1 M dithiothreitol) by heating in a boiling water bath for 5 min. Electrophorese this sample in 5% polyacrylamide gels containing 0.1% SDS by the method of Laemmli (88).

(8) Following electrophoresis treat the gel with EnHance (New England Nuclear), dry down and determine the position of the labelled proteins by autoradiography.

5. ACKNOWLEDGEMENTS

This work was supported by grants CA-29894, CA-23097 and CA-38045 from the National Institutes of Health, by grant GUKCKWF 84-6 from The Netherlands Cancer Foundation, and a grant from the Pediatric Oncology Foundation SKOG (Groningen).

6. REFERENCES

1. Solter,D. (1983) In *The Human Teratomas. Experimental and Clinical Biology.* Damjanov,I., Knowles,B.B. and Solter,D. (eds), Humana Press, Clifton, New Jersey, p. 343.
2. Damjanov,I., Knowles,B.B. and Solter,D., eds (1983) *The Human Teratoma. Experimental and Clinical Biology.* Humana Press, Clifton, New Jersey.
3. Dixon,F.J. and Moore,R.A. (1952) *Tumors of the Male Sex Organs.* Armed Forces Institute of Pathology, Washington, D.C.
4. Gonzales-Crussi,F. (1982) *Extragonadal Teratomas.* Armed Forces Institute of Pathology, Washington, D.C.
5. Mostofi,F.K. and Price,E.B.,Jr (1973) *Tumors of the Male Genital System.* Armed Forces Institute of Pathology, Washington, D.C.
6. Pugh,R.C.B. (1976) *Pathology of the Testis.* Blackwell Scientific, Oxford.
7. Scully,R.E. (1979) *Tumors of the Ovary and Maldeveloped Gonads.* Armed Forces Institute of Pathology, Washington, D.C.
8. Talerman,A. and Roth,L.M., eds (1986) *Pathology of the Testis and its Adnexa.* Churchill-Livingstone, New York.
9. Mostofi,F.K. and Sobin,L.H. (1976) *International Histologic Classification of Tumors.* WHO, Geneva.
10. Mostofi,F.K. (1980) *Cancer,* **45**, 1735.
11. Talerman,A., Fu,Y.S. and Okagaki,T. (1984) *Lab. Invest.,* **51**, 343.
12. Paiva,J., Damjanov,I., Lange,P.H. and Harris,H. (1983) *Am. J. Pathol.,* **111**, 156.
13. Damjanov,I., Fox,N., Knowles,B.B., Solter,D., Lange,P.H. and Fraley,E.E. (1982) *Am. J. Pathol.,* **108**, 225.
14. Zaloudek,C.J., Tavassoli,F.A. and Norris,H.J. (1981) *Am. J. Surg. Pathol.,* **5**, 361.
15. Miettinen,M., Virtanen,I. and Talerman,A. (1985) *Am. J. Pathol..,* **120**, 402.
16. Lee,M.C., Talerman,A., Oosterhuis,J.W. and Damjanov,I. (1985) *Arch. Pathol. Lab. Med.,* **109**, 938.
17. Damjanov,I., Clark,R.K. and Andrews,P.W. (1984) *Cell Differ.,* **15**, 133.
18. Nogales-Fernandez,F., Matilla,A., Nogales-Ortiz,F. and Galera-Davidson,H.L. (1978) *Hum. Pathol.,* **9**, 553.
19. Peters,J. (1985) *J. Biol. Chem.,* **260**, 14732.
20. Barka,T. and Anderson,P.J. (1963) *Histochemistry.* Harper and Row, New York.
21. DeLellis,R.A., ed. (1981) *Diagnostic Immunocytochemistry.* Masson, New York.
22. Pearse,A.G.E. (1983) *Histochemistry Theoretical and Applied.* Churchill-Livingston, London.
23. Shirai,T., Yoshiki,T. and Itoh,T. (1977) *Gann,* **68**, 847.
24. Hata,J., Ueyama,Y., Tamaoki,N., Akatsuka,A., Yoshimura,S., Shimuzu,K., Morikawa,Y. and Furukawa,T (1980) *Cancer,* **46**, 2446.
25. Yoshimura,S., Tamaoki,N., Ueyama,Y. and Hata,J. (1978) *Cancer Res.,* **38**, 3474.
26. Kaneko,M., Takeuchi,T., Tsuchida,Y., Saito,S. and Endo,Y. (1980) *Gann,* **71**, 14.
27. Takeuchi,T., Nakayasu,M., Hirohashi,S., Kameya,T., Kaneko,M., Yokomori,K. and Tsuchida,Y. (1979) *J. Clin. Pathol.,* **32**, 693.
28. Tveit,K.M., Fodstad,Ø, Brøgger,A. and Olsnes,S. (1980) *Cancer Res.,* **40**, 949.
29. Selby,P.J., Heyderman,E., Gibbs,J. and Peckham,M.J. (1979) *Br. J. Cancer,* **39**, 578.
30. Monoghan,P., Raghavan,D. and Munro Neville,A. (1982) *Cancer,* **49**, 683.
31. Raghavan,D. (1981) In *Early Detection of Testicular Cancer.* Skakkebaek,N.E., Berthelsen,J.G., Grigor,K.M. and Visfeldt,J. (eds), Scriptor, Copenhagen, p. 79.
32. Fogh,J. and Trempe,G. (1975) In *Human Tumor Cells In Vitro.* Fogh,J. (ed.), Plenum Press, New York, p. 115.
33. Wang,N., Perkins,P.L., Bronson,D.L. and Fraley,E.E. (1981) *Cancer Res.,* **41**, 2135.
34. Andrews,P.W., Damjanov,I., Simon,D., Banting,G.S., Carlin,C., Dracopoli,N.C. and Fogh,J. (1984) *Lab. Invest.,* **50**, 147.
35. Hogan,B., Fellous,M., Avner,P. and Jacob,F. (1977) *Nature,* **270**, 515.
36. Yamamoto,T., Komatsubara,S., Suzuki,T. and Oboshi,S. (1979) *Gann,* **70**, 677.
37. Bronson,D.L., Ritzi,D.M., Fraley,E.E. and Dalton,A.J. (1978) *J. Natl. Cancer Inst.,* **60**, 1305.
38. Bronson,D.L., Andrews,P.W., Solter,D., Cervenka,J., Lange,P.H. and Fraley,E.E. (1980) *Cancer Res.,* **40**, 2500.
39. Wang,N., Trend,B., Bronson,D.L. and Fraley,E.E. (1980) *Cancer Res.,* **40**, 796.
40. Andrews,P.W., Bronson,D.L., Benham,F., Strickland,S. and Knowles,B.B. (1980) *Int. J. Cancer,* **26**, 269.
41. Andrews,P.W., Goodfellow,P.N., Shevinsky,L., Bronson,D.L. and Knowles,B.B. (1982) *Int. J. Cancer,* **29**, 523.
42. Cotte,C.A., Easty,G.C. and Munro Neville,A. (1981) *Cancer Res.,* **41**, 1422.

43. Cotte,C.A., Raghavan,D., McIlhinney,R.A.J. and Monaghan,P. (1982) *In Vitro,* **18**, 739.
44. Bronson,D.L., Clayman,R.V. and Fraley,E.E. (1983) In *The Human Teratomas. Experimental and Clinical Biology*. Damjanov,I., Knowles,B.B. and Solter,D. (eds), Humana Press, Clifton, New Jersey, p. 267.
45. Bronson,D.L., Andrews,P.W., Vessella,R.L. and Fraley,E.E. (1983) In *Teratocarcinoma Stem Cells*. Silver,L.M., Martin,G.R. and Strickland, S. (eds), Cold Spring Harbor Laboratory Press, New York, p. 597.
46. Vogelzang,N.J., Bronson,D., Savino,D., Vessella,R.L. and Fraley,E.E. (1985) *Cancer,* **55**, 2584.
47. Sekiya,S., Kawata,M., Iwasawa,H., Inabe,N., Sugita,M., Suzuki,N., Motoyama,T., Yamamoto,T. and Takamizawa,H. (1985) *Differentiation,* **29**, 259.
48. Giovanella,B.C., Stehlin,J.S. and Williams,L.J. (1974) *J. Natl. Cancer Inst.,* **52**, 921.
49. Zeuthen,J., Nørgaard,J.O.R., Avner,P., Fellous,M., Wartiovarra,J., Vaheri,A., Rosen,A. and Giovanella,B.C. (1980) *Int. J. Cancer.,* **25**, 19.
50. Sekiya,S., Kaiho,T., Shirotake,S., Iwasawa,H., Inaba,N., Kawata,M., Higaki,K., Ishige,H., Takamizawa,H., Minamihisamatsu,M. and Kuwata,T. (1983) *In Vitro,* **19**, 489.
51. Kikuchi,Y., Momose,E., Kizawa,I., Ishida,M., Sunaga,H., Mukai,K., Seki,K. and Kato,K. (1984) *Cancer Res.,* **44**, 2952.
52. Ishiwata,I., Ishiwata,C., Soma,M., Tomita,K., Nozawa,S. and Ishikawa,H. (1985) *J. Natl. Cancer Inst.,* **75**, 411.
53. Sundström,C., Mark,J. and Westermark,B. (1980) *Acta Pathol. Microbiol. Scand. Sect. A,* **88**, 189.
54. Vogelzang,N., Andrews,P. and Bronson,D. (1983) *Proc. Am. Assoc. Cancer Res.,* **24**, 3.
55. Oosterhuis,J.W., De Jong,B., Van Dalen,J., Van der Meer, J., Visser,M., De Leij,L., Mesander,G., Collard,J.G., Schraffordt Koops,H., and Sleijfer,D.Th. (1985) *Cancer Genet. Cytogenet.,* **15**, 99.
56. Ishiwata,I., Ishiwata,C., Mukai,M., Ono,I., Ohara,K., Konoe,K. and Ishikawa,H. (1985) *Exp. Pathol.,* **27**, 143.
57. Atkin,N.B. and Baker,M.C. (1983) *Cancer Genet. Cytogenet.,* **10**, 199.
58. Oosterhuis,J.W., De Jong,B., Cornelissen,C.J., Molenaar,W.M., Meiring,A., Idenburg,V., Schraffordt Koops,H. and Sleijfer,D.Th. (1986) *Cancer Genet. Cytogenet.,* in press.
59. Kimoto,T., Ueki,A. and Nishitani,K. (1975) *Acta Pathol. Jap.,* **25**, 89.
60. Ihara,T., Ohama,K., Satoh,H., Fujii,T., Nomura,K. and Fujiwara,A. (1984) *Cancer,* **54**, 2988.
61. Kaplan,C.G., Askin,F.B. and Benirschke,K. (1979) *Teratology,* **19**, 261.
62. Oosterhuis,J.W., Suurmeyer,A.J.H., Sleijfer,D.Th., Schraffordt Koops,H., Oldhoff,J. and Fleuren,G.J. (1983) *Cancer,* **51**, 408.
63. Martin,G.R. and Evans,M.J. (1975) *Proc. Natl. Acad. Sci. USA,* **72**, 1441.
64. Reznikoff,C.A., Bertram,J.S., Brankow,D.W. and Heidelberger,C. (1973) *Cancer Res.,* **33**, 3239.
65. MacPherson,I. (1973) In *Tissue Culture Methods and Applications*. Krusse,P. and Patterson,M. (eds), Academic Press Inc., New York, p. 3.
66. Vindeløv,L.L., Christensen,I.J., Keiding,N., Spang-Thomsen,M. and Nissen,N.I. (1983) *Cytometry,* **3**, 317.
67. Vindeløv,L.L., Christensen,I.J. and Nissen,N.I. (1983) *Cytometry,* **3**, 323.
68. Vindeløv,L.L., Christensen,I.J. and Nissen,N.I. (1983) *Cytometry,* **3**, 328.
69. Vindeløv,L.L., Christensen,I.J., Jensen,G. and Nissen,N.I. (1983) *Cytometry,* **3**, 332.
70. Hedley,D.W., Friedlander,M.L., Taylor,I.W., Rugg,C.A. and Musgrove,E.A. (1983) *J. Histochem. Cytochem.,* **31**, 1333.
71. Pattillo,R.A. and Gey,G.O. (1968) *Cancer Res.,* **28**, 1231.
72. Steel,G.G., Courtenay,V.D. and Rostom,A.Y. (1978) *Br. J. Cancer,* **37**, 224.
73. Bennet,J.A., Pilon,V.A. and MacDowell,R.T. (1985) *Cancer Res.,* **45**, 4963.
74. Giovanella,B.C. and Fogh,J. (1985) *Adv. Cancer Res.,* **44**, 70.
75. Shevinsky,L.H., Knowles,B.B., Damjanov,I. and Solter,D. (1982) *Cell,* **30**, 697.
76. Andrews,P.W., Goodfellow,P.N. and Bronson,D.L. (1983) In *Teratocarcinoma Stem Cells. Cold Spring Harbor Conferences on Cell Proliferation.* Silver,L.M., Martin,G.R. and Strickland,S. (eds), Cold Spring Harbor Laboratory Press, New York, Vol.**10**, p. 579.
77. Kannagi,R., Cochran,N.A., Ishigami,F., Hakomori,S.I., Andrews,P.W., Knowles,B.B. and Solter,D. (1983) *EMBO J.,* **2**, 2355.
78. Andrews,P.W., Banting,G.S., Damjanov,I., Arnaud,D. and Avner,P. (1984) *Hybridoma,* **3**, 347.
79. Talmage,D.W. and Claman,H.N. (1967) In *Methods in Immunology and Immunochemistry*. Williams,C.H. and Chase,M.W. (eds), Academic Press, New York, Vol. **1**, Section 4, A2, p. 389.
80. Kohler,G. and Milstein,C. (1975) *Nature,* **256**, 495.
81. Herzenberg,L.A. and Herzenberg,L.A. (1978) In *Handbook of Experimental Immunology*. Weir,D.M. (ed.), Blackwell Scientific Publications, Oxford, 3rd edition, p. 22.1.
82. Kruth,H.S. (1982) *Anal. Biochem.,* **125**, 225.

83. Clark,R.K. and Damjanov,I. (1985) *Virchows Arch. (Pathol. Anat.),* **407**, 203.
84. Andrews,P.W. (1984) *Dev. Biol.,* **103**, 285.
85. Solter,D. and Knowles,B.B. (1978) *Proc. Natl. Acad. Sci. USA,* **75**, 5565.
86. Eisenbarth,G., Walsh,F.S. and Nirenberg,M. (1979) *Proc. Natl. Acad. Sci. USA,* **76**, 4913.
87. Thurin,J., Herlyn,M., Hindsgaul,O., Strömberg,N., Karlsson,K.-A., Elder,D., Steplewski,Z. and Koprowski,H. (1985) *J. Biol. Chem.,* **260**, 14556.
88. Laemmli,U.K. (1970) *Nature,* **227**, 680.

APPENDIX I

Suppliers of specialist items

Amersham International plc., White Lion Road, Amersham, Bucks. HP7 9LL, UK

Amicon Corp., 21 Hartwell Avenue, Lexington, MA 02173, USA

BDH Chemicals Ltd., Poole, Dorset BH12 4NN, UK

Bethesda Research Laboratories Inc., 2200 Wright Avenue, Richmond, CA 94804, USA

Campden Instruments, 185 Campden Hill Road, London W8 1TH, UK

Cappel Labs., 237 Lacey Street, P.O. Box 37, Westchester, PA 19380, USA

Connaught Labs. Ltd., 1755 Steels Avenue West, Willowdale, Ontario M2R 3T4, Canada

Corning Ltd., Stone, Staffs. ST15 0BG, UK

Coulter Electronics, Norwell Drive, Luton, Beds. LU3 3RH, UK

Difco Labs. Ltd., P.O. Box 143, Central Avenue, East Molesey, Surrey KT8 0SE, UK

Dynatech Diagnostics Inc., RR # 1, Inland Farm Drive, South Windham, ME 04082, USA

Eastman-Kodak Co., 343 Slate Street, Rochester, NY 14650, USA

Flow Labs. Inc., 7655 Old Spring House Road, McLean, VA 22102, USA

Gibco-Biocult, The Dexter Corp., P.O. Box 200, Pine & Mogul Streets, Chagrin Falls, OH 44022, USA

Glaxo Corp., Clarges Street, London W1, UK

Interfauna, Abbots Ripton Road, Wyton, Huntingdon, Cambs PE17 2DT, UK

Lab Tek Division, Miles Laboratories, Naperville, IL 60540, USA

Ernst Leitz, Wetzlar, GmbH, D-6330 Wetzlar, FRG

May & Baker Ltd., Liverpool Road, Barton Moss, Eccles, Manchester M30 7RT, UK

Miles Labs Inc., P.O. Box 2000, 1127 Myrtle Street, Elkhard, IN 46515, USA

Millipore Corp., Bedford, MA 01730, USA

Narashige, 27-9 Minami-Karasuyama, 4-Chome, Setagaya-Ku, Tokyo 157, Japan

New England Nuclear Research Products, 549 Albany Street, Boston, MA 02118, USA

Nikon Inc., Instrument Group, 623 Stewart Avenue, Garden City, NY 11530, USA

Olac Ltd., Shaws Farm, Blackthorn, Bicester, Oxon OX6 0TP, UK

Pharmacia Fine Chemicals AB, P.O. Box 175, S-75104 Uppsala-1, Sweden

Sera Labs. Ltd., Crawley Down, Sussex RH10 4FF, UK

Sigma Chemical Co., 3500 Dekalb Street, St. Louis, MO 63118, USA

UpJohn Diagnostics, 301 Henrietta Street, Kalamazoo, MI 49001, USA

Worthington Biochemical Corporation, P.O. Box 650, Halls Mill Road, Freehold, NJ 07728, USA

Carl Zeiss, D-7082 Oberkochen, FRG

INDEX